Contributions to Management Science

The series *Contributions to Management Science* contains research publications in all fields of business and management science. These publications are primarily monographs and multiple author works containing new research results, and also feature selected conference-based publications are also considered. The focus of the series lies in presenting the development of latest theoretical and empirical research across different viewpoints.

This book series is indexed in Scopus.

Ahmad Rafiki • Veland Ramadani
Léo-Paul Dana • Sutan Emir Hidayat
Editors

Strategic Islamic Business and Management

Solutions for Sustainability

Springer

Editors
Ahmad Rafiki
Medan Area University
Medan, Indonesia

Veland Ramadani
South East European University
Tetovo, North Macedonia

Léo-Paul Dana
Lappeenranta University of Technology
Lappeenranta, Finland

Sutan Emir Hidayat
National Committee of Islamic
Economic and Finance
University of Gunadarma
Jakarta Selatan, Indonesia

ISSN 1431-1941 ISSN 2197-716X (electronic)
Contributions to Management Science
ISBN 978-3-031-61777-5 ISBN 978-3-031-61778-2 (eBook)
https://doi.org/10.1007/978-3-031-61778-2

This Springer imprint is published by the registered company Springer Nature Switzerland AG
The registered company address is: Gewerbestrasse 11, 6330 Cham, Switzerland

Preface

In today's competitive global market, Islamic institutions or companies must implement effective business and management strategies to ensure success and sustainability. This book provides a comprehensive exploration of the various aspects of business and management in Islamic finance, from organization and planning to marketing, entrepreneurship, and innovation. It also examines the impact of globalization and dynamic markets on the performance of Islamic organizations.

Drawing on the principles and basis of Islamic religious law, the book offers a wealth of concepts, models, and frameworks that can be applied to improve the performance of Islamic organizations. It explores the history of Islamic business and management and how it can be adapted to meet the needs of modern-day Islamic organization.

The book emphasizes the importance of a just system that brings benefits to all people and promotes sustainability in the banking industry and general business. It highlights the best practices in business and management that are crucial to creating value and achieving sustainability. These practices include considering the needs of stakeholders, adopting high technology, and promoting ethical behavior.

This book also focuses on the role of digital marketing in creating sustainability for Islamic organizations. Through the use of technology and new approaches, marketing can help companies reach customers efficiently and effectively, leading to better sales and business performance. The emergence of online media platforms has enabled marketers to buy ad space on various websites and mobile apps and review live reports within hours, allowing for quick adjustments to optimize campaign performance.

By adopting strategic business and management, Islamic organizations can diversify their product lines and services, which support organizations to increase their competitiveness while staying true to the values of sustainability. This book provides an in-depth exploration of strategic Islamic business and management concepts, models, and frameworks that can be adopted by managers in any organization. It also offers best practices related to Islamic business and management.

This book is essential reading for entrepreneurs, businessmen, practitioners, policymakers, academicians, and students who are interested in developing effective strategies and future plans related to Islamic business and management and who want to develop strategies and future plans related to digital marketing for creating sustainability in business operations. It offers insights and guidance on how to improve business operations and achieve effectiveness and efficiency in Islamic organizations.

With a focus on sustainability, this book highlights how strategic Islamic business and management can help Islamic organizations and other organizations achieve their business objectives while promoting ethical and responsible practices that benefit society and the environment.

Medan/Selatan, Indonesia — Ahmad Rafiki
Tetovo, North Macedonia — Veland Ramadani
Lappeenranta, Finland — Léo-Paul Dana
Sutan Emir Hidayat

Contents

About the Editors

Ahmad Rafiki is an Associate Professor and the Dean of the Faculty of Economics and Business, Universitas Medan Area (UMA). He was with the University College of Bahrain in the Department of Business Administration as an Assistant Professor and as a Research Associate in the Mizan Research Centre, Faculty of Leadership and Management, Universiti Sains Islam Malaysia (USIM). In March 2014, he was awarded a Doctor of Philosophy from the Universiti Sains Islam Malaysia (USIM). He holds a Certified Islamic Marketing Associate (CIMA) from the International Islamic Marketing Association (IIMA). He has published many chapters of books/authored books and articles related to Islamic management, entrepreneurship, SMEs, leadership, and halal industry by international publishers such as IBA-MacMillan, IGI Global, Emerald Publishing (*Journal of Islamic Marketing, Asian Journal of Accounting Research, Journal of Investment Compliance, PSU Research Review, RAUSP Management Journal, International Journal of Organizational Analysis, Journal of Advance in Management Research, Enterprising Communities: People and Places in the Global Economy*), Springer, Routledge, etc. He also became the editorial advisory board and reviewer in reputable publishers of Emerald (*Journal of Islamic Marketing, International Journal of Islamic and Middle Eastern Finance and Management*) and Elsevier. He won the Outstanding Reviewer of Literati Awards 2020 by Emerald Publishing.

Veland Ramadani is a Professor of Entrepreneurship and Family Business at the Faculty of Business and Economics, South-East European University, North Macedonia. His primary areas of research focus on entrepreneurship, family businesses, innovation, and sustainability. Dr. Ramadani has an extensive body of work to his name, having authored or co-authored over 150 research articles, 80 book chapters, 15 textbooks, and 31 edited books. His research contributions are notably featured in prestigious academic journals, including the *Journal of Business Research, Technological Forecasting and Social Change, Technology in Society, International Journal of Entrepreneurial Behavior & Research, Review of Managerial Science,* among others. He received the Award for Excellence in 2016 for an Outstanding Paper published by Emerald Group Publishing, a testament to his outstanding contributions to the field. In addition to his research, Dr. Ramadani has been invited as a keynote speaker at numerous international conferences and as a guest lecturer at President University (Indonesia), Telkom University (Indonesia), University of Trás-os-Montes and Alto Douro (Portugal), and Universum International—UNI (Kosovo). Furthermore, his research influence is underscored by his consistent inclusion among the World's Top 2% of the most influential researchers, as determined by Elsevier and Stanford University for the last four consecutive years. Between 2017 and 2021, Dr. Ramadani served as a member of the Supervisory Board of the Development Bank of North Macedonia, where he also took on the role of Chief Operating Officer for 10 months. Dr. Ramadani has remained actively engaged in various projects, including those funded by the European Union and universities in Saudi Arabia, Kuwait, and the United Arab Emirates.

Léo-Paul Dana is a Professor at LUT School of Business and Management. He is also associated with the Chaire ETI at Sorbonne Business School. A graduate of McGill University and HEC-Montreal, he has served as a Marie Curie Fellow at Princeton University and a Visiting Professor at INSEAD. Professor at ICD Business School Paris. He has published extensively in a variety of journals including *Entrepreneurship: Theory & Practice; International Business Review; International Small Business Journal; Journal of Business Research; Journal of Small Business Management; Journal of World Business; Small Business Economics; and Technological Forecasting & Social Change.*

Sutan Emir Hidayat is an Academic of Islamic Economics (Associate Professor). Sutan Emir serves as the Director of Sharia Ecosystem Infrastructure at the National Sharia Economy and Finance Committee (KNEKS). In addition, Sutan Emir is also a member of the Expert Council at the Association of Islamic Economics Lecturers (ADESy). Sutan Emir actively writes various articles, especially on Sharia Economics and has become a national and international speaker, such as the 11th Harvard University Forum on Islamic Finance in the United States and the Gulf Research Meeting at the University of Cambridge, England, AAOIFI & WORLD Bank conference on Islamic Banking and Finance in Bahrain, and other international forums.

Driving Green Sukuk Development to Meet ESGs and Society 5.0 Aspirations Through Innovative Technology

Adlin Masood, Norfhadzilahwati Rahim, and Umi Hamidaton Mohd Soffian Lee

Abstract Three centuries post-industrial revolution, the global economy has witnessed an unparalleled expansion characterised by a fourfold rise in population and a forty-fold surge in industrial production throughout the twentieth century. Nevertheless, this expansion has resulted in considerable depletion of environmental resources throughout the production phase, hence presenting notable environmental issues in the era. The concept of Society 5.0 entails the development of a society that prioritises the needs and well-being of individuals, while leveraging advanced technologies to achieve environmental sustainability and effective governance. The rapid expansion of Green Sukuk, which involves the funding of ecologically friendly initiatives, is dependent upon the presence of favourable legal, regulatory, social, and technological contexts. The Internet of Everything (IoE), which encompasses the integration of individuals, objects, data, and operations, has the potential to augment the process of decision-making. This study conducts a literature analysis to evaluate the potential of IoE-empowered Green Sukuk in attaining Environmental, Social, and Governance (ESG) goals that are in line with the principles of Society 5.0. The results indicate that the use of private resources and technology has the potential to accelerate the achievement of environmental, social, and governance (ESG) objectives, thereby promoting a future that is both sustainable and equitable.

A. Masood (✉) · N. Rahim · Umi Hamidaton Mohd Soffian Lee
Faculty of Economics and Muamalat, Universiti Sains Islam Malaysia, Bandar Baru Nilai, Nilai, Negeri Sembilan, Malaysia
e-mail: adlin.masood@usim.edu.my; norfhadzilahwati@usim.edu.my; umihamidaton@usim.edu.my

A. Rafiki et al. (eds.), *Strategic Islamic Business and Management*, Contributions to Management Science,
https://doi.org/10.1007/978-3-031-61778-2_1

1 Introduction

Based on the findings of the United Nations Statistical Division (UN Report 2022), it is projected that around 95 million individuals will be residing in conditions of extreme poverty by the year 2022. This formidable issue of unprecedented magnitude was brought on by extreme climate change, global health crises, geopolitical tensions, warfare, and inequality, persistently elude global resolution.

In the face of mounting environmental and social challenges, the world is seeking sustainable solutions that align with Environmental, Social, and Governance (ESG) principles and foster a more equitable and prosperous Society 5.0. Society 5.0 envisions a transformative paradigm where technology is harnessed to meet human needs while actively addressing environmental, sustainable, and governance (ESG) goals. Simultaneously, the rise of Green Sukuk, Islamic bonds financing environmentally friendly initiatives, presents a promising avenue for sustainable development (Al Ansari & Alanzarouti, 2020; Ma'ruf et al., 2021). The expansion of Green Sukuk, however, relies on the presence of supportive legal, regulatory, social, and technological environments; consequently, the integration of innovative technology becomes pivotal (Aassouli & Shah, 2022).

In the face of mounting global challenges, the concept of Society 5.0 has emerged as a visionary framework for a sustainable and prosperous future. Society 5.0 envisions a human-centred society that integrates physical and digital spaces to enhance individual well-being and societal progress. At the core of Society 5.0 lies a deep commitment to Environmental, Social, and Governance (ESG) principles. ESG encompasses the interconnectedness of environmental, social, and governance factors in achieving a sustainable future. Society 5.0 emphasises the importance of environmental stewardship, promoting responsible resource utilisation, pollution reduction, and biodiversity conservation. By embracing clean technologies, circular economy principles, and renewable energy sources, Society 5.0 strives to minimise environmental impact and foster a harmonious relationship between humans and nature. Social equity and inclusiveness are fundamental pillars of Society 5.0 which underscores the importance of responsible governance, advocating for transparency, accountability, and ethical leadership. Technological advancements play a crucial role in enabling the realisation of Society 5.0. Artificial intelligence (AI), big data analytics (BDA), and the Internet of Things (IoT) are among the key technologies that can be harnessed to optimise resource utilisation, enhance service delivery, and address societal challenges.

The Islamic finance industry has gradually realised the importance of incorporating environmental, social, and governance (ESG) elements into its operations. Aassouli and Shah (2022) underlined the tremendous potential of combining Islamic finance with advanced technological achievements in boosting the worldwide goal of sustainable development in their study. Green Sukuk, an innovative financial product based on Shariah principles, has emerged as a key player in the field of sustainable finance (Alam et al., 2016). The establishment of this platform for Islamic bonds with the intention of supporting activities that promote environmental sustainability demonstrates a strong commitment to this goal. Green Sukuk is reported to have a constructive and beneficial impact on society as a whole, according to Shariah law (Fitrah & Soemitra, 2022). However, effective implementation of this programme in countries aiming for ESG necessitates a sufficient and evolving framework that includes legal, regulatory, social, and technological components (Saba et al., 2021).

The social and economic chaos caused by the COVID-19 pandemic has accelerated the adoption of Fourth Industrial Revolution (IR4.0) technology, with society more comfortable than in previous eras with the use of IR4.0 technology as a method of communications, transactions, and marketing (Masood et al., 2023). Industry 4.0 is a revolutionary shift in the industrial sector, with transformative implications for capital-market operations such as automated trading, data-driven decision-making, more effective risk management with real-time analytics, and personalised investment strategies to improve customer satisfaction. To address the challenges created by these developments, major cybersecurity measures and regulatory changes are required. Adopting Industry 4.0 innovative technology can assist financial institutions in remaining competitive in an ever-changing economy while meeting the growing demands of investors and market players.

Figure 1 diagrammatically depicts the Eleven Pillars of Industry 4.0, an extension to Nine Pillars of Industry 4.0, that has changed the manufacturing and services scene into a highly efficient, intelligent, and linked ecosystem (Erboz, 2017; Rejikumar et al., 2019).

Utilising cutting-edge technology such as distributed ledger technology (DLT), blockchains, artificial intelligence (AI), and cloud computing, green sukuk can fulfil environmental, social, and governance (ESG) requirements while also contributing to the goals of Society 5.0. The issuing of Green Sukuk and ESG compliance are both made possible by artificial intelligence. To guarantee ESG and regulatory compliance in sukuk terms, AI algorithms can automate the development of smart contracts. Hassan et al. (2023) state that artificial intelligence can sift through mountains of data in search of environmental, social, and governance (ESG) problems linked

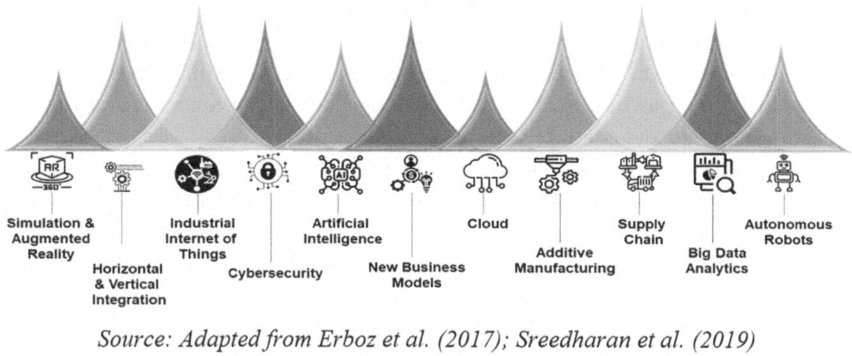

Fig. 1 Eleven pillars of IR4.0. Source: Adapted from Erboz (2017); Rejikumar et al. (2019)

to underlying projects. This helps with decision-making and lowers the risk to reputation. Investors can achieve their ESG goals and make a positive impact on sustainable development with the help of these platforms' AI-powered personalised investment recommendations. Digital ledger technology, or DLT, is a method of keeping track of financial transactions that allows for the simultaneous recording of transaction details in numerous locations. There is no centralised data repository or administrative functionality in distributed ledgers, unlike traditional databases. As a result of distributed ledger technology (DLT), stakeholders are better able to assess the environmental, social, and governance (ESG) effects of green sukuk investments since transactions involving these funds are more transparent and easier to trace. While cloud computing speeds communication and centralises data, BDA provides insights (Mohbey & Kumar, 2022).

Simultaneously, the Internet of Everything (IoE) is a highly advanced network that intimately links diverse devices, humans, and data in order to deliver fast and correct information to the relevant decision-maker, making it a very effective enabler in the future. IoE possesses the capacity to enhance the efficacy and transparency of sustainable initiatives (Langley et al. 2021). Although the Internet of Things (IoT) solely facilitates the connection known as machine-to-machine (M2M) communication, it should be noted that IoT is just one component of the broader concept of the Internet of Everything (IoE), which encompasses a wider range of interconnected entities (Fredette et al., 2012). The Internet of Everything (IoE) enables individuals to gain a deeper comprehension of the environmental and social consequences, of sponsored initiatives by collecting and analysing real-time data (Fredette et al., 2012), that are then fed to the various pillars of IoE. Consequently, this fosters a financial ecosystem characterised by openness and accountability. The integration of the IoE into the Green Sukuk ecosystem signifies a significant change in perspective, introducing a versatile framework that can effectively monitor, evaluate, and enhance the effectiveness of environmentally sustainable initiatives.

The IoE, which unites individuals, objects, information, and processes into a connected framework, stands out as a strategic partner in advancing the goals of Green Sukuk and meeting the aspirations of Society 5.0. The IoE facilitates intelligent decision-making, creating a synergy, which holds the potential to accelerate the attainment of ESG objectives. That is, the convergence of green sukuk and innovative technology represents a paradigm shift in sustainable finance, driving a more transparent, inclusive, and impactful approach to ESG investing. By harnessing DLT and AI, green sukuk can effectively meet the demands of Society 5.0, fostering a more sustainable and equitable future for all.

The principal aim of this paper is to elucidate the considerable transformative capacity inherent in the model under consideration. It seeks to unravel how the symbiotic relationship between Green Sukuk and IoE can function as a formidable force in accelerating progress towards ESGs attainment as desired by Society 5.0. The crux of this argument lies in the cultivation of innovation, inclusion, and sustainability, positioning the combination of Islamic finance and innovative technology as a pivotal driver for societal progress. By intricately interweaving private sector dynamism with technological prowess, this research contends that a profound and positive impact can be achieved, transcending the conventional boundaries of financial instruments and sustainability initiatives.

The research begins with a brief review of Green Sukuk and IoE before delving into the potential benefits of IoE towards empowering Green Sukuk ecosystem. It presents earlier research papers on how digitalisation and the Internet of Things (IoT), which is a component of IoE, are already being utilised to improve the Green Sukuk market. In doing so, it contributes significantly to continuing research on the critical role of Green Sukuk in achieving global sustainability goals. The sections that follow examine the theoretical underpinnings, methodology, and literature findings that underline the need of imbuing Green Sukuk with the IoE in order to effectively fulfil the ESGs goals. Finally, the study discusses the policy consequences and future research objectives.

2 Research Methodology

2.1 Methodology

The concept of Society 5.0 encompasses the establishment of a societal framework that places paramount importance on the welfare and interests of individuals, all the while harnessing cutting-edge technologies to attain ecological sustainability and efficient governance. The proliferation of Green Sukuk, a financial instrument dedicated to supporting environmentally conscious endeavours, hinges upon the existence of propitious legal, regulatory, social, and technological frameworks. The Internet of Everything (IoE), an all-encompassing concept that entails the seamless integration of individuals, objects, data, and operations, possesses the inherent capacity to enhance the intricate process of decision-making. This scholarly investigation undertakes a comprehensive examination of existing literature to assess the prospective capabilities of Internet of Things (IoE)-enabled Green Sukuk in achieving Environmental, Social, and Governance (ESG) objectives that align harmoniously with the fundamental tenets of Society 5.0.

The primary objective of this study is to conduct a comprehensive analysis of key concepts and explore the potential utilisation of cutting-edge Industry 4.0 technology, such as the integration of the IoE, within the Green sukuk ecosystem. This integration aims to ensure the fulfilment of ESG criteria and ultimately align with the overarching goals of Society 5.0. The forthcoming data shall be extracted from bibliographic databases in order to unveil the primary areas of inquiry in previous studies pertaining to the Green Sukuk ecosystem and the application of the IoE within the context of Islamic Finance.

The forthcoming analysis will expound upon the discoveries and proffer counsel and initiatives to be undertaken by relevant parties in order to guarantee the worldwide proliferation of Green Sukuk, thereby fostering Maqasid Syariah and societal welfare. Furthermore, it will delineate the trajectory for future research endeavours.

2.2 Literature Review

Green Sukuk industry is still in its infancy, thereby few literature reviews have been conducted worldwide and author found the same with regard to Society 5.0. This study attempted to use bibliographic database Scopus which yielded only two articles and even then it was on Islamic Finance and not Green Sukuk, whilst Google Scholar yielded only seven articles but even then, not on Green Sukuk which is the main focus in this chapter. Thus, authors decided to abandon performing SLR and opted for performing literature review in a comprehensive manner and focused on the keywords chosen.

2.2.1 Sukuk

The Islamic financial system uses sukuk, a phrase that means "a legal instrument, deed, or cheque" in Arabic (from the plural of "صك" (sakk)), as a certificate of financial worth, similar to conventional bonds. Islamic finance is based on the principle that all financial transactions must be free of interest (riba) and uncertainty (gharar). To meet this specific need, sukuk were established as a practical way to raise funds that adhere to Shari'a law. It is absolutely forbidden under Islamic law to charge or receive interest, or to participate in a transaction defined by extreme uncertainty and unclear results, sometimes known as "gharar" or gambling (El Shazly & Tripathy, 2013).

The Securities Commission of Malaysia (SC) and the respected Shariah Advisory Council (SAC) have agreed that sukuk, which are certificates of equal value, are proof of undivided ownership or investment in assets. Securities of uniform value representing individual ownership interests in a collection of present or future assets that fulfil the specified eligibility requirements are referred to as sukuk according to the Accounting and Auditing Organisation for Islamic Financial Institutions (AAOIFI) (2003: pp. 298). Scholarly works such as Ahmed and Elsayed (2019), Asutay and Hakim (2018), and Hassan et al. provide strong support for the previously given concept.

Therefore, sukuk are a type of Islamic financial instrument that meet the requirements of Islamic Shariah while also providing investors with the opportunity to earn a return on their capital. This works in a manner analogous to asset-backed securities. It can be structured in a way that yields a fixed rate of return, similar to the interest on a bond. In contrast to bonds, however, Sukuk holders actually get a piece of the pie when the assets or company become funded. Sukuk do not pay interest but instead depend on the value of the assets they are backed by.

Based on the country's Islamic finance ecosystem, regulatory climate, and the acceptance of Sharia-compliant services, Fitch Ratings predicts that Islamic banks will continue to outpace conventional banks in terms of growth. An all-time high of 13% was achieved in 2022 by Islamic financing, marking a significant growth from the previous 5 years. The sustained increase in working-capital loans and the stability of the household sector have been the main drivers of the robust economic recovery, which in turn has contributed to this expansion. Due to the aforementioned event, Malaysia is now the third biggest Islamic banking market in the world. From 38% at the end of 2021 to 41% at the end of 2022, the proportion of Islamic finance in the overall portfolio of loans made by the banking system rose steadily. With sukuk making up over 64% of the total local outstanding issuance as of November 2022, Malaysia's sukuk market is well-known worldwide for its long-standing reputation and substantial impact..

2.2.2 Green Sukuk

Emerging economies have experienced remarkable growth in green bond markets, providing funding for environmentally friendly infrastructure such as renewable energy, eco-friendly housing, and low-carbon transportation (CBI, 2018). Islamic scholars, financial practitioners, and environmental think tanks assert that Islamic finance aligns naturally with green finance, as it aims to direct investments towards environmental advantages. To maintain the natural balance of the world, it is essential to steer clear of extravagance, opulence, and dishonesty, as demanded by Wasatiyyah. Islam strictly prohibits activities that promote disorder and engage in unethical transactions such as interest/usury, uncertainty or deceitful contracts, and gambling (RFI, 2018).

This framework necessitates that enterprises meet the needs of individuals within a sustainable and efficient economy. Therefore, actions that result in chaos, especially harm to the environment, are considered unlawful (Obaidullah, 2018; RFI, 2018). According to Islamic scholars, Islamic finance is crucial in addressing climate change, environmental degradation, and humanitarian disaster, and in promoting environmental conservation and climate action (Obaidullah, 2018). Green sukuk offers a potential solution to the challenges faced by green finance, thanks to its unique structure and ability to incorporate economic, environmental, and Islamic principles.

Since 2016, the World Bank Group Knowledge and Research Hub, Bank Negara Malaysia, and the Securities Commission have collaborated to promote the Malaysia Green Finance Programme, focusing on the development of green Islamic finance markets. The main objective of this programme is to enhance funding for projects that promote environmental sustainability. In 2017, thanks to the programme, a groundbreaking event took place as the first Green Sukuk was issued. This initiative successfully raised an impressive US$ 59 million to support the development of a solar power plant in Malaysia. By launching innovative Green Sukuk projects, Malaysia has established itself as a trailblazer in Green Sukuk issuance (Liu & Lai, 2021). These efforts help direct sukuk investments towards addressing societal and environmental concerns. Green Sukuk are crucial for connecting infrastructure development with economically sustainable practices that prioritise environmental concerns in accordance with Shariah Law. Figure 2 illustrates the Green Sukuk ecosystem in Malaysia.

The World Bank states that the adoption of Green Sukuk can contribute significantly to achieving various Sustainable Development Goals. These goals include promoting good health and well-being, providing quality education, ensuring clean

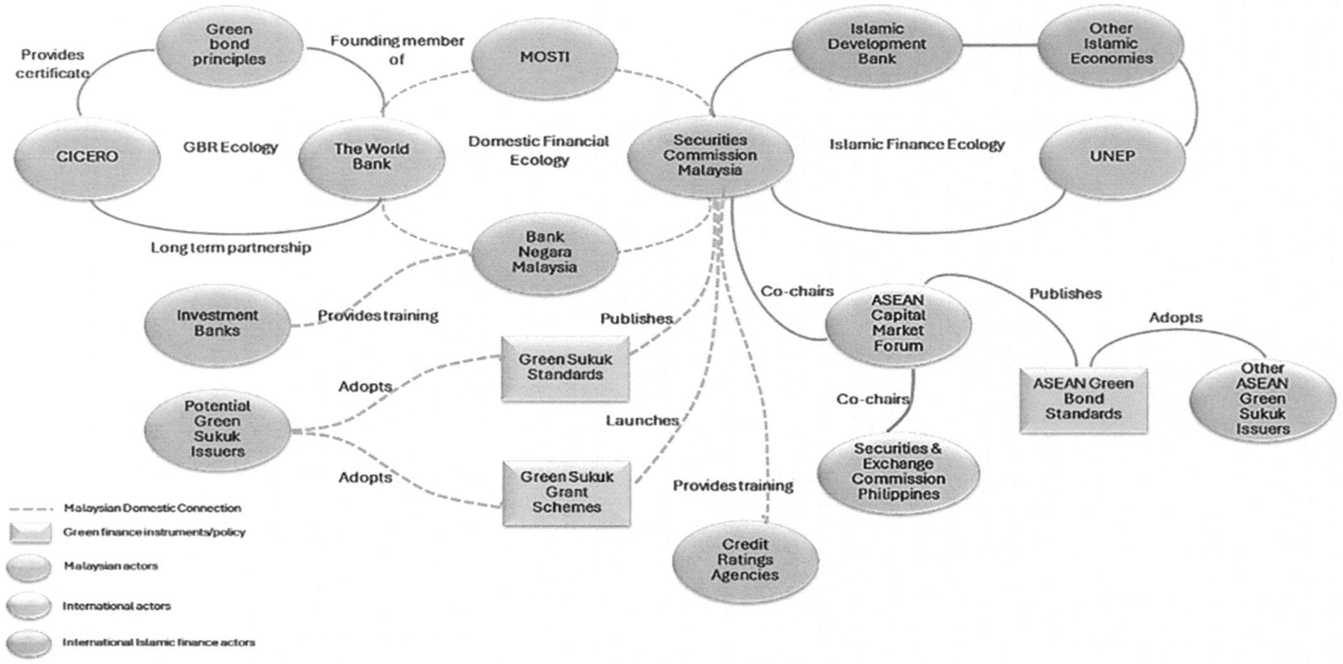

Fig. 2 The financial ecologies of green sukuk in Malaysia. Source: Adapted from Liu and Lai (2021)

water and infrastructure, promoting affordable and clean energy, fostering decent work and economic growth, emphasising industry, innovation, and infrastructure, creating sustainable cities and communities, advocating for responsible consumption and production, and addressing climate action (Araminta & Timur, 2022). Furthermore, Bank Negara Malaysia Assistant Governor, in a Bloomberg report (March 2022) reiterated that, "*We see Islamic finance with the potential to play a significant part in supporting the sustainable development goal (SDG) commitments. A large financing gap needs to be filled to realize the SDGs in developing nations, which is estimated to be between US$2.5 trillion to US$3 trillion annually. OIC countries alone need about US$1 trillion annually to implement the SDGs. In Malaysia, green and SRI sukuk issuances have served as a bridge between Islamic finance and sustainable investment. As of February 2021, a total of US$2 billion green and SRI sukuk has been issued since its launch in 2014*".

The report titled "The Global Green Bond Market: Size & Forecast with Impact Analysis of COVID-19 (2021–2025)" emphasises the growing global demand for green, social, and sustainability bonds. The increase in demand can be linked to a range of factors, including the rising importance of renewable energy, access to clean drinking water, and improved sanitation (Abdullah & Keshminder, 2022; Delle Foglie & Keshminder, 2022). In addition, there is a growing concern regarding CO2 emissions, a heightened awareness of forest conservation, and the rapid expansion of urban populations worldwide (Rahman et al., 2023). There has been a significant rise in Islamic green financing as more and more investors are choosing to invest in sustainable business models. Fitch Ratings reported significant growth in the second quarter of 2023 for ESG sukuk, reaching a total outstanding value of US$ 30.5 billion. Green Sukuk is gaining popularity as part of ESG Sukuk. There has been a significant increase of 22% compared to the previous quarter. According to Fitch Ratings 2023, the percentage of ESG sukuk is expected to exceed 7.5% of the total outstanding sukuk worldwide in the next 5 years. In the first half of 2023, it reached 3.8%. The growth we have observed follows a consistent pattern that was also observed in the previous 5-year period. Malaysia has established itself as a prominent player in sukuk issuance, consistently ranking among the top three countries worldwide.

2.2.3 Environmental, Social, and Governance (ESG) Concerns

The Environmental, Social, and Governance (ESG) framework serves as a comprehensive tool for evaluating an organisation's operational strategies and overall performance in relation to a wide range of sustainability and ethical concerns. Furthermore, it offers a means by which one can assess the potential hazards and prospects pertaining to various business domains. The primary function of an ESG programme is to uphold a sense of responsibility and oversee the execution of various frameworks and methodologies aimed at effectively managing a company's influence, encompassing critical aspects such as its carbon emissions and the manner in which it interacts with its workforce, suppliers, and other vested parties. ESG initiatives, in addition to their inherent value, serve as integral components of comprehensive business sustainability endeavours. These endeavours are designed to establish companies on a trajectory towards enduring prosperity, grounded in conscientious corporate governance and astute business strategies.

Green Sukuk perfectly correlates with the increasing focus on Environmental, Social, and Governance (ESG) standards in the financial industry. The concept of ESG, which focuses on a company's duty to promote social well-being and create sustainable wealth for stakeholders, has been increasingly popular due to changing societal demands regarding production and consumption practices (Jamali et al., 2017; Nishitani et al., 2021). Green Sukuk enables the funding of eco-friendly initiatives by integrating Environmental, Social, and Governance (ESG) concepts. It specifically tackles issues like curbing greenhouse gas emissions and fostering sustainable progress. The combination of Green Sukuk and ESG not only encourages ethical and responsible investment, but also supports the sustainable growth of the global economy (Li et al., 2021). ESG factors are now widely acknowledged as crucial for companies aiming to secure more affordable capital and improved corporate governance, demonstrating a broader dedication to generating beneficial social and environmental effects in the financial market.

2.2.4 Society 5.0 Lifestyle

Society 5.0 envisions a future innovative technology which seamlessly integrates into our daily lives, thereby foster a harmonious coexistence between humans and machines and ultimately improving human well-being. The Japanese were the first to propose the idea of incorporating modern analytical technological innovations like Big Data analytics, Artificial Intelligence (AI), and the Internet of Things (IoT) into various industries and societal segments. It represents a significant upgrade from the current societal structure, promising to enhance human existence significantly.

Therefore, it is essential to pay close attention to the parts of the idea that concern the expansion of industry within a flexible framework. To solve societal problems while living in harmony with nature, Society 5.0 seeks to harness the power of digital change and promote economic growth (Fukuyama, 2018). Its unique selling points include diversity, decentralisation, resilience, sustainability, problem-solving, value creation, and environmental harmony. "A human-centered society that balances economic advancement with the resolution of social problems through a system highly integrating cyberspace and physical space" (Federation, 2016) describes Society 5.0, an environment modelled after Industry 4.0 that is focused on people. Unlike in earlier societal iterations (up to Society 4.0), members of Society 5.0 will be able to live according to their own set of beliefs and principles, unrestricted by a number of restrictions. In an effort to address the many challenges facing modern society, the idea of Society 5.0 has gained traction in recent years. By maximising the potential of information and communication technologies, it foresees a future super-smart society in which all members live luxurious, carefree lifestyles connected to both the digital and physical realms (Fig. 3). There is a strong connection

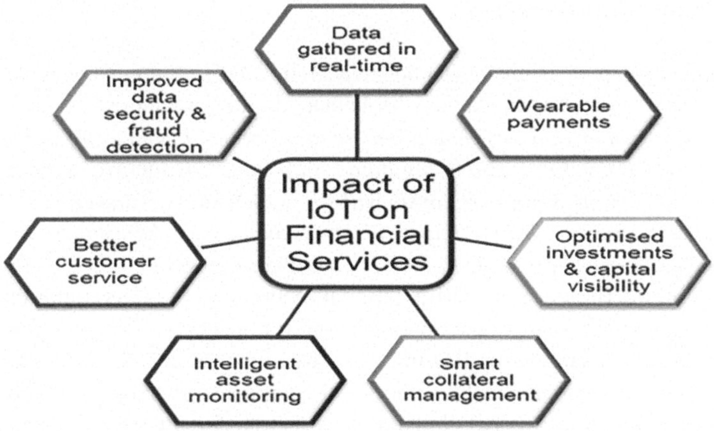

Fig. 3 Top applications of IoT in the financial service industry. Source: Adapted from Barrett et al. (2015); Maiti and Ghosh (2021)

between this idea and the belief that promoting harmonious interactions between humans and technology is essential for long-term progress in many areas of society, such as healthcare, education, democracy, and the economy.

In order to improve people's lives, Society 5.0 plans for the seamless integration of technical advancements like the Internet of Things (IoT), Big Data analytics, and Artificial Intelligence (AI) into everyday life's essential components. By providing extra value, this integration aims to strengthen the supply chain and pave the way for the formation of Society 5.0. Using these methods can simplify market scenario analysis, says Morella (2021). Society 5.0 involves the collection and storage of massive amounts of sensor data in cyberspace. In this setting, AI is about to revolutionise the integration of IoT and Big Data, giving rise to a new kind of intelligence that will reach every corner of society. By supplying people with the commodities and services they need at the exact moment they need them, this evolution seeks to improve people's lives by making them more sustainable and comfortable. By sifting through this mountain of data stored in cyberspace, AI provides humans in physical space with a variety of forms of feedback. Morella (2021) asserts that improving the efficiency of the global food supply chain is highly dependent on stakeholders and market conditions. Economic development would result from attaining Society 5.0, which centres on people and emphasises the role of technology in effectively tackling societal problems and raising the general level of living (Carayannis et al., 2022).

Modifying technical resources and bringing new adjustments to raise society's standard of living is the main goal of Society 5.0. According to this theory, happiness and long-term sustainability should be the top priorities (Gladden, 2019). Society 5.0, the domain of social and cultural shifts, is intrinsically related to the Economy 5.0 economic development. Thus, Society 5.0 can be seen as a meeting point for digitisation, technological progress, and information flow; it aims to promote economic and social sustainability. At its core, Society 5.0 is defined by the traits of a smart society that has undergone digital transformation (Sharp, 2020).

2.2.5 Innovative Technology

According to report from Bloomberg (March 2022), Bank Negara Malaysia Assistant Governor, Adnan Zaylani Mohamad Zahid was quoted as saying that, "*Islamic financial institutions should step up the adoption of emerging technologies such as artificial intelligence and big data to tap into new opportunities in underserved markets. Bank Negara Malaysia will strengthen key infrastructures to ensure the ecosystem remains vibrant for digitalization, including enhancements to regulatory sandbox to accelerate time to market for digital innovation*".

According to Barrett et al. (2015), digital technology presents opportunities at every stage of value creation and appropriation. It impacts both the operational and strategic levels of a company, allowing it to create new value propositions and fulfil its purpose. In modern financial institutions, the impact of technology affects a multitude of operational facets, including information transfer, processing, and management across multiple layers (Fig. 3). The integration of devices via the IoT enables smooth functioning and the exchange of data (Maiti & Ghosh, 2021).

Hyperconnectivity, as defined by Fredette et al. (2012), is the expansion of the IoT concept into the Internet of Everything through a variety of constantly available, readily accessible, information-rich, and interactive communication and interaction methods. An expanded version of the IoT, the concept of the IoE links data, people, and business processes together (Charmonman & Mongkhonvanit, 2015). Hence, it incorporates supplementary connectivity-based frameworks like the Industrial Internet (II), the IoT, and the Internet of People (Di Martino et al., 2018). In this view, the term "Internet of Everything" (IoE) refers to a vast system of networked computers, phones, tablets, and other electronic devices that allows for the instantaneous transfer of information and data. According to Bandara and Ioras (2016) and Langley (2021), the IoE serves as a means of connecting individuals, objects, data, and processes in order to enhance user experiences and facilitate intelligent decision-making (Fig. 4). With the rise of digital platforms, distant work and communication have become more accessible, as seen by the current epidemic. In his study, Murali Raman predicted that the metaverse, when properly integrated into business strategies and social principles, will have a lifespan similar to that of the "dot-com to dot-gone" e-commerce phenomenon. To achieve comprehensive connection, expanded intelligence, and advanced cognition, the IoE must be implemented.

This is necessary since the Islamic Finance ecosystem integrates multiple components, such as the Green Sukuk ecosystem, the interactive and interconnected aspect of IoE holds great potential for addressing numerous global restrictions. So, it's perfect for protecting Green Sukuk ventures from fraud. Consequently, this promotes an environment in the financial sector that is defined by transparency and responsibility. A paradigm shift has occurred with the introduction of the IoE into the Green Sukuk ecosystem, which provides a flexible framework for the efficient monitoring, evaluation, and improvement of ecologically friendly projects.

Green Sukuk involves numerous domestic and international entities, including regulators, which increases the urgency of finding creative uses for technology, even

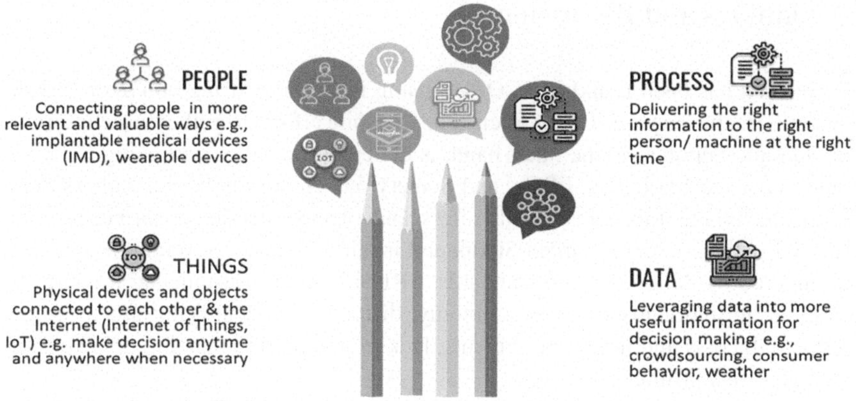

Fig. 4 Pillars of IoE. Source: Adapted from Mohanty et al. (2020)

if financial institutions already know how important it is to maintain a healthy macro, meso, and micro ecosystem. The integration of cutting-edge technologies like blockchains (BC), distributed ledger technology (DLT), Artifical Intelligence (AI), big data analytics (BDA), cloud computing (CC), and the Internet of Things (IoT) system would greatly benefit the Green Sukuk ecosystem. The issuing of Green Sukuk and ESG compliance are both made possible by AI. To guarantee ESG and regulatory compliance in sukuk terms, AI algorithms can automate the development of smart contracts.

Hassan et al. (2023) state that AI can sift through mountains of data in search of ESG problems linked to underlying projects. This helps with decision-making and increases risks mitigation. Investors can achieve their ESG goals and make a positive impact on sustainable development with the help of these platforms' AI-powered personalised investment recommendations. Green sukuk investments and their ESG impacts can be better assessed with the use of DLT, a kind of blockchain that improves the visibility and auditability of Green Sukuk transactions and monitors the progress and results of projects in real time. CC streamlines communication and centralises data, while BDA give more precise insights (Mohbey & Kumar, 2022).

However, implementing IoE also presents significant challenges for firms. These include, but are not limited to, developing system-to-system interoperability, dealing with long-standing industry partners who refuse to work with new innovations, dealing with path-dependent legacy processes and transactions, issues with liability and contracts, security concerns, loss of control, and privacy issues stemming from the deluge of data collected and utilised by businesses and their smart devices. This is why it's critical for companies to study how the Internet of Things (IoT) will influence value generation in service ecosystems and how smart things will change current business models (Langley et al., 2021). The setup and capabilities of the smart items would have a significant impact on this.

3 Findings and Discussion

No research has been found in SCOPUS using this search string on Green Sukuk, innovative technology, ESGs, or Society 5.0, despite efforts to gather more information. Google Scholar, on the other hand, is deficient in studies pertaining to Green Sukuk, novel technologies, ESGs, and Society 5.0. Accordingly, the authors draw the conclusion that this field of study offers a novel angle that has so far been underexplored in the context of Green Sukuk and Islamic Finance. Several obstacles are encountered by ASEAN green standards, SRI Sukuk, and sustainable investment, including an unorganised green taxonomy, difficulty in identifying green project assets, high costs and time requirements, lack of compelling benefit, and exposure to a higher risk profile.

With the Internet of Everything (IoE) seamlessly integrated into the Green Sukuk ecosystem, various integrity and communication issues in Islamic finance, like the timely sharing of data and information among diverse stakeholders, might be resolved. A network of physical items that can connect to each other and exchange data automatically, without human intervention, is called the IoT (Shafiq et al., 2022). This network is made possible by sensors, software, and other technological components. Using data collected in real time from the IoT makes it easier to run business operations from afar. The benefit of Machine-to-Machine (M2M) communication is the effective transmission of data packets via a network, which leads to time and cost savings, when information is exchanged among IoT devices. In contrast to the People-to-Things (P2T) and Things-to-Things (T2T) facets of the IoT, the IoE refers to the interconnection of all entities, including data, processes, and physical objects. By creating seamless connections among the four cornerstones of the IoE, as shown in Fig. 5, the author has suggested a simplified connection paradigm for the IoE, which has the ability to improve both individual experiences and decision-making.

The Islamic Finance ecosystem, which encompasses Green Sukuk, is defined by its extensive range, encompassing both domestic and international companies. Furthermore, it functions inside a regulatory framework supervised by pertinent institutions, guaranteeing that transactions and communications comply with the norms and standards mandated by Shariah Law. However, integrating the Internet of Everything (IoE) and other advanced IR4.0 technologies into the Islamic Finance ecosystem would ensure the efficient and reliable execution of these processes, thereby improving the Green Sukuk ecosystem and facilitating the attainment of Environment Sustainable Governance (ESGs) and Society 5.0 goals.

4 Conclusion and Implications

By including the perspective of Environmental, Social, and Governance (ESG) factors and the concept of Society 5.0, the argument for using technology in the Green Sukuk ecosystem becomes even more compelling.

Alignment with ESG Principles: The utilisation of data analytics and technology may enhance the efficiency of resource allocation, reduce unnecessary waste, and

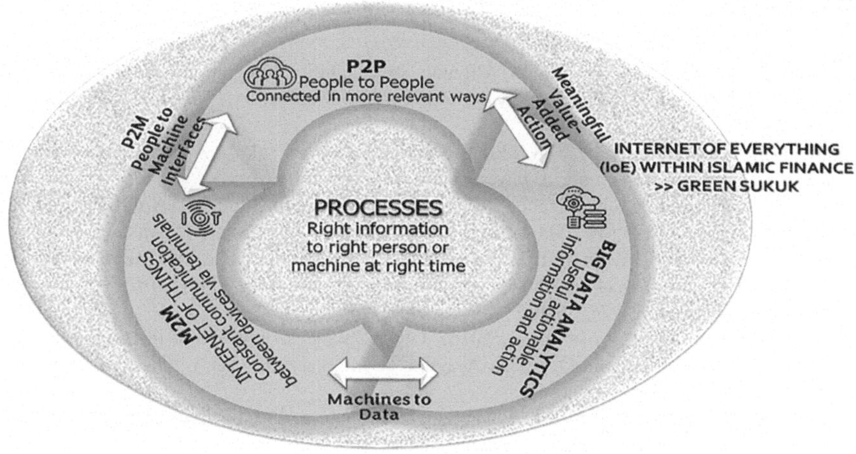

Fig. 5 Proposed model of internet of everything incorporated into Islamic finance. Source: Authors own

monitor the ecological consequences in real time. This, in turn, contributes to advancing Green Sukuk initiatives by promoting environmental sustainability, which is a fundamental aspect of ESG principles. Moreover, the utilisation of technology can enhance transparency and inclusivity, thereby empowering all participants involved in the Green Sukuk value chain. This, in turn, can foster equitable labour practices, community involvement, and responsible economic growth, aligning with the social aspects of ESG principles. Automated compliance checks and safe data management using blockchain and AI may effectively maintain ethical standards, address greenwashing, and improve regulatory compliance. This will boost governance practices within the Green Sukuk ecosystem, aligning with the governance aspect of ESG.

Embracing Society 5.0 Aspirations: Adhering to ethical and responsible use of technology, while safeguarding data privacy and security, is in line with the goal of Society 5.0, which aims for human-centred development by utilising technology to empower individuals and enhance well-being. The utilisation of advanced analytics and artificial intelligence can harness the full potential of Big Data to enhance intelligent decision-making across the whole lifecycle of Green Sukuk. This, in turn, contributes to the realisation of a highly intelligent society as envisioned by Society 5.0. The utilisation of technology enables data-driven insights and transparent governance, which can facilitate the emergence of Green Sukuk as a significant catalyst for sustainable economic growth. This aligns with the overarching objectives of Society 5.0 in the long run.

Through the adoption of technology and adherence to ESG and Society 5.0 principles, Green Sukuk has the potential to evolve from a financial tool into a potent driver for a future that is more sustainable, fair, and prosperous. Collectively, these influences have the potential to shape a future in which the preservation of the environment, societal obligations, and advancements in technology are interconnected, leading us towards a more sustainable and equitable global society.

References

Aassouli, D., & Shah, M. E. (2022). Aligning sustainability ıntegration, digitalization and ınclusivity for green, sustainable recovery. *Journal of Infrastructure Policy and Management, 5*(1), 39–47.

Abdullah, M. S., & Keshminder, J. S. (2022). What drives green sukuk? A leader's perspective. *Journal of Sustainable Finance & Investment, 12*(3), 985–1005.

Accounting and Auditing Organization for Islamic Financial Institutions (AAOIFI). (2003). Shari'a Standards, 1424-5H/2003-4, Shari'a Standard No. 17. Investment Sukuk, (p. 298).

Ahmed, H., & Elsayed, A. H. (2019). Are Islamic and conventional capital markets decoupled? Evidence from stock and bonds/sukuk markets in Malaysia. *The Quarterly Review of Economics and Finance, 74*, 56–66.

Alam, N., Duygun, M., & Ariss, R. T. (2016). Green sukuk: an innovation in Islamic capital markets. In A. B. Dorsman, Ö. Arslan-Ayaydin, & M. B. Karan (Eds.), *Energy and finance* (pp. 167–186). Springer.

Al Ansari, R., & Alanzarouti, F. (2020). ESG and Islamic finance: An ethical bridge built on shared values. *Journal of Islamic Financial Studies, 6*(1).

Araminta, D. V., & Timur, Y. P. (2022). The role of green sukuk in realizing the sustainable development goals 2030 agenda. *Jurnal Ekonomi & Bisnis Islam, 8*(2).

Asutay, M., & Hakim, A. (2018). Exploring international economic integration through sukuk market connectivity: A network perspective. *Research in International Business and Finance, 46*, 77–94.

Bandara, I., & Ioras, F. (2016). The evolving challenges of internet of everything: enhancing student performance and employability in higher education. In *INTED2016 proceedings* (pp. 652–658). IATED (pdf).

Barrett, M., Davidson, E., Prabhu, J., & Vargo, S. L. (2015). Service innovation in the digital age. *MIS Quarterly, 39*(1), 135–154.

Bloomberg. (2022, March). Malaysia seeks sustainable lead to drive Islamic finance growth. Retrieved from: https://www.bloomberg.com/news/articles/2022-03-24/malaysia-seeks-sustainable-lead-to-drive-islamic-finance-growth?embedded-checkout=true

Carayannis, E. G., Dezi, L., & Gregori, G. (2022). Smart environments and techno-centric and human-centric innovations for industry and society 5.0: A quintuple helix innovation system view towards smart, sustainable, and inclusive solutions. *Journal of Knowledge Economics, 13*, 926–955.

Charmonman, S., & Mongkhonvanit, P. (2015). Special consideration for big data in IoE or internet of everything. In *2015 13th international conference on ICT and knowledge engineering (ICT & knowledge engineering 2015)* (pp. 147–150). IEEE.

Climate Bond Initiative (CBI). (2018). Bonds and climate change: state of the market 2018. Available at: https://www.climatebonds.net/resources/reports/green-bonds-state-market-2018

Delle Foglie, A., & Keshminder, J. S. (2022). Challenges and opportunities of SRI sukuk toward financial system sustainability: a bibliometric and systematic literature review. *International Journal of Emerging Markets*. https://doi.org/10.1108/IJOEM-04-2022-0601

Di Martino, B., Li, K. C., Yang, L. T., & Esposito, A. (2018). Trends and strategic researches in internet of everything. In *Internet of everything: Algorithms, methodologies, technologies and perspectives* (pp. 1–12). Springer.

El Shazly, M. R., & Tripathy, P. (2013). Sukuk structures, profiles and risks. *Paper of Columbia College* (pp. 1–18). (pdf)

Erboz, G. (2017). How to define industry 4.0: Main pillars of industry 4.0. In *Managerial trends in the development of enterprises in globalization era* (pp. 761–767).

Federation, K. J. B. (2016). Toward realization of the new economy and society—Reform of the economy and society by the deepening of "Society 5.0". Retrieved January 7, 2024, from https://www.keidanren.or.jp/en/policy/2016/029_outline.pdf.

Fitch Ratings. *ESG-Sukuk strengthening as pivot towards sustainability grows*. Retrieved from https://www.fitchratings.com/research/non-bank-financial-institutions/esg-sukuk-strengthening-as-pivot-towards-sustainability-grows-08-08-2023

Fitrah, R., & Soemitra, A. (2022). Green Sukuk for sustainable development goals in Indonesia: a literature study. *Jurnal Ilmiah Ekonomi Islam, 8*(1), 231–240.

Fredette, J., Marom, R., Steiner, K., & Witters, L. (2012). The promise and peril of hyperconnectivity for organizations and societies. *The Global Information Technology Report, 2012*, 113–119.

Fukuyama, M. (2018). Society 5.0: Aiming for a new human-centered society. *Japan Spotlight, 27*(5), 47–50.

Gladden, M. E. (2019). Who will be the members of Society 5.0? Towards an anthropology of technologically posthumanized future societies. *Social Sciences, 8*, 148.

Hassan, P., Passing, F., & Goméz, J. M. (2023). ESG fingerprint: How big data and artificial intelligence can support investors, companies, and stakeholders? In R. Schmidpeter & R. Altenburger (Eds.), *Responsible artificial intelligence. CSR, sustainability, ethics & governance*. Springer. https://doi.org/10.1007/978-3-031-09245-9_11

Jamali, D., Karam, C., Yin, J., & Soundararajan, V. (2017). CSR logics in developing countries: Translation, adaptation and stalled development. *Journal of World Business, 52*(3), 343–359.

Langley, D. J., Doorn, J. V., Ng, I. C. L., Stieglitz, S., Lazovik, A., & Boonstra, A. (2021). The internet of everything: Smart things and their impact on business models. *Journal of Business Research, 122*, 853–863. ISSN 0148-2963. Retrieved from: https://doi.org/10.1016/j.jbusres.2019.12.035.

Li, T. T., Wang, K., Sueyoshi, T., & Wang, D. D. (2021). ESG: Research progress and future prospects. *Sustainability, 13*(21), 11663.

Liu, F. H., & Lai, K. P. (2021). Ecologies of green finance: Green sukuk and development of green Islamic finance in Malaysia. *Environment and Planning A: Economy and Space, 53*(8), 1896–1914.

Maiti, M., & Ghosh, U. (2021). Next generation internet of things in fintech ecosystem. *IEEE Internet Things Journal*. https://doi.org/10.1109/JIOT.2021.3063494

Ma'ruf, A., Mahomed, Z., & Mohamad, S. (2021). Sustainable finance and a Sharī' analysis of environmental, social and governance (ESG) criteria. In M. K. Hassan, M. Saraç, & A. Khan (Eds.), *Islamic finance and sustainable development: A sustainable economic framework for Muslim and non-Muslim countries* (pp. 193–217). Springer.

Masood, A., Rahim, A. A., & Lee, U. H. M. S. (2023). Harnessing internet of everything (IoE) for sustainability of halal cosmetics ecosystem. In A. Rafiki (Ed.), *Digitalization in halal management* (pp. 29–45). Springer.

Mohanty, S. P., Yanambaka, V. P., Kougianos, E., & Puthal, D. (2020). PUFchain: A Hardware-Assisted Blockchain for Sustainable Simultaneous Device and Data Security in the Internet of Everything (IoE). *IEEE Consumer Electronics Magazine, 9*(2), 8–16. https://doi.org/10.1109/MCE.2019.2953758

Mohbey, K. K., & Kumar, S. (2022). The impact of big data in predictive analytics towards technological development in cloud computing. *International Journal of Engineering Systems Modelling and Simulation, 13*(1), 61–75.

Morella, P., Lambán, M. P., Royo, J., & Sánchez, J. C. (2021). Study and analysis of the implementation of 4.0 technologies in the agri-food supply chain: A state of the art. *Agronomy, 11*(12), 2526.

Nishitani, K., Nguyen, T. B. H., Trinh, T. Q., Wu, Q., & Kokubu, K. (2021). Are corporate environmental activities to meet sustainable development goals (SDGs) simply greenwashing? An empirical study of environmental management control systems in Vietnamese companies from the stakeholder management perspective. *Journal of Environmental Management, 296*. https://doi.org/10.1016/j.jenvman.2021.113364

Obaidullah, M. (2018). Managing climate change: The role of Islamic finance. *IES Journal Article, 26*(1).

Rahman, A., Islam, J., Kundu, D., Karim, R., Rahman, Z., Band, S. S., & Kumar, N. (2023). Impacts of blockchain in software-defined internet of things ecosystem with Network Function

Virtualization for smart applications: Present perspectives and future directions. *International Journal of Communication Systems*. https://doi.org/10.1002/dac.5429
Rejikumar, G., Sreedharan, V., Arunprasad, P., Persis, J., & Sreeraj, K. M. (2019). Industry 4.0: key findings and analysis from the literature arena. *Benchmarking: An International Journal, 26*(8), 2514–2542. https://doi.org/10.1108/BIJ-09-2018-0281
Responsible Finance Institute (RFI). (2018). *Environmental impact in Islamic finance*. Available at: http://report.rfi-foundation.org/download.html
Saba, I., Khan, A., & Jawed, H. (2021). Islamic finance and SDGs: Connecting dots. In M. K. Hassan, M. Saraç, & A. Khan (Eds.), *Islamic finance and sustainable development: A sustainable economic framework for Muslim and non-Muslim countries* (pp. 55–76). Springer.
Shafiq, M., Gu, Z., Cheikhrouhou, O., Alhakami, W., & Hamam, H. (2022). The rise of "internet of things": Review and open research issues related to detection and prevention of IoT-based security attacks. *Wireless Communications and Mobile Computing*. https://doi.org/10.1155/2022/8669348. Article ID 8669348, 12 pages.
Sharp, L. (2020). Society 5.0: A brave new world. *Impact,* 2–3.

Adlin Masood is a Senior Fellow in the Faculty of Economics and Muamalat, Universiti Sains Islam Malaysia. Having proven dynamic 15 years track record in the hospitality industry with special focus in the field of marketing, public relations and communications has given Adlin a totally different perspective and outlook in the academic world. This has ultimately resulted towards a practitioner's approach in educating students, research, and consultancy. Preferring to look at the dynamics and framework of halal systems in various fields in research, from marketing extending to other functional business areas, even including education through experiential learning. Currently involved in simulation game innovation for halal training and consultancy in Taiwan and Zhengzhou, China and exploring other avenues with network in Korea and Uzbekistan. Other areas of interest include SDGs and ESG of Asia Pacific nations through halal economy and green sukuk as well as application of IR4.0 technologies.

Norfhadzilahwati Rahim is a Senior Lecturer in the Faculty of Economics and Muamalat at Universiti Sains Islam Malaysia (USIM). She received her Master of Science (Finance) from Universiti Utara Malaysia (UUM) and a Ph.D. in Economics and Muamalat Administration from Universiti Sains Islam Malaysia. Her primary research interests include financial management, corporate finance, and Islamic finance perspectives. She taught at UNITEN for 7 years and served on several organising committees, including symposiums, conferences, and seminars. She worked on various finance research projects at UNITEN as a researcher/consultant. She and her colleagues were awarded a silver medal in the poster competition at the International Invention, Innovation, and Technology Exhibition (ITEX) 2014 and the National Symposium and Exhibition on Business and Accounting (NSEBA) 2014. She has held several positions at UNITEN, including Head of Unit (Unit Islamic Finance), Department of Finance and Economics, and Head of Academic, College of Business Management and Accounting.

Umi Hamidaton Mohd Soffian Lee is a Senior Lecturer in the Faculty of Economics & Muamalat at Universiti Sains Islam Malaysia (USIM) and possesses a bachelor's and master's degree in human science from IIUM, as well as a Ph.D. in Islamic Civilization from UKM. She is an accomplished author, having written a book on ethical leadership, multiple articles in scholarly publications, and book chapters, as well as presenting talks at seminars and conferences in her subject field. Appointed as the halal programme coordinator at the Graduate School of Muamalat (USIM) and Research Fellow at the Institute of Fatwa and Halal (IFFAH). Dr. Umi is also a journal and research grant reviewer, a guest lecturer both locally and abroad, and a primary investigator and member of numerous research programmes and initiatives. Her research interests are mostly in management, with an emphasis on organisational management and behaviour, the halal business, and Islamic philanthropy.

Strategies and Applications of Blockchain Technology of Zakat Collection and Distribution in Malaysia and Indonesia

Khairil Faizal Khairi, Ahmad Rafiki, Sari Nuzullina Rahmadhani, and Rana Fathinah Ananda

Abstract This chapter aims to elaborate and discuss on application of blockchain technology of zakat collection and distribution in Malaysia and Indonesia. The study adopts content analysis technique where various sources are referred including research database named Scopus, reports, official documents, and research articles. The chapter also explains on the development, the strength of zakat collection using blockchain, the research of zakat practices in Malaysia and Indonesia. The challenges and other issues on the adoption of blockchain technology in zakat are explained. It is expected that the blockchain technology in zakat will give more solutions, bring efficiency and transparency, which affect the national economy. The information revealed in this study could be used as a reference for future studies. Additionally, it is important to be discussed by practitioners, academicians, and other stakeholders, which assumed will be impacted to the economic development of the countries.

1 Introduction

Blockchain is a digital transaction record based on its structure where individual records called blocks are linked together in a registration called a chain. Islamic social finance and Fintech (digital finance) offer new opportunities for Muslims around the world to explore these technological fields in an Islamic context, especially fields related to the use of blockchain. The use of Blockchain is not only limited to digital finance, but extends to technology that requires transparency in

K. F. Khairi
Universiti Sains Islam Malaysia, Nilai, Negeri Sembilan, Malaysia
e-mail: khairil@usim.edu.my

A. Rafiki (✉) · S. N. Rahmadhani · R. F. Ananda
Universitas Medan Area, Medan, Indonesia

A. Rafiki et al. (eds.), *Strategic Islamic Business and Management*, Contributions to Management Science,
https://doi.org/10.1007/978-3-031-61778-2_2

transactions, where this technology offers high security in data storage. This can fulfill the need to provide confidence in Islamic finance by providing a more transparent, trustworthy, and reliable method. Several technology experts have expressed the opinion that future transactions will not require a third party and transactions will be carried out more quickly. A PwC report published in 2022 showed that 84% of 600 executives across companies agreed that blockchain technology is more reliable because it offers data security, transparency, and traceability to their business processes.

It is also expected that more than 3 trillion dollars of business in 2030 will be blockchain technology-based businesses. Parts of Islamic Social Finance, for example Zakat, will benefit from blockchain technology because it has end-to-end transaction characteristics that have data fragmentation and functions as a gatekeeper for automatic data access. Future zakat transactions will be able to guarantee traceability, transparancy, and trust in zakat institutions and will subsequently be able to increase the amount of zakat collection at the state and national levels. This is the first system to use a blockchain platform regarding the collection and distribution of zakat funds where zakat payers can track the flow of zakat to recipients transparently.

Blockchain technology has the potential to significantly impact zakat management by enhancing transparency, efficiency, and trust in the collection and distribution of zakat (Nordin et al., 2021). Several studies and articles discuss the benefits of integrating blockchain into zakat management, highlighting its ability to provide a transparent and reliable environment for zakat transactions, as well as to increase the effectiveness and trust in zakat collection (Beik et al., 2021). The use of smart contracts attached to zakat transactions can guarantee effectiveness, transparency, and trust, leading to increased zakat collection at the state and country level (Owoyemi, 2020). Additionally, blockchain technology offers opportunities for zakat institutions to re-establish trust in the sector, reduce transaction costs, and provide transparency to zakat payers, ultimately leading to increased support and collection (Shaukat & Zhu, 2021). Furthermore, the use of blockchain technology in zakat management is seen as a way to overcome trust issues, increase transparency, and make the zakat collection system more efficient and accessible (Bin-Nashwan et al., 2021). While the potential benefits are evident, the adoption of blockchain in zakat management may face challenges such as technical knowledge and budget constraints (Saad et al., 2023). Overall, the application of blockchain technology in zakat management is believed to be necessary for improving capital optimization, transactional transparency, and collaboration in socio-economic development, ultimately promoting the Sustainable Development Goals (SDGs) of a country (Mohammed et al., 2021a, 2021b).

The main objective of this research is: "To analyze the effectiveness factors of zakat collection and distribution using blockchain in Malaysia and Indonesia." The results of the analysis found will later be socialized and discussed with related parties so that zakat collection and distribution is more effective using blockchain.

It is hoped that this research will increase knowledge regarding the development of the zakat sector, especially in the use of blockchain technology and provide opportunities for academic researchers to conduct further studies. The use of blockchain technology can increase public access to the zakat system, ensure zakat is distributed to those in need more effectively, and raise awareness and education among the public about the importance of zakat and blockchain technology. The government can carry out better monitoring and supervision by implementing blockchain technology to monitor and supervise the distribution of zakat more efficiently and transparently.

Meanwhile, innovation in zakat management is needed by applying blockchain in the collection, management, and distribution of zakat, increasing efficiency and transparency and can encourage the growth of the zakat services sector and blockchain technology in the sharia financial industry.

2 Literature Review

2.1 Development of Zakat

The potential for national zakat funds is estimated at hundreds of trillions of rupiah. Even though the potential is large, the funds raised to date have not been optimally realized. Therefore, public trust in zakat management institutions needs to be increased by, among other things, transparent and targeted management. According to BAZNAS data in 2022, zakat, infaq and alms (ZIS) funds collected nationally amounted to IDR 22.43 trillion. The role of zakat can contribute greatly to efforts to overcome poverty and improve community welfare. But not according to expectations, from such a large potential, the zakat collected is only around 4% of the potential zakat available. The following is the national zakat collection data in Indonesia for 2013–2023 and Malaysia in 2015–2022 (Fig. 1 and Table 1):

For muzakki, having a BAZ or LAZ will help distribute the zakat that must be paid to mustahiq more easily. However, some muzakki (zakat obligators) still doubt the existence of BAZ or LAZ, in terms of distributing zakat to those entitled to it, in addition to the many muzakki's desire to give zakat directly to those entitled to it. This shows that the majority of muzakki still want better zakat management, namely that zakat management must have professionalism, transparency in reporting, and distribution that is right on target.

In this accountability issue, zakat management institutions are strongly required to be accountable regarding zakat management. The issue of accountability is related to how the OPZ is able to provide administrative reports, collection and distribution of zakat that are accountable and managed by professional responsible persons (Al-Bawwab, 2023). The curisiosity toward public sector organizations is more caused by the information gap between management who have direct access to information and institutions or the public who are outside management (Owoyemi, 2020).

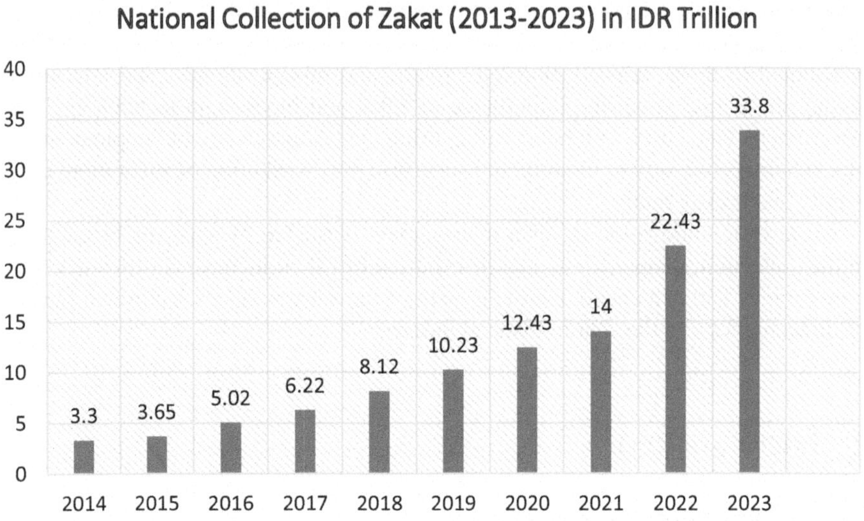

Fig. 1 National Collection of Zakat (2013–2023). Source: Badan Amil Zakat Nasional (2023)

Table 1 Zakat collection in Malaysia 2015–2022

	2015	2016	2017	2018	2019	2020	2021	2022
Zakat collection (in million)	2483.38	2846.56	2923.1	2854.82	3175.72	3365.55	3164.52	4042.69

Source: PPZ-MAIWP Malaysia (2023)

The lack of synergy between zakat managers is very visible in the lack of cooperation between BAZNAS and LAZ (Aziz et al., 2020). The cause is the egoism that appears on both sides of the zakat management. These conditions should not occur considering that all zakat managers are essentially institutions that are oriented toward the benefit of the people, especially muzakki and mustahiq.

The concepts of accountability and accessibility occupy very important criteria related to the organization's responsibility in presenting, reporting, and disclosing all activities as well as the extent to which financial reports contain all the relevant information needed by users and how easily this information can be accessed by the public (Omar & Khairi, 2021).

With the growing evolution of Islamic finance over the last few decades, zakat has taken part in the practical side by being present through a number of zakat management institutions in both countries. Very similar to ordinary charities from an operational and systematic point of view, these institutions contribute greatly to improving social welfare and alleviating poverty (Akhyar Adnan & Barizah Abu Bakar, 2009). In fact, they aim to develop a system of collecting alms from individuals who have fulfilled the requirements of the obligation to pay zakat and distributing the collected donations to poor people and those who fall into the zakat category (Masarif al-zakat) of society who need social support (Alshater et al., 2021).

Fintech is taken from the abbreviation used to refer to the use of technology in the current financial system. This seems to be developing very rapidly at the moment and is considered a determining factor in the success of the Industrial Revolution 4.0. Among the services involving financial institutions that have gone through digital systems are payment systems, e-wallets, cryptocurrency, crowdfunding, blockchain, artificial intelligence (AI), and so on (Zulfikri et al., 2023). A lot of data shows an increase every year and this illustrates that the digitalization of the previously more traditional financial system is starting to be implemented in Indonesia and Malaysia (Mawardi et al., 2023).

Islamic social finance (waqf, zakat, baitulmal, and infaq) and its digital economy offer opportunities for Muslims and non-Muslims to build transparency, trust, and confidence in the financial system (Hudaefi et al., 2022). The tech world is embracing IR 4.0 with a vision of a world without intermediaries, while the rest of the tech world simply wants faster and more efficient transaction methods. Zakat institutions as important institutions that manage the collection and distribution of zakat money in Indonesia and Malaysia must take the opportunity to explore and adapt Fintech in existing zakat management.

Zakat institutions have benefited from blockchain technology because it provides a peer-to-peer transaction system that is based on a fully distributed system through autonomous processes, especially the implementation of smart contracts. Blockchain is essentially a database record or logbook for all transactions or digital events that have been implemented and shared among participants. Each ledger transaction will be confirmed with the approval of the majority of system participants (Nazeri et al., 2023). Once the data or information is entered, it cannot be deleted. Blockchain contains specific records of every transaction that has been carried out. Through the blockchain system, participants do not need to know or trust each other to interact where electronic transactions can be verified automatically and recorded by network nodes via cryptographic algorithms, without human intervention, centralized authorities, third parties (e.g., government, banks, financial institutions, or others). The use of a zakat collection blockchain platform can increase the transparency of the collection process where all zakat collection transactions can be followed and investigated (Mawardi et al., 2023; Mohammed et al., 2021a, 2021b).

Paying zakat is one of the obligations of Muslims, especially toward the end of Ramadan. As the largest Muslim country in the world, Indonesia also collects a fantastic amount of zakat every year. The National Zakat Amil Agency (Baznas) noted that the collection of zakat, infaq, alms (ZIS) and other religious social funds (DSKL) reached IDR 22.43 trillion in 2022. This value increased by 58.90% compared to the previous year. The increase mainly came from zakat payments on malls amounting to 22.11% and zakat on sacrificial animals of up to 400.95%. However, this realization has only reached 86.29% of the target set for 2022 of IDR 26 trillion.

2.2 *Strengths of the Zakat Collection Blockchain System*

In reporting accountability, the main principles that must be used are transparency and honesty. With this principle, zakat institutions strive to provide information on activity reports and reports on the collection and distribution of zakat funds in a clear, honest, and trustworthy manner (Shaukat & Zhu, 2021; Muneeza et al., 2022). Below are some of the important principles:

1. Transparency—The system developed provides robustness, traceability, and transparency of transactions so that it can increase the efficiency of BAZNAS in providing services to zakat payers and better guarantee the achievement of zakat goals through the use of this system.
2. Data and information security—Digital ledgers cannot be altered or modified once the data is entered. Dishonesty in reporting can be controlled and more easily seen. This can be aligned with existing policies at BAZNAS through the ICT Governance Unit that has been formed. Apart from that, BAZNAS can convey information related to zakat collection easily and can be accessed anywhere.
3. Real-time information—When information is updated, it is updated for everyone on the network at the same time and is convenient for monitoring purposes. As a result, searching for the data needed at a certain time becomes faster, easier, and more accurate, thereby saving BAZNAS time and costs (Figs. 2 and 3).

However, there are legal implications of using blockchain technology for zakat management, which include (Nazeri & Nor, 2023):

1. *Regulatory Compliance:* Organizations must comply with local and international laws and regulations governing the use of blockchain technology in zakat management.

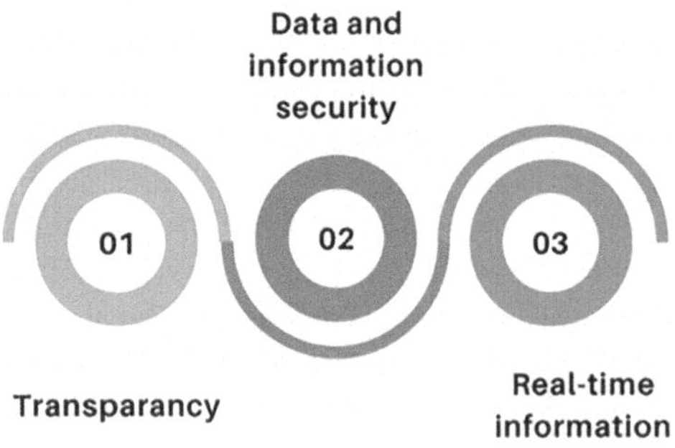

Fig. 2 Principles of blockchain system. Source: Adapted from Shaukat and Zhu (2021); Muneeza et al. (2022) and drawn by authors

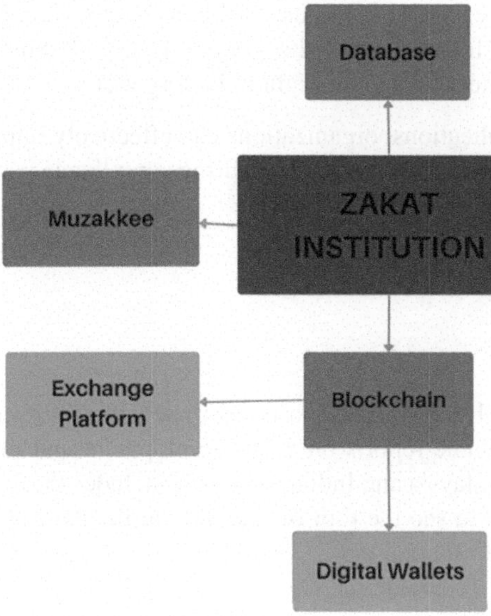

Fig. 3 Proposed framework of Zakat Blockchain. Source: Proposed by Authors

2. *Data Protection:* Organizations must ensure that the data stored on the blockchain is protected and complies with data protection laws, such as the General Data Protection Regulation (GDPR) and the Personal Data Protection Act (PDPA).
3. *Contractual Obligations:* Smart contracts used in blockchain-based zakat management must be legally binding and comply with contractual obligations.
4. *Taxation:* Organizations must consider the tax implications of using blockchain technology for zakat management, including the taxation of cryptocurrency transactions and the potential for tax evasion through the use of blockchain.
5. *Liability:* Organizations must consider the liability implications of using blockchain technology for zakat management, including the potential for legal disputes and the allocation of liability in the event of a breach or failure of the blockchain system.
6. *Jurisdiction:* Organizations must consider the jurisdictional implications of using blockchain technology for zakat management, including the potential for cross-border disputes and the need for international cooperation in the event of a breach or failure of the blockchain system.
7. *Intellectual Property Rights:* Organizations must consider the intellectual property rights implications of using blockchain technology for zakat management, including the potential for infringement of patents, trademarks, or copyrights.

8. *Privacy:* Organizations must ensure that the use of blockchain technology for zakat management does not infringe upon the privacy rights of individuals, including the right to data protection and the right to be forgotten.

By addressing these legal implications, organizations can effectively implement blockchain technology for zakat management while ensuring compliance with local and international laws and regulations.

3 Method

This study adopts the content analysis approach based on review of past researches and publications as well as reputable reports on zakat development and management system in both countries (Malaysia and Indonesia). This includes the search to the reputable database of Scopus to see the root of research on the field of blockchain or financial technology and zakat.

4 Discussion

As shown in Table below, which counts the number of articles related to zakat and blockchain in the two countries, which is very minimum except the recent 4 years since 2020. This indicates that the initiative of adopting blockchain is still progressing and thus the implementation of blockchain technology in zakat collection and distribution must be considered seriously (Table 2).

From the Table below, the highest cited documents on zakat and blockchain is written by Alaeddin O. et al. in 2021, which emphasized on implementing the

Table 2 Publications on Zakat and Blockchain in 10 years by Scopus

Year	Documents
2011	1
2012	0
2013	0
2014	0
2015	0
2016	0
2017	0
2018	0
2019	0
2020	3
2021	7
2022	4
2023	5

blockchain technology in Islamic financial institutions, though it is not merely focus on zakat but the study could be used as a reference to enhance the blockchain technology in other instruments like zakat. The second highest cited document is by Tajudin M. et al., which also touched on financing technology, while the third highest cited document by Alam A. et al. discussed on perception of generation Z on paying zakat which is associated with the use of technology. These information leads to the opportunities of adopting technology (blockchain) in managing fund (collecting and distributing) like zakat or other Islamic social finance instruments (Table 3).

All selected articles in this study discuss the factors related to blockchain technology of zakat collection and distribution. The above Table shows the articles specifically discussed on blockchain technology of zakat collection and distribution in Malaysia and Indonesia, which have the highest number of citations. Several citations indicate the quality and reputation of the papers that are being acknowledged by scholars or researchers around the world. The above chosen 20 articles have a range of 0–10 citations which are dominated source by the Journal of Islamic Marketing (2), while other varies. The list consists of quantitative and qualitative methods which indicate that blockchain technology of zakat collection and distribution is being explored widely.

Blockchain technology has the potential to improve management effectiveness in various areas, including supply chain management and human resource management. In supply chain management, blockchain can enhance transparency, traceability, and reduce administrative costs (Laaper & Fitzgerald, 2017). It can also increase supply chain resilience, viability, and trust across a complex ecosystem (Difrancesco et al., 2023). In human resource management, blockchain can aid in effective decision-making by reducing the time, money, and effort required by potential recruiters and HR professionals (Ramachandran et al., 2023). However, there are still challenges to consider, such as the lack of a generalized view and an unclear understanding of the coordination aspects, impact, and performance of blockchain's application in supply chains (Pal et al., 2021; Khairi et al., 2023). Despite these challenges, blockchain technology has the potential to transform and disrupt supply chains and improve management effectiveness.

The trend in zakat management is toward digitization, transparency, and accountability. This includes the use of technology to improve the collection and distribution process, as well as the establishment of more well-organized and professional zakat management institutions. In the digital era, zakat institutions are collaborating with various online platforms to make zakat management more accessible and transparent to the public. There are four aspects elaborated in the below Table on the trend analysis of zakat management system in Indonesia and Malaysia (Table 4):

A phenomenon challenge in adopting blockchain technology in zakat collection and distribution is resistance to change. The usage of FinTech is still not made by all as some group of people yet to believe in the system. They still do the traditional approach in paying zakat through the counter. Among others, due to the lack of communication and involvement. If team members are not adequately informed about the new technology or not given the opportunity to provide feedback, they

Table 3 20 Highest Citation Articles on Zakat and Blockchain

Authors	Title	Year	Source title	Cited by
Alaeddin O. et al.	Implementing the Blockchain Technology in Islamic Financial….	2021	J. of Information Technology Management	10
Tajudin M. et al.	Financing with heart and intelligence: Augmenting intimacy and …	2020	Intl. J. of Advanced Science and Technology	9
Alam A. et al.	Generation Z perceptions in paying Zakat, Infaq, and Sadaqah…	2022	Investment Management and Financial Innovations	6
Ascarya; Sakti A.	Proposing new Islamic microfinance model for sustainable…	2021	Islamic Finance and Sustainable Development: A Sustainable Economic Framework for Muslim and Non-Muslim Countries	4
Ashurov S. et al.	Developing a Conceptual Framework for Zakat Collection and Distribution…	2022	Research Anthology on Measuring and Achieving Sustainable Development Goals	4
Ahmad K.; Yahaya M.H.	Islamic social financing and efficient zakat distribution: impact of fintech…	2023	J. of Islamic Marketing	4
Rosele M.I. et al.	The Digitalized Zakat Management System in Malaysia and the Way Forward	2022	Al-Ihkam: Jurnal Hukum dan Pranata Sosial	3
Ahmed M.U.; Kasri N.S.B.	Zakat Calculation Software for Corporate Entities	2021	Islamic FinTech: Insights and Solutions	2
Amilahaq F. et al.	Digital platform of zakat management organization…	2021	Advances in Intelligent Systems and Computing	2
Mukit M.M.H. et al.	Contributions of Professor M. Kabir Hassan to the Islamic Finance ….	2021	Jurnal Ekonomi Malaysia	2
Beik I.S. et al.	The Optimization of Blockchain for Greater Transparency…	2021	Islamic FinTech: Insights and Solutions	1
Ashurov S. et al.	Developing a Conceptual Framework for Zakat Collection and…	2020	Impact of Zakat on Sustainable Economic Development	1
Yulianita Gitaharie B. et al.	Research on firm financial performance and consumer behavior	2020	Research on Firm Financial Performance and Consumer Behavior	1

(continued)

Table 3 (continued)

Authors	Title	Year	Source title	Cited by
Raza Rabbani M. et al.	A Fintech-Based Zakat Model Using Artificial Intelligence	2022	FinTech in Islamic Financial Institutions: Scope, Challenges, and Implications in Islamic Finance	1
Abdullah N.	The Innovative Technique of AD/AC and ZP/ZR Appraising of….	2023	J. of Information Technology Management	0
Rahim D.A.A.; Mahmud R.	Intention to Pay Zakat on Digital Money….	2023	Contributions to Management Science	0
Beik I.S.; Arsyianti L.D.	Digital Technology and Its Impact on Islamic Social….	2021	Islamic FinTech: Insights and Solutions	0
Andespa R. et al.	A systematic review of customer Sharia compliance behaviour in Islamic….	2023	J. of Islamic Marketing	0
Nugroho B. et al.	Digital Technology-Enhanced ZISWAF Management: ….	2023	Proceedings - 2023 tenth Intl. Conference on Computer, Control, Informatics and its Applications: Exploring the Power of Data: Leveraging Information to Drive Digital Innovation, IC3INA 2023	0
Abdul Manan H. et al.	An appraisal by its entrepreneurial asnaf (Usahawan Asnaf) on Lembaga Zakat ….	2011	2011 IEEE Colloquium on Humanities, Science and Engineering, CHUSER 2011	0

Table 4 Trend analysis of the Zakat management system in Indonesia and Malaysia

Aspect	Trends
Awareness	Knowledge and awareness of the community are good enough about zakat; however, the communities need to learn and well accept the use of technology in zakat management.
R&D and Technology	A strategic study center has been established by zakat institutions in both countries and some of them have been adopted digital platforms in collecting zakat fund. However, optimizing the use of technology applications in zakat management, collection and distribution should be considered seriously, which also make an efficient coordination among the zakat institutions.
Regulation	Regulations must be standardized (standard in zakat management and system) and well-enforced among the zakat institutions. It is associated with the need of a transparent audit as well as an initiative to strengthen the synergy among various zakat-related institutions.
HR	Lack of competencies and professionalism in managing zakat fund. Quality and quantity of human resources on zakat management are still need to be enhanced.

Source: Author's compilation

may resist the change. Moreover, new technology can be intimidating, and employees may fear that they will lose their jobs or be unable to adapt. Employees more familiar with old systems thus may be comfortable with the existing technology and resist the change to a new system. Blockchain technology in fact able to highlight areas of inefficiency, analyse rational costs and resource allocation and to increase value-added output.

5 Conclusion

The chapter elaborates the development, potential, and trend of blockchain technology on zakat collection and distribution in Malaysia and Indonesia. The comparison of applications in both countries will show the adaptability of the technology on this kind of Islamic social finance instrument (zakat). The research on blockchain technology associated with zakat was not many to be referred, thus the research on the topic still being explored and studied mostly in Muslim emerging countries, i.e. Turkey, Brunei Darussalam, United Arab Emirates, Qatar, Saudi Arabia, and Oman. Each country has uniqueness in managing zakat fund. Through World Zakat Forum, many countries have agreed on certain platform of collecting and distributing zakat but yet some potentials need to be enhanced.

It is an interesting topic whereby blockchain technology would be a solution for an effective zakat management which may discover by researchers in terms of best practices related to the zakat collection and distribution. It involves government commitment, rules and regulations, manpower, facilities and infrastructure, and decision support systems. Meanwhile, understanding zakat management with an adoption of blockchain technology is crucial to do acceleration in developing national economy especially Muslim countries.

Based on the issues in those papers that have been explored by authors from these three countries, there are some suggestions; *firstly* is to strengthen coordination among the policy makers and other stakeholders on an effective policy for zakat management. It is found that some policies are redundant and overlapping each other. *Secondly,* to continue holding training and support. If employees do not feel prepared to use the new technology, they may resist its implementation. Making zakat as supporters of the halal industry from the supply side by becoming a backbone for MSME players in the halal industry.

The conclusion from the search results is that blockchain technology has the potential to significantly improve the management and transparency of zakat collection and distribution. The use of blockchain in zakat management can increase public trust, facilitate transactions, and provide a more comprehensive and efficient system for managing zakat funds. This can lead to a more effective distribution of zakat funds and ultimately contribute to economic growth and poverty reduction. However, the implementation of blockchain in zakat management requires careful consideration and collaboration between various stakeholders, including zakat institutions, regulators, and technology experts. This study is expected to give valuable information and insights for future research in mapping out more comprehensive framework mainly on zakat management and blockchain technology. Initiatives should be made by stakeholders to explore potentials of blockchain technology in collecting and distributing zakat fund.

References

Al-Bawwab, R. A. (2023). Zakat: Changing the framework of giving. *Islamic Economic Studies, 30*(2), 86–103. https://doi.org/10.1108/IES-08-2021-0026

Akhyar Adnan, M., & Barizah Abu Bakar, N. (2009). Accounting treatment for corporate zakat: A critical review. *International Journal of Islamic and Middle Eastern Finance and Management, 2*(1), 32–45.

Alshater, M. M., Saad, R. A. J., Abd Wahab, N., & Saba, I. (2021). What do we know about zakat literature? A bibliometric review. *Journal of Islamic Accounting and Business Research, 12*(4), 544–563.

Aziz, Y., Mansor, F., Waqar, S., & Haji Abdullah, L. (2020). The nexus between zakat and poverty reduction, is the effective utilization of zakat necessary for achieving SDGs: A multidimensional poverty index approach. *Asian Social Work and Policy Review, 14*(3), 235–247.

Beik, I. S., Zaenal, M. H., & Saoqi, A. A. Y. (2021). *The optimization of blockchain for greater transparency in zakat management* (pp. 281–297). Insights and Solutions.

Bin-Nashwan, S. A., Abdul-Jabbar, H., & Aziz, S. A. (2021). Does trust in zakat institution enhance entrepreneurs' zakat compliance? *Journal of Islamic Accounting and Business Research, 12*(5), 768–790.

Difrancesco, R. M., Meena, P., & Kumar, G. (2023). How blockchain technology improves sustainable supply chain processes: A practical guide. *Operations Management Research, 16*(2), 620–641.

Hudaefi, F. A., Caraka, R. E., & Wahid, H. (2022). Zakat administration in times of COVID-19 pandemic in Indonesia: A knowledge discovery via text mining. *International Journal of Islamic and Middle Eastern Finance and Management, 15*(2), 271–286.

Khairi, K. F., Laili, N. H., Sabri, H., Ahmad, A., Pham, V. H., & Tran, M. D. (2023). The development and application of the zakat collection blockchain system. *Journal of Governance & Regulation., 12*, 294.

Laaper, S., & Fitzgerald, J. (2017). *Using blockchain to drive supply chain transparency and innovation*. Accessed Jan 21, 2019, from www2.deloitte.com/us/en/pages/operations/articles/blockchainsupply-chain-innovation.html

Mawardi, I., Widiastuti, T., Al Mustofa, M. U., & Hakimi, F. (2023). Analyzing the impact of productive zakat on the welfare of zakat recipients. *Journal of Islamic Accounting and Business Research, 14*(1), 118–140.

Mohammed, M. O., Robbana, A., & Bedoui, H. (2021a). *Zakat digital management techniques* (pp. 299–317). Insights and Solutions.

Mohammed, M. O., El Amri, M. C., & Shabani, R. M. (2021b). Zakat on wealth and asset: Lessons for SDGs. In *Islamic wealth and the SDGs: Global strategies for socio-economic impact* (pp. 375–392). Springer International Publishing.

Muneeza, A., Bin-Nashwan, S. A., Moshin, M. I. A., Mohamed, I., & Al-Saadi, A. (2022). Zakat payment from cryptocurrencies and crypto assets. *International Journal of Islamic and Middle Eastern Finance and Management, 16*(3), 482–497.

Nazeri, A. N. N., Nor, S. M., Rahman, A. A., Majid, M. A., & Hamid, S. N. A. (2023). Exploration of a new zakat management system empowered by Blockchain Technology in Malaysia. *ISRA International Journal of Islamic Finance, 15*(4), 127–147.

Nazeri, A. N. N., & Nor, S. M. (2023). The implementation of blockchain technology in the Zakat management system: A conceptual research. *Journal of Information Systems and Digital Technologies, 5*(2), 287–300.

Nordin, N., Nordin, N., Ab Kadir, M. S., Fauzi, M. A., Hamran, M., & Meor Hamdan, M. M. K. (2021). The acceptance of blockchain system among zakat users: Case of Pengkalan Chepa, Kelantan. In *The importance of new technologies and entrepreneurship in business development: In the context of economic diversity in developing countries: The impact of new technologies and entrepreneurship on business development* (pp. 581–593). Springer International Publishing.

Omar, N., & Khairi, K. F. (2021). Zakat and blockchain: A review. *International Journal of Islamic Economics and Finance Research, 4*(2 December), 60–66.

Pal, A., Tiwari, C. K., & Haldar, N. (2021). Blockchain for business management: Applications, challenges and potentials. *The Journal of High Technology Management Research, 32*(2), 100414.

Ramachandran, R., Babu, V., & Murugesan, V. P. (2023). The role of blockchain technology in the process of decision-making in human resource management: A review and future research agenda. *Business Process Management Journal, 29*(1), 116–139.

Saad, R. A. J., Ahmi, A., Sawandi, N., & Abdul Aziz, N. M. (2023). Zakat administration reformation towards an effective and efficient zakat revenue generation. *Journal of Islamic Accounting and Business Research, 14*(8), 1232–1260. https://doi.org/10.1108/JIABR-05-2021-0151

Shaukat, B., & Zhu, Q. (2021). Finance and growth: Particular role of zakat to levitate development in transition economies. *International Journal of Finance & Economics, 26*(1), 998–1017.

Owoyemi, M. Y. (2020). Zakat management: The crisis of confidence in zakat agencies and the legality of giving zakat directly to the poor. *Journal of Islamic Accounting and Business Research, 11*(2), 498–510.

Zulfikri, Z., Kassim, S., & Othman, A. H. A. (2023). A conceptual framework of the Blockchain technology adoption for zakat institution in Indonesia. *European Journal of Islamic Finance, 10*(1), 16–23.

Khairil Faizal Khairi is an Associate Professor at Universiti Sains Islam Malaysia (USIM) since 2004, is a distinguished subject matter expert specializing in Islamic Social Finance, Fintech, and Blockchain for Islamic Finance. Currently serving in the Faculty of Economics and Muamalat at USIM, Dr. Khairil has been an invaluable contributor to the academic landscape. His expertise extends to the development of innovative solutions, notably leading a team in the creation of the groundbreaking MySIF Blockchain. This cutting-edge technology is designed for Islamic Social Finance instruments, focusing on Zakat, Waqf, and Sadaqah. Dr. Khairil's commitment to advancing the field has earned him several awards, recognizing his outstanding contributions and pioneering work in Islamic social finance fintech modeling. Dr. Khairil's academic journey includes earning a Ph.D. in Management from Macquarie University of Sydney, an MBA from Universiti Utara Malaysia, and a Bachelor of Banking and Finance from Philadelphia University, Jordan. Despite his numerous achievements, Dr. Khairil remains grounded and dedicated to the pursuit of knowledge, continually striving to make meaningful contributions to the field of Islamic Finance.

Ahmad Rafiki is currently as an Associate Professor in Faculty of Economics and Business, the Dean of Faculty of Economics and Business, and the Head of Islamic Management and Halal Industry Center at Universitas Medan Area, Indonesia. He obtained his Bachelor Degree in Business Administration with a major of Marketing from the MARA University of Technology (UiTM) and Masters in Management from the International Islamic University of Malaysia (IIUM). He completed his thesis for the award of Doctor of Philosophy from the Islamic Science University of Malaysia (USIM). He is holding Certified Islamic Marketer (CIMA) from the International Islamic Marketing Association (IIMA). His research interests include the Islamic entrepreneurship, management, marketing, and leadership. He has published chapters and papers in international publishers such as IBA-MacMillan, IGI Global, Emerald Publishing, Springer, IntechOpen, Routledge etc. He also became the editorial advisory board and reviewer in reputable publishers of Emerald and Elsevier.

Sari Nuzullina Rahmadhani is an assistant professor at the Faculty of Business and Management, Universitas Medan Area. She graduated for her Master of Accounting from Universitas Gadjah Mada and has been serving for 6 years in the said Faculty. Her research area focuses on Management Accounting and Auditing and she has published several articles and book chapters in the mentioned fields. She has also participated in several International Conferences and Seminars.

Rana Fathinah Ananda is an assistant professor at the Faculty of Business and Management, Universitas Medan Area. She graduated for her Master of Science from Universitas Sumatera Utara with 3 years teaching experience. Her research area focuses on Auditing, Management Accounting, and Financial Accounting and she has published several books and book chapters in the said fields. She has also participated in several Conferences and Seminars.

Blockchain Integration: Transforming Islamic Frozen Estate Management in Malaysia

Nur Hidayah Laili, Khairil Faizal Khairi, and Azuan Ahmad

Abstract This chapter provides a comprehensive exploration of the intricate challenges involved in managing frozen estates in Malaysia. It meticulously dissects the complexities surrounding legal disputes, ambiguous legislation, and resource deficiencies, offering innovative solutions through the integration of blockchain technology. The chapter begins by delving into the intricacies of the Malaysian legal landscape, focusing on entities such as the High Court, Syariah Court, Amanah Raya Berhad, and the Land Office. It sheds light on the challenges posed by contested wills, adherence to Islamic inheritance laws, and the absence of specific legislation tailored for frozen estates. The resulting legal complexities, often leading to protracted legal battles lasting months or even years, underscore the urgent need for nuanced reforms in legal frameworks. Recognizing the inadequacies within the current estate management paradigm, the authors propose blockchain technology as a transformative remedy. The decentralized ledger system is explored, emphasizing its features, including immutability, transparency, and decentralized consensus. The chapter illustrates how blockchain can bring transparency to estate management processes in Malaysia, providing a singular, trusted data source that diminishes disputes and ensures meticulous record-keeping. Security emerges as a critical facet, with the authors delving into the decentralized nature of blockchain, cryptographic algorithms, and smart contracts. These elements fortify data against tampering and unauthorized access, ultimately reducing administrative overhead and expediting asset transfers by obviating the need for intermediaries. Furthermore, the chapter explores the potential of blockchain technology to revolutionize estate management by offering real-time tracking and monitoring of frozen estates. This ensures precise and current information for all relevant parties, mitigating disputes and fostering efficient asset transfers. Ongoing research and collaborative efforts between the legal and technology sectors are highlighted, indicating a positive trajectory toward the adoption of blockchain in estate management practices. In conclusion, this

N. H. Laili · K. F. Khairi (✉) · A. Ahmad
Universiti Sains Islam Malaysia, Bandar Baru Nilai, Nilai, Negeri Sembilan, Malaysia
e-mail: khairil@usim.edu.my

A. Rafiki et al. (eds.), *Strategic Islamic Business and Management*, Contributions to Management Science,
https://doi.org/10.1007/978-3-031-61778-2_3

chapter offers a comprehensive examination of the challenges in frozen estate management in Malaysia and proposes a transformative solution through blockchain integration. The chapter serves as a valuable resource for legal professionals, administrators, beneficiaries, and stakeholders interested in navigating the complex landscape of frozen estates with efficiency, transparency, and trust.

1 Introduction

The intricate task of handling frozen estates presents a multifaceted challenge that extends beyond its surface complexities. Frozen estates, a phenomenon triggered by legal disputes, concerns regarding estate administration, or challenges to will validity, cast a shadow on the seamless distribution of assets, creating a web of difficulties for beneficiaries and the broader estate management process. In Malaysia, a nation grappling with similar challenges as others, the management of frozen estates is entangled in a web of issues, ranging from transparency and security to efficiency. A closer examination reveals that legal complexities amplify the difficulties. Ambiguities in legislation, overlapping jurisdictions, and procedural hurdles contribute to the intricacies faced by estate managers. When an estate is entangled in legal disputes, the ensuing legal proceedings are not merely cumbersome but can also stretch over extensive periods, impeding the timely resolution of matters. Such prolonged legal battles often arise from familial conflicts, disputes concerning asset distribution, or challenges questioning the validity of the will. It is worth noting that the duration required for resolving these legal intricacies can extend to months, or even years, casting a prolonged shadow on the administration and distribution of the estate (Rahman, 2019). This perspective underscores the need for nuanced reforms in legal frameworks and procedures to streamline the management of frozen estates, ensuring a more efficient and equitable resolution of these complex issues.

Another significant issue in the effective management of frozen estates in Malaysia is the lack of expertise and resources. Individuals or administrators appointed to handle frozen estates may lack the necessary expertise, leading to inefficiencies and potential mismanagement. The management of frozen estates demands a comprehensive understanding of probate law, property rights, and relevant regulations (Nasrul et al., 2021). Weak land administration systems, legal complexities, and a lack of expertise and resources contribute to difficulties in administering and distributing assets within these estates. Blockchain technology offers several potential advantages in the management of frozen estates. Blockchain is a distributed ledger technology that enables secure and transparent transactions without the need for intermediaries. Its key features, such as immutability, transparency, and decentralized consensus, make it suitable for enhancing the efficiency, security, and accountability of frozen estates management.

Blockchain technology emerges as a potential remedy for the complexities associated with managing frozen estates. Functioning as a distributed ledger technology, blockchain facilitates secure and transparent transactions sans intermediaries. Its salient features, including immutability, transparency, and decentralized consensus, render it well-suited for enhancing the efficiency, security, and accountability of frozen estates management. Blockchain proffers a decentralized and transparent solution to these challenges. Transparency stands out as a pivotal benefit of employing blockchain in managing frozen estates. This technology enables transparent and unalterable record-keeping, streamlining estate management processes in Malaysia. By recording estate-related transactions and information on a blockchain, all stakeholders gain access to a singular, trusted data source. This transparency diminishes disputes, ensures meticulous record-keeping, and furnishes a dependable audit trail for estate administrators, beneficiaries, and legal authorities.

Security constitutes another critical facet that blockchain introduces to the management of frozen estates. The decentralized nature of blockchain ensures the distribution of estate data across multiple nodes, fortifying it against tampering and unauthorized access. Additionally, cryptographic algorithms employed in blockchain furnish robust data encryption, safeguarding sensitive estate information and upholding stakeholders' privacy (Chhabra et al., 2021). The utilization of smart contracts, which are self-executing agreements coded on the blockchain, obviates the need for intermediaries, diminishing paperwork, expediting asset transfers, and reducing administrative overhead. Furthermore, blockchain-based systems facilitate real-time tracking and monitoring of frozen estates, delivering precise and current information to all pertinent parties.

Blockchain technology harbors the potential to significantly enhance estate management in Malaysia by providing transparency, security, and efficiency. Through transparent record-keeping, streamlined asset ownership verification, automated execution through smart contracts, heightened security and privacy, and facilitated dispute resolution, blockchain has the capacity to revolutionize estate management practices in the country. Ongoing research and collaborative efforts between the legal and technology sectors indicate that the adoption of blockchain in estate management is poised to usher in positive transformations.

2 Blockchain Technology: Mechanisms, Characteristics, and Implications

Blockchain technology, heralded as a transformative innovation, has not only ascended to prominence but has become a cornerstone across diverse sectors, showcasing a myriad of advantages, especially in finance, supply chain, and healthcare. The transparency and security inherent in blockchain have propelled its adoption in various industries, fundamentally altering the way we handle transactions and data. In the financial realm, blockchain's impact goes beyond facilitating secure transactions; it challenges the traditional role of intermediaries like banks. Swan (2015) notes that blockchain enables direct peer-to-peer transactions, reducing the need for a trusted third party. Moreover, the introduction of smart contracts exemplifies the technology's potential in finance. This innovation automates and ensures the integrity of contractual agreements, streamlining processes and minimizing the risk of disputes. In supply chain management, blockchain's decentralized nature addresses long-standing challenges. By providing an immutable and transparent record of a product's journey from manufacturing to delivery, blockchain enhances traceability. This not only bolsters efficiency but also reduces the risk of fraud by ensuring the authenticity of products. For instance, a consumer can trace the origin of a product, such as organic produce, to verify its authenticity and quality.

The impact of blockchain in healthcare is equally profound. The security and interoperability features of blockchain offer a solution to the long-standing issue of patient data management. Through blockchain, patients and healthcare providers can access medical records securely. Each interaction with the patient's data is recorded and encrypted, ensuring privacy and facilitating seamless collaboration among healthcare professionals, as suggested by Yaqoob et al., 2022. As blockchain continues to evolve, its transformative influence expands beyond specific sectors. It represents a paradigm shift toward decentralized and transparent systems, challenging established norms and fostering innovation. The versatility of blockchain technology, as evident in its applications across finance, supply chain, and healthcare, underscores its potential to reshape the way we conduct transactions and manage data in our interconnected world. To comprehend the extensive applications and implications of blockchain technology, it is crucial to explore its underlying mechanisms. At its core, blockchain functions as a decentralized and distributed ledger system. In contrast to conventional centralized databases, it relies on a network of computers, known as nodes, which collaboratively validate and record transactions. Each transaction forms a "block," and these blocks are chronologically linked to create the blockchain. Significantly, the information within each block is secured through cryptographic hashing, ensuring immutability. The blockchain system revolves around a distributed ledger maintained by multiple network participants (nodes). This intricate process operates through interconnected blocks, each encapsulating a set of transactions or data. The essence of blockchain lies in its decentralized nature, eliminating the need for a central authority to authorize transactions. Instead, network participants collectively uphold ledger integrity through a consensus mechanism, fostering a more democratic and resilient system (Yadav et al., 2023).

To fully grasp the vast applications and potential implications of this revolutionary technology, an exploration of its foundational mechanisms becomes essential. At its core, blockchain functions as a decentralized and distributed ledger system, setting itself apart from traditional centralized databases by operating on a network of computers known as nodes. These nodes work collectively to validate and record transactions, encapsulating them into blocks that are linked chronologically to construct the blockchain. Within each block, there are multiple transaction records, each uniquely identified by a hash—an alphanumeric string of fixed length generated from a character string using a mathematical function. Upon the creation of a block, it acquires a "stamp" in the form of a hash, connecting it to the preceding block and forming a chain. Before each block is seamlessly incorporated into the chain, a meticulous verification process, facilitated by a consensus mechanism among network participants, becomes imperative. This careful process theoretically allows for an infinite number of blocks. The security and immutability of information within each block are upheld through cryptographic hashing. This hash intricately details specific transaction information, encompassing sender, recipient, participants, date, time, and other details encoded into the blockchain algorithm (Brito & Castillo, 2013; Salmon & Myers, 2019).

As blockchain technology advances, the foundational principles of decentralization, transparency, and security remain at the forefront. The decentralized nature, facilitated by distributed ledger technology, ensures that no single entity holds control, fostering a democratic and resilient infrastructure. This approach not only reduces the risk of a single point of failure but also safeguards against the abuse of power. Participants, known as nodes, collectively validate transactions, creating a distributed consensus that enhances security (Yaqoob et al., 2022). Attempts to compromise the system would require manipulating a majority of the network, making it highly impractical and resistant to unauthorized alterations. This shift toward decentralization not only reshapes traditional power dynamics but also cultivates a more inclusive and accessible ecosystem, empowering individuals and organizations in a transparent and trustless environment. As blockchain technology progresses, its impact on reshaping industries and fostering a decentralized paradigm is set to become even more pronounced.

Transparency is inherent in blockchain, where every transaction is recorded on a tamper-resistant ledger accessible to all participants. This feature fosters trust and accountability, as the immutability of blockchain records ensures that once a transaction is added, it cannot be altered without network consensus. This transparency instills confidence among stakeholders, streamlines auditing processes, and reduces the likelihood of fraud and errors. In industries like supply chain management, consumers can trace product journeys, ensuring authenticity and ethical sourcing (Sedlmeir et al., 2022). In financial systems, blockchain transparency can mitigate fraud risks and enhance regulatory compliance. As blockchain technology matures, its transparent nature becomes a cornerstone for building robust and accountable systems across various sectors, paving the way for more efficient and trustworthy interactions in the digital age.

Security in blockchain is fortified by cryptographic algorithms and consensus mechanisms, safeguarding data integrity and preventing malicious attacks. Cryptography plays a pivotal role in securing transactions through techniques like encryption. Digital signatures authenticate participant identities and verify data integrity. Consensus mechanisms, like proof-of-work or proof-of-stake, govern transaction validation, making it challenging for a single entity to compromise the system. The decentralized and distributed nature of blockchain further enhances security, eliminating central points vulnerable to attacks. In the face of evolving cyber threats, the robust combination of cryptographic techniques and consensus mechanisms positions blockchain as a resilient solution. It provides a secure foundation for applications ranging from financial transactions to sensitive data management and beyond (Zeba et al., 2022).

The ongoing evolution of blockchain technology promises to reshape industries and instigate a paradigm shift toward decentralized, transparent, and secure systems. The foundational principles of decentralization, transparency, and security remain steadfast as blockchain progresses, with its impact spanning across finance, supply chain, healthcare, and beyond. The decentralized nature, collaborative validation of transactions, transparency, and robust security mechanisms collectively position blockchain as a transformative force, challenging traditional norms and fostering innovation. As blockchain technology matures, its potential to enhance efficiency, trustworthiness, and inclusivity becomes increasingly apparent, paving the way for a future where decentralized and transparent systems redefine how we conduct transactions and manage data in our interconnected world.

3 Diverse Blockchain Models: Characteristics, Applications, and Trade-Offs

Blockchain technology has undergone remarkable advancements, giving rise to a myriad of blockchain models that exhibit diverse features and functionalities. These models have been tailored to address the specific needs and requirements of various industries and use cases. A comprehensive understanding of the distinctive characteristics and nuances among blockchain models is imperative for making informed decisions in selecting the appropriate model for specific applications.

The term "blockchain models" predominantly refers to the public permissionless blockchain, a paradigm exemplified in well-established cases such as Bitcoin and Ethereum. In the public permissioned model, writing privileges are exclusively granted to authorized nodes, while commitment can be orchestrated by either the entire network or a subset of authorized nodes. In the context of closed systems, specific actions may be exclusively carried out by a designated party. The enduring challenge of transparency and auditability in a closed blockchain arises from the legal and moral obligations of participants, particularly organizations or corporations, to uphold confidentiality. Even in cases where sensitive information is deliberately withheld, it remains feasible to discern or reasonably estimate a participant's identity through transaction patterns or alternative methods (Biryukov et al., 2014; Meiklejohn et al., 2013). To address this challenge, certain academics, such as Kosba et al. (2016), advocate for the utilization of public blockchains incorporating zero-knowledge proofs. This approach enables validators to verify transactions without the necessity of disclosing sensitive information.

The private blockchain model confines participation to a specific group of entities, typically controlled by a single entity or consortium. Access to the network and transaction validation are permissioned, ensuring a delicate balance between privacy and control over the blockchain. Prioritizing scalability and efficiency, private blockchains find suitability in applications within organizations or industries with stringent data privacy requirements (Siegfried et al., 2020). In the realm of private blockchains, paramount considerations include enhanced privacy and confidentiality. Sensitive data is meticulously kept confidential and shared solely among authorized participants, allowing organizations to harness the benefits of blockchain technology while safeguarding sensitive information (Christidis & Devetsikiotis, 2016). Private blockchains often exhibit superior scalability and performance compared to their public counterparts, thanks to restricted participation. This restriction facilitates higher transaction processing speeds and throughput, making private blockchains suitable for applications with high transaction volumes and low latency requirements (Lu et al., 2019).

Consortium blockchains navigate a middle ground between public and private models by permitting a select group of participants to collectively validate transactions and maintain the blockchain network. This model provides enhanced privacy, scalability, and control while fostering collaboration and shared governance among trusted entities. Restricted participation in consortium blockchains involves a limited set of participants known and trusted by consortium members. Unlike public blockchains, consortium blockchains operate on a permissioned basis, controlling and restricting access to authorized participants. This characteristic ensures that consensus and decision-making within the consortium are orchestrated by trusted entities (Yang et al., 2022). The increased privacy in consortium blockchains is attributed to participation being confined to known entities, enabling secure sharing of sensitive data within the network. This feature holds particular significance in industries where confidentiality and regulatory compliance are paramount, such as healthcare, finance, and supply chain management (Aslam et al., 2021).

Hybrid blockchains offer a distinctive amalgamation of public and private blockchain characteristics, providing organizations with heightened flexibility and control over their blockchain networks. This model facilitates the integration of public and private networks, enabling the simultaneous benefits of transparency, decentralization, and data privacy. Comprising both public and private components, the hybrid blockchain model ensures transparency, immutability, and decentralized consensus through its public facet, allowing for secure transactions and open participation. Conversely, the private component offers control, privacy, and confidentiality over sensitive data, limiting access exclusively to authorized participants (Wylde et al., 2022). A pivotal feature of hybrid blockchains is the ability to delineate the level of data visibility and control. Organizations can selectively determine which data is stored and shared on the public network, while safeguarding sensitive information within the private network. This characteristic empowers organizations to strike a delicate balance between transparency and privacy based on their specific requirements (Agrawal et al., 2021). Hybrid blockchains foster interoperability between public and private networks, facilitating seamless data exchange and collaboration. This characteristic enables organizations to leverage the benefits of public networks for functions like authentication, verification, or consensus while retaining the privacy and control provided by private networks (Casino et al., 2019).

The evolution of various blockchain models has significantly broadened the horizons and applications of blockchain technology. Public, private, consortium, and hybrid blockchain models offer distinct features and characteristics, catering to a diverse array of use cases and industry requirements. A nuanced comprehension of the intricate trade-offs associated with each model is paramount for selecting the most suitable blockchain model for specific applications.

4 Mechanisms of Islamic Estate Management and Inheritance Distribution

The estates management from an Islamic standpoint is deeply rooted in foundational texts such as the Qur'ān and Sunnah, alongside legal enactments concerning estate distribution. Key to this framework are instruments like farā'id, wasiyyah (bequest or will), hibah (gift), and muafakat (takharuj). Among these, farā'id holds paramount importance, serving as the primary mechanism for estate distribution. Islamic inheritance, as governed by the Qur'ān and Sunnah, establishes a system of regulations dictating the fair distribution of a deceased individual's assets among their heirs. Farā'id, defined by specific regulations outlined in the Qur'ān and Sunnah, mandates the entitlements of legal heirs and prescribes the distribution of inheritance within Islamic law (Anuar et al., 2022). Scholarly discourse has elaborated upon the nuances of farā'id, defining it as the prescribed division of assets among heirs as established by Allah. This system is considered the superior method of estate distribution due to its fairness, efficiency, and orderliness. Legal heirs are identified and entitled to inherit according to the stipulations of farā'id, with distribution ratios specified in the Qur'ān (Lahsasna, 2016). The reliance on farā'id within Muslim communities underscores its significance, albeit sometimes without full comprehension of its Qur'ānic origins. The Qur'ān delineates specific inheritance fractions in verses such as those found in Surah An-Nisa, alongside additional guidelines dispersed throughout various chapters. Hadith literature further elucidates aspects of Islamic inheritance, contributing to a comprehensive understanding of estate distribution.

Wasiyyah, or testamentary bequests, reaffirms an individual's autonomy over their property posthumously, subject to conditions delineated by Shariah. Wasiyyah constitutes a contractual arrangement for estate distribution upon the testator's demise, limited to one-third of the total estate for non-legal heirs (Nor Muhamad, 2017). The formulation of a valid wasiyyah requires adherence to established criteria and pillars, ensuring its enforceability posthumously. In contrast, hibah serves as an alternative means of managing wealth, encompassing voluntary acts of giving informed by Qur'ānic principles, Sunnah, and juristic reasoning. While Malaysia lacks specific legislation for hibah among Muslims, the practice finds legitimacy within Islamic jurisprudence (Kamarudin & Nor Muhamad, 2018). Hibah encompasses both charitable alms and gifts, differentiated by their intent and purpose. Consensus, or takharuj, offers a mechanism for peaceful inheritance distribution through deliberation and agreement among heirs. This approach, applied within Malaysia's Small Estate Distribution Section, facilitates conflict resolution and equitable asset allocation (Ahmad, 2017). Consensus methods involve the voluntary withdrawal of heirs from their entitlements in exchange for compensatory arrangements, promoting harmony and fairness in estate distribution.

5 Navigating Frozen Estates: Legal, Administrative, and Technological Perspectives in Malaysia

The administration and distribution of assets following an individual's passing is a process governed by a legally binding document known as a will. This document outlines the deceased's wishes regarding the division and allocation of their estate among beneficiaries, providing clarity for the distribution process. However, in some instances, the smooth execution of a will encounters obstacles, leading to what is commonly referred to as frozen estates. These frozen estates occur when legal disputes, challenges to the will's validity, unresolved financial obligations, or other complications hinder or delay the normal administration and distribution of assets. Frozen estates present a significant challenge in the resolution of a deceased person's affairs, often causing distress and frustration among beneficiaries and the executor of the will. Navigating the complex legal landscape surrounding frozen estates within the context of wills is crucial to understanding the intricacies involved in the administration of an estate.

Effectively managing frozen estates in Malaysia related to wills requires a thorough understanding of the legal framework and procedures involved in estate administration. Disputes, especially in cases of contested wills, can lead to delays in asset distribution, posing challenges for beneficiaries and executors alike. These disputes commonly arise due to conflicts over entitlement, asset valuation, or disagreements about the deceased's intentions as expressed in the will (Ayuni et al., 2022). Key individuals involved, such as executors, administrators, and beneficiaries, must possess a thorough knowledge of legal provisions and requirements.

The regulation of frozen estates in Malaysia encompasses a comprehensive legal framework and enforcement mechanisms designed to govern the management and preservation of assets subject to freezing or confiscation. Various institutions, including the High Court, Syariah Court, Amanah Raya Berhad, and the Land Office, collaborate to ensure transparency, accountability, and facilitate asset recovery processes (Nasrul et al., 2021).

The High Court in Malaysia plays a significant role in the distribution of assets after an individual's passing. With jurisdiction over civil matters, including estate administration and inheritance disputes, the High Court oversees the distribution process in accordance with either the Distribution Act 1958 (for non-Muslims) or Islamic inheritance laws (for Muslims) through the High Court's probate division. When a person passes away, their estate needs to be administered and distributed to the rightful beneficiaries. If the deceased left a valid will, the executor named in the will can apply to the High Court for a Grant of Probate. The Grant of Probate establishes the authenticity of the will and the executor's authority to administer the estate. The High Court examines the will, verifies its validity, and if satisfied, grants the probate (Mohd Alam Shah et al., 2022). In cases where there is no valid will or appointed executor, the High Court may grant Letters of Administration to an appropriate person, usually a close relative or next of kin. This allows the appointed person to administer the estate and distribute the assets according to the law of intestacy. Once the Grant of Probate or Letters of Administration are obtained, the High Court oversees the distribution of assets.

Operating in parallel with the civil court system, the Syariah Court holds jurisdiction over matters related to Islamic law, including the regulation of frozen estates in cases involving Muslims. Islamic inheritance laws, based on the Quran and Hadith, govern the distribution of assets among Muslim heirs. The Syariah Court has the power to issue freezing orders or injunctions based on Islamic principles, ensuring the preservation and management of assets in accordance with Islamic law. If the deceased individual was a Muslim and had not made a valid will, the Syariah Court has the authority to distribute the assets based on the principles of Faraid, the Islamic law of inheritance. The court examines familial relationships, calculating the shares of each eligible heir according to the prescribed rules of Faraid. The Syariah Court oversees the administration of the deceased individual's estate, ensuring that the assets are managed and distributed in accordance with Islamic law.

Amanah Raya Berhad, a government-appointed trust corporation, takes on the responsibility of managing and administering frozen estates in Malaysia. Acting as the custodian of frozen assets, the corporation safeguards and preserves their value until they are released or disposed of according to court orders or legal provisions. Governed by the Amanah Raya Berhad Act 1995 (Act 533), the corporation provides professional estate management services, including asset valuation, financial management, and the distribution of assets to beneficiaries. Amanah Raya Berhad can also act as an executor or trustee for individuals who appoint them in their wills or through trust instruments. In cases where no executor is appointed, the deceased person's family or beneficiaries can engage Amanah Raya Berhad to administer the estate. The corporation handles necessary legal processes, including obtaining the Grant of Probate or Letters of Administration from the High Court, collecting and safeguarding assets, settling debts, and distributing the remaining assets to the rightful beneficiaries.

Under the jurisdiction of state governments, the Land Office plays a crucial role in regulating frozen estates involving immovable properties, such as land and real estate. The office ensures proper registration, transfer, and management of frozen properties, including imposing restrictions on their disposal or transfer. Collaborating with other institutions, the Land Office facilitates the preservation and recovery of frozen assets.

The regulation of frozen estates in Malaysia poses several challenges that hinder effective asset management and distribution. The multifaceted regulatory framework involves multiple laws, regulations, and court decisions, resulting in legal complexities (Nasrul et al., 2021). The interpretation and application of these laws can be intricate, leading to uncertainties and delays in asset management and distribution. The overlapping jurisdiction of different authorities, such as the High Court, Syariah Court, Amanah Raya Berhad, and Land Office, further complicates the regulation of frozen estates. Each authority may have its set of rules and procedures, creating ambiguity and potential conflicts in the distribution of assets.

The absence of specific legislation tailored to comprehensively address frozen estates is another challenge. While relevant acts, such as the Administration of Estates Act 1965, Probate and Administration Act 1959, Civil Law Act 1956, and Land Titles Act 1960, exist, they primarily focus on general estate administration rather than providing comprehensive guidance on handling frozen estates. As a result, the legal framework may not adequately address the unique challenges and complexities associated with frozen estates, leading to further delays and legal disputes. Efficient coordination among multiple agencies is crucial for the streamlined management of frozen estates. Regulating these estates requires collaboration among various agencies, including law enforcement bodies, regulatory authorities, and judicial institutions. Ensuring effective collaboration and information sharing is challenging, particularly when multiple agencies are involved in the investigation and management of frozen estates. The lack of coordination often results in inefficiencies and delays in resolving disputes related to frozen estates (Abdullah et al., 2020). The limited sharing of information among agencies further obstructs a comprehensive understanding of frozen estates, leading to delays and inadequate decision-making (Ahmad & Laluddin, 2010).

In the face of these challenges, the integration of blockchain technology emerges as a potential solution to address the complexities in the management of frozen estates in Malaysia. Leveraging inherent features such as decentralization, transparency, and immutability, blockchain holds promise in overcoming issues related to information sharing, trust, and accountability. By enabling secure and tamper-proof record-keeping, blockchain ensures transparency in the management of estate-related transactions, ownership transfers, and asset preservation.

Blockchain technology, with its transformative components like smart contracts, introduces an opportunity to revolutionize estate management processes. Smart contracts, intrinsic to blockchain, possess the capability to automate and enforce agreements, thereby fostering an efficient and transparent execution of estate management tasks. Concurrently, the incorporation of blockchain-based identity verification mechanisms enhances trust among the various stakeholders involved in frozen estate management in Malaysia. By leveraging blockchain solutions, the country has the potential to significantly streamline the intricate processes associated with managing frozen estates, leading to heightened efficiency and the safeguarding of stakeholder interests.

6 Blockchain Technology's Transformative Role in Frozen Estate Management

The intricate realm of managing frozen estates is fraught with challenges, ranging from legal complexities to coordination issues and the ever-pressing need for transparency and accountability. In this context, blockchain technology emerges as a powerful solution, offering unique features such as decentralization, transparency, immutability, and smart contract functionality. This chapter explores how blockchain has the potential to revolutionize the management of frozen estates, addressing key challenges and providing a comprehensive framework for transparency, traceability, automation, and security.

Blockchain technology introduces transparency and traceability as fundamental pillars in addressing issues inherent in the management of frozen estates. Through its decentralized and immutable nature, blockchain ensures transparency throughout the entire process. Every transaction and activity related to the frozen estate is meticulously recorded on a shared ledger visible to authorized participants. This transparency not only minimizes the risk of fraudulent activities but also fosters trust among stakeholders (Azmin & Khairi, 2022). All parties involved, including administrators, legal representatives, and beneficiaries, can access the blockchain to verify and audit transactions and decisions made during the management process.

The traceability feature of blockchain creates an immutable audit trail of asset ownership, transfers, and transactions. Each change or transfer of ownership is recorded as a permanent and time-stamped entry on the blockchain, providing a clear and transparent view of the estate's management history (Andoni et al., 2019). This comprehensive and auditable history ensures accountability and prevents disputes or discrepancies in ownership claims. By leveraging these transparency and traceability features, blockchain technology addresses the complexities of managing frozen estates, creating a robust foundation for trust and accountability.

Smart contracts, encoded within the blockchain, play a pivotal role in revolutionizing the management of frozen estates. These self-executing contracts with predefined rules and conditions automate various processes, mitigating challenges such as delays, inefficiencies, and disputes. The execution of estate-related processes becomes automated, ensuring the efficient and accurate implementation of predefined rules (Bailis et al., 2017). This automation reduces reliance on manual processes and intermediaries, minimizing human error and improving the overall speed and accuracy of estate management.

The transparency and auditability provided by smart contracts contribute to preventing disputes and ensuring that all parties involved have access to the same information. This not only promotes fairness in the management process but also enhances trust among stakeholders (Casino et al., 2019). Smart contracts, as an integral part of blockchain technology, become a catalyst for efficiency and accountability, reshaping the landscape of frozen estate management.

The decentralized architecture and cryptographic algorithms inherent in blockchain enhance the integrity and security of data in frozen estate management.

Blockchain's immutability and tamper-resistant nature ensure that once a transaction or record is added to the blockchain, it cannot be altered or deleted without consensus from the network participants (Nakamoto, 2012). This feature significantly reduces the risk of fraud, tampering, and unauthorized access to estate-related information.

The cryptographic techniques employed by blockchain technology secure data and verify the authenticity of transactions. Each transaction recorded on the blockchain is encrypted and linked to previous transactions through cryptographic hashes, creating a chain of trust (Swan, 2015). This cryptographic security ensures that the data remains secure and cannot be tampered with, providing increased confidence in the integrity and reliability of estate management processes.

Thus, blockchain technology holds immense potential in addressing the challenges faced in the management of frozen estates. By leveraging its transparency, smart contract functionality, and enhanced security, blockchain can streamline processes, reduce disputes, and enhance accountability in frozen estate management. The transparent and traceable nature of blockchain addresses issues of fraud and trust, while smart contracts revolutionize the automation of estate-related processes. The decentralized architecture and cryptographic security ensure the integrity and security of data, creating a robust foundation for efficient and trustworthy frozen estate management. As we navigate the complexities of estate management, blockchain technology stands as a transformative force, reshaping traditional practices and opening avenues for a more secure and transparent future.

7 Blockchain Integration for Enhanced Frozen Estate Management

The introduction of a blockchain-based system for frozen estate management seeks to harness the unique features of blockchain technology to elevate transparency, security, efficiency, and coordination among various entities involved in the management process. The system architecture comprises several interconnected components designed to collectively enhance the management of frozen estates. The system architecture consists of the following components (Fig. 1):

Blockchain Network: A pivotal element of the system is the use of a private or permissioned blockchain network, ensuring privacy and controlled access for authorized participants. Asset records within this network are digital records stored on the blockchain, meticulously tracking ownership and transaction history. This approach establishes a decentralized and immutable ledger, recording transactions in a transparent and secure manner. The distributed ledger ensures tamper-proof and fraud-resistant asset records. Critical information, such as ownership details, transaction timestamps, and associated metadata, is stored across multiple nodes within the network. This distribution makes it challenging for any single entity to alter records without consensus from the majority of network participants. The blockchain

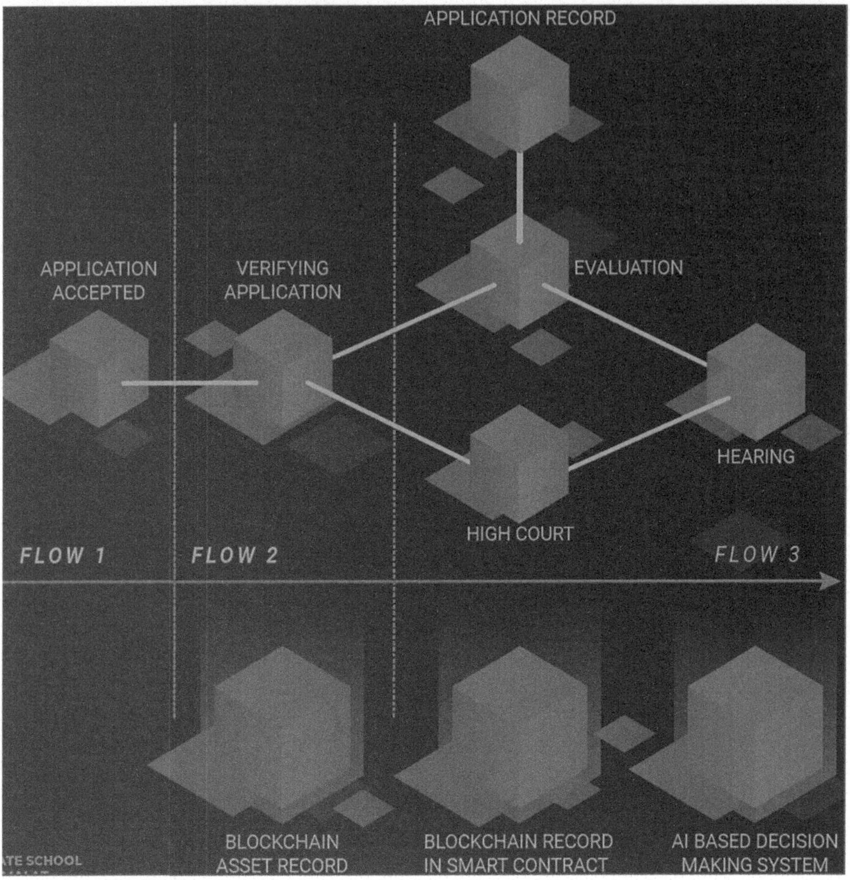

Fig. 1 The blockchain architecture for frozen estate. Source: Author's own concept

network spans relevant agencies, including the High Court, Syariah Court, Land Office, and Amanah Raya, with each node maintaining a transparent copy of the blockchain ledger.

User Interfaces: The system prioritizes user-friendly interfaces tailored to diverse stakeholders, including administrators, legal representatives, and beneficiaries. These interfaces enable seamless interaction with the blockchain network, allowing stakeholders to access estate-related information, initiate transactions, and monitor ongoing processes. Dashboards, forms, and notifications are integrated into the interfaces to facilitate efficient communication and collaboration among stakeholders.

Coordination Mechanisms: Facilitating coordination among multiple agencies involved in frozen estate management is a primary objective of the architecture system. The blockchain network serves as a shared platform, enabling secure information exchange, transaction validation, and collaborative decision-making. Consensus mechanisms embedded within the blockchain network ensure unanimous agreement on transaction validity, minimizing disputes and fostering cooperation among participating agencies.

Smart Contracts: Central to the system's functionality are smart contracts, playing a pivotal role in automating and enforcing estate-related processes and agreements. A set of predefined smart contracts govern tasks such as application records and the evaluation of applications. The blockchain record of a smart contract encapsulates details from the application record, involved parties, specified terms and conditions, and the evaluation process. This record is distributed across multiple nodes, ensuring consensus and preventing tampering or alteration of the contract's history. Network participants can verify the validity and integrity of a contract's execution by examining the blockchain record, thereby eliminating the need for trust in a central authority.

Identity Verification: To bolster security, the system incorporates robust identity verification mechanisms, ensuring that only authorized individuals or entities participate in the estate management process. This verification process may involve the integration of digital signatures, biometric authentication, or other secure identification methods. Robust identity verification not only prevents fraud and unauthorized access but also ensures the integrity of transactions and decision-making within the system.

The system architecture strategically leverages the transparency, security, and automation capabilities of blockchain technology to address the multifaceted challenges encountered in frozen estate management. By integrating smart contracts, user interfaces, identity verification, and coordination mechanisms, the system aims to streamline processes, enhance trust among stakeholders, and elevate overall efficiency in managing frozen estates.

8 Advancing Frozen Estate Management Through Blockchain Technology

The provided flowchart system architecture outlines a comprehensive blueprint for a blockchain-based system designed to address the challenges inherent in managing frozen estates. The complexities associated with coordination among multiple agencies, asset preservation, transparency, accountability, and security are key considerations in the proposed solution. This discussion explores the intricacies of the flowchart, elucidating the various components and stages involved in the implementation of a blockchain-based system tailored for frozen estate management. The flowchart's initiation underscores the establishment of a blockchain network customized for the specific requirements of frozen estate management. This ensures privacy and controlled access for authorized participants. The network comprises multiple nodes distributed among pertinent agencies such as the High Court, Syariah

Court, Land Office, and Amanah Raya, each maintaining decentralized copies of the blockchain ledger. The utilization of blockchain networks forms the bedrock of the proposed system, fostering decentralization and ensuring a transparent and tamper-proof transaction history.

Smart contracts, highlighted in the flowchart, emerge as a central component in automating and enforcing estate-related processes. These self-executing contracts, governed by predefined rules and conditions, alleviate the need for manual intervention and enhance accuracy. The automated execution of tasks such as asset preservation, ownership transfer, and distribution to beneficiaries is facilitated through the seamless integration of smart contracts, contributing to increased efficiency within the estate management process (Ullah & Al-Turjman, 2023).

User interfaces are meticulously designed to cater to the diverse needs of stakeholders, including administrators, legal representatives, beneficiaries, and auditors. These interfaces serve as gateways for stakeholders to interact with the blockchain network, providing access to estate-related information, transaction initiation, and real-time monitoring of ongoing processes. The emphasis on user-friendly interfaces aligns with the goal of fostering efficient communication and collaboration among stakeholders.

Identity verification mechanisms play a crucial role in ensuring the integrity and security of the estate management process. Robust verification methods, such as digital signatures or biometric authentication, are integrated into the system to thwart fraudulent activities and restrict participation to authorized individuals or entities. This layer of security complements the overall robustness of the proposed blockchain-based system.

The flowchart underscores the significance of secure data management through the application of cryptographic techniques. Estate-related data, encompassing transactions, ownership transfers, and asset preservation activities, is encrypted and stored within the blockchain. The immutable and tamper-proof nature of the blockchain ensures a transparent and auditable record of all estate-related activities, mitigating concerns related to data integrity and security.

Facilitating coordination among multiple agencies is a pivotal aspect of the architecture system, as highlighted in the flowchart. The blockchain network serves as a shared platform for secure information exchange, validation of transactions, and collaborative decision-making. Consensus mechanisms embedded within the blockchain network ensure unanimous agreement on transaction validity, minimizing disputes and fostering cooperation among participating agencies.

Thus, the flowchart system architecture for a blockchain-based system presents a structured and comprehensive approach to address the challenges faced in frozen estate management. By leveraging the unique features of blockchain technology, including decentralization, transparency, security, and automation, the proposed system aims to streamline processes, enhance trust, and improve overall efficiency in managing frozen estates.

The potential transformative impact of blockchain technology on frozen estate management is evident through the analysis of challenges and the exploration of blockchain-based solutions. Streamlining processes, enhancing transparency, and improving efficiency are achievable outcomes with the adoption of blockchain in this industry. While acknowledging these potential benefits, it is crucial to recognize the need for careful consideration of legal, regulatory, and technical aspects, along with stakeholder collaboration and adoption. As the technology evolves, further research and practical implementations are imperative to fully realize the potential of blockchain in streamlining frozen estate management. With the right infrastructure, governance frameworks, and adoption strategies, blockchain has the capacity to reshape the frozen estate management landscape, ushering in a more efficient and transparent future.

Acknowledgements This research was supported by Fundamental Research Grant Scheme (FRGS) provided by the Ministry of Education of Malaysia with grant code: FRGS/1/2021/SS01/USIM/02/4. Special thanks to the professionals and experts in the field of blockchain technology and frozen estate management who generously shared their knowledge and experiences with us. Their willingness to engage in discussions and provide feedback greatly enriched our understanding of the subject matter.

References

Abdullah, M. M., Nasir, N. M., Muhamad, N. H., Aziz, M. R., & Awang, A. B. (2020). A literature review on Islamic estate planning from year 2014 to 2019. *Library Philosophy and Practice (e-Journal)*, 4290.

Agrawal, T. K., Kumar, V., Pal, R., Wang, L., & Chen, Y. (2021). Blockchain-based framework for supply chain traceability: A case example of textile and clothing industry. *Computers & Industrial Engineering, 154*. https://doi.org/10.1016/j.cie.2021.107130

Ahmad, M. Y. (2017). Amalan Muafakat dan Aplikasi Takharuj Dalam Pengagihan Pusaka Menurut Perspektif Undang-undang Pusaka Islam. *Islamiyyat, 39*(1), 21–28. https://doi.org/10.17576/islamiyyat-2017-3901-03

Ahmad, M. Y., & Laluddin, H. (2010). *Pengurusan Harta Pusaka: Permasalahan Sikap Masyarakat Islam di Malaysia Dan Penyelesaiannya Menurut Perspektif Islam*. Shariah Law Report.

Andoni, M., Robu, V., Flynn, D., Abram, S., Geach, D., Jenkins, D., McCallum, P., & Peacock, A. (2019). Blockchain technology in the energy sector: A systematic review of challenges and opportunities. *Renewable and Sustainable Energy Reviews, 100*, 143–174.

Anuar, S. B., Md Azmi, F. A., & Sidek, S. N. S. (2022). Exploring the potential and suggesting a new conceptual framework of blockchain technology in estate distribution. *International Journal of Advanced Research in Technology and Innovation, 4*(1), 35–43.

Aslam, J., Saleem, A., Tariq Khan, N., & Bae Kim, Y. (2021). Factors influencing blockchain adoption in supply chain management practices: A study based on the oil industry. *Journal of Innovation & Knowledge, 6*(2), 124–134.

Ayuni, N., Wahab, N., Salleh, R. M., Yusoff, R. M., Atan, S. M., & Sainan, Z. (2022). An analysis on the issues of the frozen estates in Kampong Bharu, Malaysia. *International Journal of Academic Research in Business and Social Sciences, 12*(9), 1977–1985.

Azmin, U. S., & Khairi, K. F. (2022). *Review of blockchain-based governance framework for the frozen estate in Malaysia*. Paper presented at the 10th Islamic banking, accounting, and finance international conference (iBAF 2022), Nilai, Negeri Sembilan, Malaysia.

Bailis, P., Narayanan, A., Miller, A., & Han, S. (2017). Research for practice: Cryptocurrencies, blockchains, and smart contracts; hardware for deep learning. *Communications of the ACM, 60*(5), 48–51.

Biryukov, A., Khovratovich, D., & Pustogarov, I. (2014). *Deanonymization of clients in bitcoin P2P network*. Paper presented at the Proceedings of the 2014 ACM SIGSAC conference on computer and communications security, Scottsdale, Arizona, USA. https://doi.org/10.1145/2660267.2660379

Brito, J., & Castillo, A. (2013). *Bitcoin: A primer for policymakers*. Mercatus Center. https://books.google.co.id/books?id=yC-nAwAAQBAJ

Casino, F., Dasaklis, T. K., & Patsakis, C. (2019). A systematic literature review of blockchain-based applications: Current status, classification and open issues. *Telematics and Informatics, 36*, 55–81. https://doi.org/10.1016/j.tele.2018.11.006

Chhabra, A., Saha, R., Kumar, G., & Kim, T. H. (2021). A survey of privacy concerns in blockchains and information retrieval. In *CEUR workshop proceedings (CEUR-WS.org)* (pp. 1–29).

Christidis, K., & Devetsikiotis, M. (2016). Blockchains and smart contracts for the internet of things. *IEEE Access, 4*, 2292–2303. https://doi.org/10.1109/ACCESS.2016.2566339

Kamarudin, M. K., & Nor Muhamad, N. H. (2018). Islamic estate planning and management: Malaysian experience. *International Journal of Management Applications, 1*(1), 8–16. https://www.researchgate.net/publication/322975181

Kosba, A., Miller, A., Shi, E., Wen, Z., & Papamanthou, C. (2016). *Hawk: The blockchain model of cryptography and privacy-preserving smart contracts*. Paper presented at the 2016 IEEE symposium on security and privacy (SP).

Lahsasna, A. (2016). *Shariah estate planning* (1st ed.). Islamic Banking and Finance Institute Malaysia (IBFIM).

Lu, H., Huang, K., Azimi, M., & Guo, L. (2019). Blockchain technology in the oil and gas industry: A review of applications, opportunities, challenges, and risks. *IEEE Access, 7*, 41426–41444. https://doi.org/10.1109/ACCESS.2019.2907695

Meiklejohn, S., Pomarole, M., Jordan, G., Levchenko, K., McCoy, D., Voelker, M. G., & Savage, S. (2013). *A fistful of bitcoins: Characterizing payments among men with no names*. Paper presented at the Proceedings of conference on Internet Measurement Conference. Barcelona, Spain. https://doi.org/10.1145/2504730.2504747

Mohd Alam Shah, N. A., Drs Nasrul, M. A., Halim, A. H., & Abdul Hak, N. (2022). Administration of estates: Resolution of disputes in achieving sustainability among the deceased's family. *IIUM Law Journal, 30*(2).

Nakamoto, S. (2012). *Bitcoin: A peer-to-peer electronic cash system*. Retrieved from www.bitcoin.org

Nasrul, M. A. D., Abdul Manaf, Z. I., Syafril, S., & Fathurrohman, M. (2021). An overview of the inheritance legal system in Malaysia and Indonesia: Issues FACED BY BOTH COUNTRY. *Journal of Shariah Law Research, 6*(2), 181–200.

Nor Muhamad, N. H. (2017). Wasiat Sebagai Instrumen Perancangan Harta Islam: Prosedur Dan Pelaksanaan. *Jurnal Hadhari: An International Journal, 9*(1), 17–32. https://doi.org/10.17576/JH-2017-0901-02

Rahman, M. H. (2019). Faraid distribution in Malaysia: Issues and challenges. In *Pemerkasaan Filantropi Islam di Malaysia. Institut Kajian Zakat (IKAZ); Akademi Pengajian Islam Kontemporari (ACIS)*. Universiti Teknologi Mara (pp. 212–223). ISBN 987-967-1329-86-3.

Salmon, J., & Myers, G. (2019). Blockchain and associated legal issues for emerging markets. https://openknowledge.worldbank.org/handle/10986/31202

Sedlmeir, J., Lautenschlager, J., Fridgen, G., & Urbach, N. (2022). The transparency challenge of blockchain in organizations. *Electronic Markets, 32*, 1779–1794. https://doi.org/10.1007/s12525-022-00536-0

Siegfried, N., Rosenthal, T., & Benlian, A. (2020). Blockchain and the industrial internet of things. *Journal of Enterprise Information Management*. https://doi.org/10.1108/jeim-06-2018-0140

Swan, M. (2015). *Blockchain: Blueprint for a new economy*. O'Reilly Media.

Ullah, F., & Al-Turjman, F. (2023). A conceptual framework for blockchain smart contract adoption to manage real estate deals in smart cities. *Neural Computing and Applications, 35*, 5033–5054.

Wylde, V., Rawindaran, N., Lawrence, J., Balasubramanian, R., Prakash, E., Jayal, A., Khan, I., Hewage, C., & Platts, J. (2022). Cybersecurity, data privacy and blockchain: A review. *SN Computer Science, 3*, 127. https://doi.org/10.1007/s42979-022-01020-4

Yadav, A. S., Singh, N., & Kushwaha, D. S. (2023). Evolution of blockchain and consensus mechanisms & its real-world applications. *Multimedia Tools and Applications, 82*, 34363–34408. https://doi.org/10.1007/s11042-023-14624-6

Yang, Y., Lin, T., Liu, P., & Xiao, S. (2022). UCBIS: An improved consortium blockchain information system based on UBCCSP. *Blockchain: Research and Applications, 3*(2). https://doi.org/10.1016/j.bcra.2022.100064

Yaqoob, I., Salah, K., Jayaraman, R., & Al-Hammadi, Y. (2022). Blockchain for healthcare data management: opportunities, challenges, and future recommendations. *Neural Computing and Applications, 34*, 11475–11490.

Zeba, S., Amjad, M., & Rizvi, D. (2022). Sustainable paradigm for computing the security of wireless internet of things: Blockchain technology. In M. Afshar Alam, M. A. Ahad, S. I. Ali, S. Zafar, & D. Mehta (Eds.), *Smart and sustainable approaches for optimizing performance of wireless networks* (pp. 51–66). Wiley.

Nur Hidayah Laili is currently a member of the Faculty of Economics and Muamalat of USIM, where she brings a humble dedication to her academic pursuits. Her academic foundation includes a Bachelor's degree in Accounting from Universiti Utara Malaysia, followed by a Master's in Accounting from UiTM. Driven by a sincere curiosity, she pursued and earned her Ph.D. in Management from the Macquarie University, a testament to her commitment to continuous learning. In the realm of research, she has taken on the role of Principal Investigator for the project titled "developing a blockchain-based governance framework for waqf organisations toward a sustainable economic wellbeing," which underscores her humble yet purposeful approach to contributing knowledge within the field. Additionally, Dr. Nur Hidayah serves as a Co-Researcher in the project "design and development of a prototype corporate waqf blockchain-based traceability using eos private smart contract platform for islamic social finance." Her collaborative efforts reflect a simple yet profound commitment to exploring the intersection of blockchain technology and Islamic social finance. In the spirit of shared learning, she has contributed to publications emanating from these projects, demonstrating a down-to-earth approach to disseminating insights on blockchain, corporate Waqf, and Islamic social finance.

Khairil Faizal Khairi, an Associate Professor at Universiti Sains Islam Malaysia (USIM) since 2004, is a distinguished subject matter expert specializing in Islamic Social Finance, Fintech, and Blockchain for Islamic Finance. Currently serving in the Faculty of Economics and Muamalat at USIM, Dr. Khairil has been an invaluable contributor to the academic landscape. His expertise extends to the development of innovative solutions, notably leading a team in the creation of the groundbreaking MySIF Blockchain. This cutting-edge technology is designed for Islamic Social Finance instruments, focusing on Zakat, Waqf, and Sadaqah. Dr. Khairil's commitment to advancing the field has earned him several awards, recognizing his outstanding contributions and pioneering work in Islamic social finance fintech modeling. Dr. Khairil's academic journey includes earning a Ph.D. in Management from the Macquarie University of Sydney, an MBA from Universiti Utara Malaysia, and a Bachelor of Banking and Finance from Philadelphia University, Jordan. Despite his numerous achievements, Dr. Khairil remains grounded and dedicated to the pursuit of knowledge, continually striving to make meaningful contributions to the field of Islamic Finance.

Azuan Ahmad is a seasoned professional specializing in the dynamic intersection of information security, blockchain, artificial intelligence, malware research, IoT, green IT, software development, and ICT training. With a solid academic foundation, he earned his Ph.D. and MSc in Computer Science (with a focus on Cloud Security and Information Security, respectively) from Universiti Teknologi Malaysia (UTM). Dr. Azuan's commitment to excellence is evident through his BSc

(Hons) in Information Security and Assurance from Universiti Sains Islam Malaysia (USIM). In addition to his academic prowess, Dr. Azuan holds esteemed certifications, including Certified Ethical Hacker, Certified CompTIA A+, Nav6 IPv6 Engineer, MSCP, MCP, and Smart/Green IT. These certifications underscore his comprehensive expertise in various facets of the ever-evolving technology landscape. Dr. Azuan Ahmad's dedication to knowledge dissemination is evident through his role as a certified trainer. He imparts his wealth of knowledge to diverse sectors, including military, government, and private organizations, contributing significantly to the enhancement of cybersecurity practices and IT proficiency within these domains. With a humble and focused demeanor, Dr. Azuan Ahmad continues to make meaningful contributions to the fields of information security and technology. His academic achievements, coupled with extensive industry experience, position him as a respected authority in the realms of cybersecurity and emerging technologies.

Sustaining Halal Industry Through Synergising IR4.0 Technologies and Data Analytics: *Exploring the Integration of AR, VR, MR, and Consumer Insights*

Adlin Masood, Umi Hamidaton Mohd Soffian Lee, and Nurul Adilah Hasbullah

Abstract The Halal cosmetics sector is actively embracing the digital age by leveraging the vast potential of Industry 4.0 technologies and data analytics. This article delves into the intricate exploration of the incorporation of IR4.0 technologies including Augmented Reality (AR), Virtual Reality (VR), and Mixed Reality (MR) technologies, in conjunction with the profound utilisation of data analytics, within the domain of Halal cosmetics. Moreover, it delves into the intricate ways in which these ground-breaking advancements are moulding the perceptions and inclinations of consumers, thus ushering in a paradigm shift within the industry. Through a meticulous analysis of the interplay between these cutting-edge technologies and data-derived insights, this article effectively underscores their pivotal role in augmenting consumer experiences, amplifying the transparency of supply chains, and propelling the trajectory of strategic decision-making to enable sustainability of Halal cosmetics industry.

1 Introduction

To maximise profits and advance sustainable development, a country's primary goal is to provide steady, comprehensive economic growth that is inclusive in the sense that it ensures economic stability and increases job opportunities for the general public. One industry deserving attention in contributing to this goal is the halal industry, which has experienced significant growth in recent years. The development of the industry can be attributed to the rising global demand for halal goods and services due to the increasing percentage of educated and technologically savvy

A. Masood (✉) · Umi Hamidaton Mohd Soffian Lee · N. A. Hasbullah
Faculty of Economics and Muamalat, Universiti Sains Islam Malaysia, Nilai, Negeri Sembilan, Malaysia
e-mail: adlin.masood@usim.edu.my; umihamidaton@usim.edu.my; n.adilahasbullah@usim.edu.my

A. Rafiki et al. (eds.), *Strategic Islamic Business and Management*, Contributions to Management Science,
https://doi.org/10.1007/978-3-031-61778-2_4

Muslims, projected to reach 31% of the global population by 2060 (Bhutto et al., 2022). Non-Muslim demand also plays a significant role (Firdaus et al., 2022). One of the promising areas in the Halal industry is the cosmetic market. In the halal cosmetics market, a similar trend is observed due to the growing consumer awareness and demand for halal beauty care products, particularly in regions with a large population of Muslim consumers (Azam & Abdullah, 2020; Naseri & Raja Abdullah, 2022; Ngah et al., 2020; Shahid et al., 2018; Shamsuddin & Mohd Yusof, 2020).

According to SGIER 2021/2022, the global market for halal cosmetics is expected to grow at a Compound Annual Growth Rate (CAGR) of 7.4% over the next 4 years, reaching $93 billion by 2025. The implementation of Industry Revolution 4.0 (IR4.0) technologies, coupled with advanced data analytics by emerging halal cosmetics manufacturers, is partially accountable for the rise in popularity of halal cosmetics. Additionally, Muslim spending on cosmetics was estimated at US$ 84 billion in 2022 and is forecast to reach US$ 129 billion by 2027 (SGIER, 2023).

Muslim consumers are the primary force behind the growth of the Halal cosmetics industry, but remarkably, the Halal brand has witnessed a notable upswing in appeal among consumers who identify as non-Muslim, given that Halal certification mandates that products adhere to principles of safety, hygienic practices, fair trade, and animal welfare (Rezai et al., 2015). This trend has compelled many multinational cosmetics companies to produce halal-certified goods. Some companies offer cruelty-free, organic, or vegan cosmetics as an additional endeavour to attract a wider range of consumers. Businesses, entrepreneurs, and economists have all taken notice of the increased demand for Halal cosmetics, and this growth has also piqued the curiosity of the scientific community. Masood et al.'s (2023) comprehensive literature review showed a significant upsurge in Halal cosmetics-related research during the past 15 years. The existing body of literature has endeavoured to ascertain a comprehensive comprehension of the principal determinants and motivations behind the usage and expansion of Halal cosmetics. Research conducted in countries such as Pakistan (Bhutto et al., 2022; Ertz, 2022), China (Ali et al., 2021), India (Shahid et al., 2018, 2022), and Indonesia (Rizkitysha, 2020; Widyanto & Sitohang, 2022) focusing on Millennials drew on the theoretical framework of theory of planned behaviour (TPB).

Every industrial revolution centred around boosting productivity. IR4.0, which could ultimately be referred to as the fourth industrial revolution, is a highly complex framework that has been commonly debated and discovered. It has a significant impact on the industrial sector because it introduces relevant improvements related to smart and future factories. This developing IR4.0 concept is an umbrella term for a new industrial paradigm that includes Cyber-Physical Systems (CPS), the Internet of Things (IoT), the Internet of Services (IoS), Robotics, Big Data, Cloud Manufacturing, Augmented Reality, etc. (Oztemel & Gursev, 2020).

According to Zaid (2020), the sustainability of the halal industry is significantly influenced by the complexity of technology. Given that the industry is still in its nascent stage, this chapter endeavours to enhance the current understanding of the impact of the incorporation of IR4.0 technologies including Augmented Reality (AR), Virtual Reality (VR), and Mixed Reality (MR) technologies, together with the utilisation of data analytics, within the domain of Halal cosmetics. Moreover, it delves into the intricate ways in which these ground-breaking advancements are moulding the perceptions and inclinations of consumers, thus ushering in a paradigm shift within the industry of behaviours regarding Halal cosmetics products. The aforementioned impacts will be thoroughly examined and analysed within the context of this research. The findings of this research will not only advance the theoretical knowledge but also provide practical implications. It is believed that decision makers and marketers, equipped with this knowledge, may effectively strategise, synchronise, and oversee both the macro- and micro-systems of the Halal cosmetics business. This will ultimately contribute to the long-term viability and continuity of the Halal cosmetics and strive towards achieving inclusive economic expansion, thereby benefiting society.

2 Literature Review

As the halal cosmetics market continues to evolve, the adoption of innovative IR4.0 technologies will play a crucial role in ensuring the long-term sustainability. The convergence of AR, VR, MR, and data analytics provides a unique opportunity to reshape the industry's landscape, aligning with the preferences and values of modern consumers. This chapter focuses on the interplay between these technologies, shedding light on their potential applications and benefits.

To establish a theoretical framework, an initial step involved conducting a comprehensive literature review. According to Lambert, a literature review can be defined as a comprehensive evaluation of existing knowledge pertaining to the subject of inquiry, encompassing the associated themes and diverse viewpoints presented on the matter. Meredith proposed that it aids in the advancement of theoretical frameworks as literature review can ascertain the conceptual substance of the domain being examined.

To get insight into and understand the factors that drive the consumer or customer to purchase a product, it is important to look at their purchasing behaviour. Purchasing behaviour refers to the consumer's search, purchase, usage, appraisal, and disposal of product, service, or ideas offered to them to satisfy their needs (Lin & Niu, 2021). Basically, it differs from one individual to another, and one generation to another.

2.1 *The Purity Pursuit: Unveiling Halal Cosmetics and Consumer Behaviour*

In the fabric of human needs, the search of beauty is intertwined with that of religion. Halal cosmetic goods are regarded as an innovation and revolution in the cosmetic sector, providing high-quality items that adhere to halal regulations and meet demanding scientific criteria. Despite these characteristics, halal cosmetic firms face hurdles in presenting themselves and determining how to encourage customers to make purchases. Halal cosmetics are items made from halal materials and designed to beautify, cleanse, protect, and change the appearance of specific body parts (Sugibayashi et al., 2019). The fundamental distinctions between halal cosmetics and regular ones centre on three key factors: the cosmetic materials utilised, the manufacturing techniques involved, and the marketing strategies used. For Muslim customers, the importance of Halal cosmetics extends beyond religious observance. These goods must go through thorough scientific testing to assure purity, safety, and the usage of high-quality natural ingredients. Notably, they avoid adding forbidden ingredients such as alcohol or meat into their manufacture. Furthermore, they frequently promote ethical and cruelty-free activities. As a result, the Halal label is not only an assurance for Muslims, but also a global emblem of superior quality and a conscious lifestyle choice (Wilson & Liu, 2011). This allure extends to non-Muslims, causing them to explore using halal cosmetics. According to the State of Global Islamic Economy study, the halal cosmetics business is growing, notably with halal-certified items. According to statistics, Muslims are expected to spend US$ 129 billion on halal cosmetics by 2027, up from US$ 84 billion in 2022.

2.1.1 The Pillars of Purity

According to Wilson (2019), the Pillars of Purity are often linked with halal research. He points out that halal branding includes ingredient branding as well as putting a halal logo on the package, giving purchasers full confidence. In keeping with the tenets of purity and ethics inherent in the Halal concept, Malaysia has placed an emphasis on halal branding through the use of certified halal logos and permitted ingredients. Cleanliness is a spiritual principle embedded in the Quran (2: 222), as both Siraj (2011) and Soltani et al. (2022) point out. Every Muslim has a need to pursue purity, and this idea includes the beauty ritual products they use, as this verse echoes. That is why halal cosmetics encompass more than just cruelty-free beauty products. Many global companies rely on preservatives derived from pork or alcohol, which has made Muslim consumers wary of these brands' claims (Abdulkadir, 2022). This incident exemplifies why halal cosmetics are becoming more popular among Muslim people. All the way from the ingredients used to make the product to its final packaging, halal products take a comprehensive approach. According to Shahid et al. (2022), it is imperative that no materials that are considered Haram, such as blood, carrion, or animal by-products, be present on this journey. The principle of '*halalan thoyyiban*', which promotes clean and high-quality products for Muslim consumers, is based on this dedication to purity (Hashim & Hashim, 2012). As per Islamic law, Muslims must only eat halal food and must abstain from anything that is considered haram or questionable. Consequently, Muslim buyers are picky about what they buy. Instead of simply looking for a product standardisation certification, they prefer to buy products that have the International Organisation for Standardisation (ISO) mark, which indicates that they are Halal-certified (Ariffin et al., 2019).

2.1.2 Beyond Religion: A Scientific Symphony

The appeal of Halal cosmetics, however, goes beyond simple religious considerations, as Sugibayashi et al. (2019) rightly note. To guarantee the effectiveness and safety of its products, the industry requires thorough scientific investigation. In the crucibles of halal cosmetics research and development, science and religion come together to create skin-nourishing products that are in line with Islamic beliefs. The careful balancing act makes sure that Muslim consumers aren't put in a position where they have to choose between their faith and the effectiveness of their beauty routines. This is because health factors can also impact the consciousness of using halal cosmetics. Fearing for their skin, customers are becoming more selective in the cosmetics they buy, thanks to the rise of illegal and harmful products. Muslim consumers can reap spiritual and material benefits from using halal cosmetics since they are free of impurities and harmful substances (Mohezar et al., 2016). These products also score highly in terms of kindness, health, safety, and beauty. Accordingly, Islam forbids the use of and trade in anything that is deemed evil, filthy, or useless. Because of the potential dangers they pose, it is against the law to consume them. The profusion of halal-certified pharmaceuticals shows that pharmaceutical products have progressed considerably more technically than cosmetic items. However, in order to meet the increasing global demand, further scientific study into making halal cosmetics is required. Many Muslim customers place a high value on knowing where their cosmetic products come from and how they are made (Briliana & Mursito, 2017).

2.1.3 Decoding Consumer Choices

Both local and imported products related to cosmetics and healthcare are available in Malaysia, including those from regions outside of Muslim-majority countries. This raises concerns among consumers, particularly within the Muslim community, regarding the true compliance of these items with halal standards. Consumers today have expressed significant concerns about whether beauty and personal care items comply with halal and Sharia laws, given the intricate nature of international trade. Consumer behaviour, as Belch and Belch (2004) eloquently describe, is a complex interplay between needs, desires, and external influences. In the case of Halal cosmetics, this takes on a unique rhythm, driven by both faith and worldly considerations. Panitapu (2013) underlines the multidimensional nature of consumer behaviour, highlighting its critical role in informing marketing strategies. Gogoi (2013) further sheds light on the internal and external forces that shape purchase decisions. Studies by Hong et al. (2019), Ishak et al. (2019), Kamaljeet et al. (2018), and Masood et al. (2021) offer valuable insights into these intricate motivations, revealing that Halal consumers prioritise factors like product quality, certification, brand reputation, social influence, and even packaging aesthetics.

2.1.4 Generation Z: The Tech-Savvy Seekers

Bilgen and Kanata (2016) predicted that by 2023 and 2030, Muslims would account for 26 and 30% of the global population, respectively. A sizable section of this cohort are tech-savvy millennials. As a result of this development, a globally integrated halal industry has emerged, providing its own unique set of issues and considerations that are especially pertinent to the halal market. Enter Generation Z, the digital natives born between 1997 and 2012. According to Ghosh (2016), their objectives are worldwide, their communication styles are technologically advanced, and their curiosity is unbounded. Hays (2017) describes them as a generation filled with entrepreneurial zeal, self-sufficient problem-solving abilities, and a deep understanding of cause-driven decisions. These characteristics pique their interest in Halal cosmetics, making them sophisticated consumers who want not only purity and quality, but also brands that reflect their beliefs and resonate with their technologically savvy environment. The demand for halal cosmetics has expanded due to the younger generation's careful consumption habits.

2.2 Technology Acceptance Model (TAM)

Figure 1 displays the technology adoption model (TAM), which states that technology adoption is predicted by users' behavioural intentions, which are defined by the perception of technology usefulness in accomplishing the task and perceived ease of use (Davis, 1989). Furthermore, TAM typically serves as a foundation for study into how to improve user acceptability through the careful selection of system design elements, as well as why people reject or accept technology. This finding is consistent with previous research by Rizkitysha and Hananto (2022), who observed that people who have a strong and positive sense of the effectiveness of systems use them more than those who have a weak or negative perspective. According to Mohd Amir et al. (2020), adoption of technology is an important component in influencing an individual's behavioural intention, as well as in customer purchasing power, as discovered by Juaneda-Ayensa et al. (2016). Furthermore, according to Forbes Technology Council (2016), 'if something makes customers' lives easier, more fun,

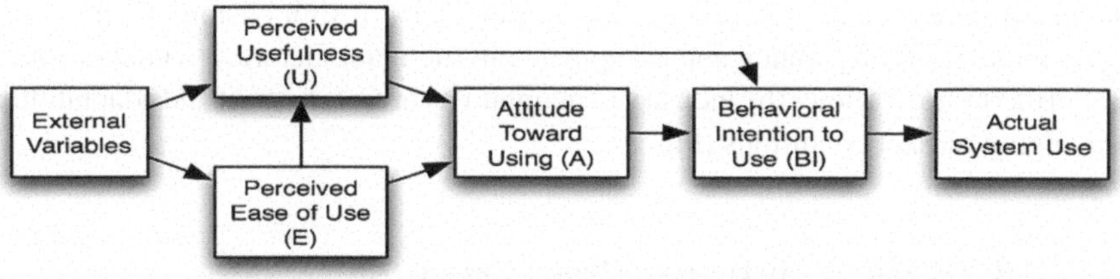

Fig. 1 Technology acceptance model. Source: Adapted from Davis (1989), 'Perceived usefulness, perceived ease of use, and user acceptance of information technology'

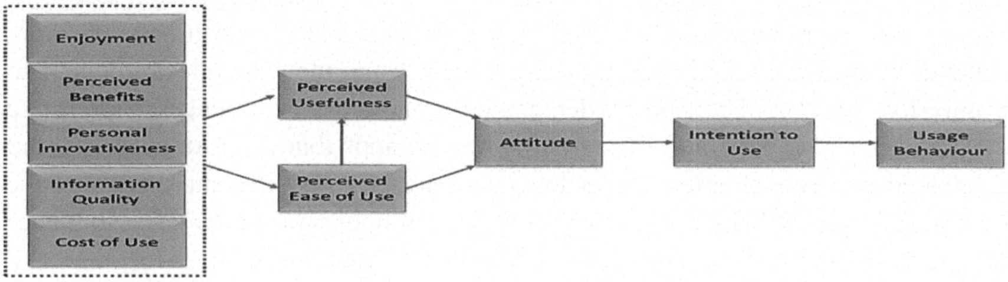

Fig. 2 Proposed AR acceptance model. Source: Adapted from Leue et al. (2014) of a theoretical model of augmented reality acceptance

or more interesting, customers will adopt it and prompt similar technology to be built'.

The particular opportunities and constraints of consumer adoption in this domain can be better understood when augmented reality (AR) is integrated into the TAM, which is often called TAM-AR or TAM2-AR. According to Dieck (2014), TAM-AR or TAM2-AR adds new aspects to the original UX/UI framework, which include: enjoyment, perceived benefits, personal innovativeness, information quality, costs of use, and perceived utility and simplicity of use. Here is the model, as seen in Fig. 2. The model's capacity to grasp user perceptions and behaviour in the ever-changing IR4.0 technology ecosystem is improved by this addition. In addition, by including elements directly related to the AR experiences, the model can handle the complexities of user acceptability.

2.3 *Beauty Reimagined: How AR, VR, MR, and Data Analytics Are Transforming Halal Cosmetics*

In this era of IR4.0, the convergence of cutting-edge technologies is transforming many industries, including the beauty sector. A ground-breaking change is taking place in the Halal cosmetics industry, which is deeply rooted in religious and ethical concerns, thanks to the integration of mixed reality (MR), augmented reality (AR), and virtual reality (VR) with advanced data analytics. In addition to enhancing the customer experience and ensuring a bright future for Muslim customers, this synergy guarantees transparency and compliance in the manufacture of Halal cosmetics. Furthermore, it allows the industry to maintain its inventive spirit and adaptability to changing market conditions (Fig. 3).

2.3.1 AR/VR/MR: A Mirror on a Digital Canvas

The days of getting by with just product descriptions and swatches are over. Makeup, hairstyles, and skincare routines may all be virtually tried on, tested for impact, and verified to be halal with the help of augmented reality, virtual reality, and MR technology. Verizon Media found in 2020 that sixty per cent of consumers are more inclined to buy from brands that produce content using innovative technology, and seventy-five per cent of consumers demand new interactive experiences. Brands like MAC and Sephora have introduced online try-on apps that use virtual reality and augmented reality technology. This lets customers try on different shades of eye shadow and lipstick from the comfort of their own homes or on the go. Using augmented reality technology, New Kinpo Group scans customers' faces for flaws and then suggests skincare items that will fix those issues. To help customers along the way, Olay and Clinique have released a new app that addresses hair and skin issues. To further aid customers with application techniques, numerous brands also give real-time support in the form of product combinations and cosmetic videos (Thomas, 2019). Using augmented reality (AR) in real-time marketing has given consumers a fun, interactive, and useful experience.

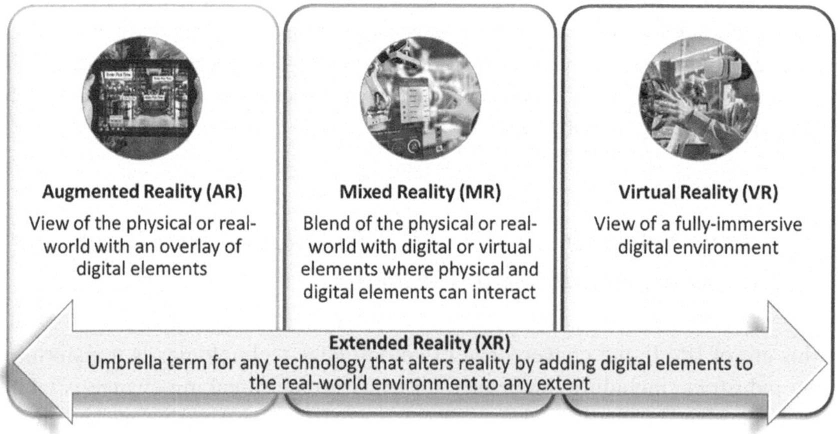

Fig. 3 Representation of current AR, MR, VR, and XR technologies. Source: Authors own concept

Augmented Reality (AR) in the Halal Cosmetics Experience

A new platform for mobile technology called Augmented Reality (AR) has come out as an engaging and dynamic way to show visual information about products (Vonkeman & Van Dolen, 2017). Presenting virtual, three-dimensional product information in real-time within a consumer market context is the core of augmented reality (Javornik, 2016). By providing a full 360-degree picture of virtual products, augmented reality builds upon 3D technology. By superimposing digital features onto the user's physical environment, this idea merges the physical and digital worlds (Cawood & Fiala, 2008). Sugibayashi et al. (2019) found that augmented reality (AR) in halal cosmetics lets buyers digitally investigate and confirm the halal status of items. Products' ingredient lists, production details, storage info, packaging details, and shipping details can all be accessed with a simple scan. By letting users get emotionally invested in the brands they purchase from and conduct more thorough product evaluations, this immersive technology improves both the shopping experience and the value of the products itself (Vilkina & Klimovets, 2019). Brands must quickly expand their augmented reality (AR) ecosystems if they want to capture a larger portion of the market, as consumers are increasingly demanding more tailored and interactive experiences from brands (Deloitte, 2021).

As a result, augmented reality is revolutionising the way people engage with Halal beauty products. Customers may virtually try on different things through smartphone applications and smart mirrors, creating a personalised and immersive experience with augmented reality. According to Vonkeman et al. (2017), augmented reality makes it easier for customers to visually examine things in their own homes. By incorporating augmented reality into the purchasing experience, Halal-conscious consumers may expect a more engaging and powerful adventure.

Virtual Reality (VR) for Immersive Exploration

According to Cawood and Fiala (2008), the main difference between AR and VR is the way they handle the technology. AR adds virtual components to users' real-world contexts, whereas VR completely immerses users in virtual worlds. Virtual features can be superimposed onto real-world objects using augmented reality technology, allowing users to interact with and manipulate them in new ways. But virtual reality goes further and beyond by letting customers virtually investigate the making of Halal cosmetics. Virtual tours allow customers to see where Halal-certified foods are sourced and provide insight into the stringent quality control procedures used. Customers can rest easy knowing that Halal cosmetics are genuine because of this openness. Brands may connect with the Halal-conscious community on a deeper level using virtual reality by sharing their commitment to sustainable and ethical practices.

Mixed Reality (MR) Bridging the Physical and Virtual Worlds

By integrating aspects of augmented reality and virtual reality, MR brings a fresh perspective to the Halal cosmetics sector. MR technologies possess the capacity to evoke emotions by rendering the invisible visible and imbuing our physical environment with additional depths of significance. It facilitates the emergence of novel possibilities associated with establishing meaningful connections between the physical and digital realms, between what is immediately observable and what remains to be discovered, and between our tangible environment and our creative faculties (Malinverni et al., 2018). Consequently, it permits concurrent interaction between tangible and digital components for consumers in order to superimpose digital content onto corporeal products. This practice not only improves the overall purchasing experience but also guarantees that consumers are well-informed when it comes to selecting products that align with their Halal preferences.

One could envision themselves virtually entering a salon that has obtained Halal certification, engaging in a consultation with an avatar stylist, and virtually experimenting with various styles in real-time, all from the convenience of their own residence. This facilitates well-informed decision-making and not only increases convenience but also decreases product returns and waste. In addition to virtual try-on capabilities, MR has the potential to significantly transform product education, envision the capability of a holographic display to reveal the ingredients, ethical procurement practices, and Halal certifications of a lipstick tube upon scanning it. This level of transparency fosters confidence and enables consumers to make well-informed decisions that are consistent with their personal values. The beauty and cosmetics sector has experienced a significant enhancement in efficiency, flexibility, and responsiveness through the incorporation of technological advancements into its fundamental value chain.

2.3.2 Data Analytics: The Invisible Alchemist

The power of data analytics is revolutionising how Halal cosmetics brands understand and cater to their consumers. Data analytics, the unseen magician, wields its power behind the scenes. By analysing consumer preferences, purchase patterns, and online reviews, Halal beauty brands can personalise product recommendations, predict trends, and develop innovative formulas that cater to diverse skin types and needs thereby meet the diverse needs of the Halal-conscious market.

Imagine a Halal cosmetics app that analyses your skin tone and texture, recommends personalised skincare routines based on your preferences and religious requirements, and tracks your progress over time. This level of customisation fosters a deeper connection between brands and consumers, creating a sense of community and shared values.

Ensuring Compliance and Integrity

Rapid product innovation is required in the cosmetics and personal care business due to the high level of competition and the rapid pace of the market. An increasing number of people are looking for all-natural, halal cosmetics that respect their religious and cultural customs, which could lead to a surge in the market. Therefore, in order to conform to halal standards, cosmetics companies that make halal products are forced to use innovative technologies in all stages of the product life cycle, including development, formulation, and manufacturing. This change is being driven by the growing number of Muslim consumers who are more conscious of the chemicals and non-halal components in the cosmetics they buy. A product is considered 'Halal' if its production, storage, packaging, and distribution adhere to Sharia law. This includes not only the components, but also the manner in which they are prepared and delivered. In addition to ingredients, halal covers a wide range of consumables, such as medicines, cosmetics, personal care products, toiletries, and more.

According to research by Othman et al. (2023), halal cosmetics producers face challenges such as sourcing halal ingredients, dealing with unhelpful suppliers, keeping products halal throughout production, and creating halal application documentation. Fortunately, data analytics provide a solution to these problems by ensuring the authenticity and conformance of halal cosmetics. Using data analytics, one can trace the whole halal cosmetics supply chain, beginning with the sourcing of raw materials and ending with the finished product. By following this approach, halal cosmetics can be more easily made without the use of any non-halal materials or methods.

Moreover, data analytics may be used to keep an eye on the halal cosmetics quality. By conducting quality control tests, analysing consumer comments, and reviewing products, manufacturers can identify issues with product quality. This guarantees that the halal cosmetics are of high quality and conform to all regulations. In addition, producers can be more open and honest with customers about the halal cosmetics information by using data analytics technologies. In the Halal cosmetics industry, where honesty is paramount, this openness fosters devotion and trust. Lastly, it has the potential to strengthen the credibility of the brand.

2.3.3 The Ethical Equation

However, amidst the technological frenzy, navigating the ethical landscape remains crucial. Data privacy and security are paramount, and ensuring that sensitive user information is collected, stored, and utilised responsibly is non-negotiable. Transparency and user consent are key, while AI algorithms must be free from bias and discrimination to uphold the core values of inclusivity and fairness that are central to the Halal philosophy.

3 Conclusion and Implications for the Future

The influence of AR, VR, MR, and data analytics on Halal cosmetics is just starting to unfold. With the continuous advancement of these technologies, it is likely that virtual consultations with dermatologists who specialise in Halal-certified products will become more common. Additionally, personalised cosmetic formulas may be tailored to individual genetic profiles and religious preferences. The boundaries between the physical and digital worlds will blend together, resulting in a one-of-a-kind and captivating beauty experience that goes beyond geographical limits and embraces diversity.

Ultimately, the combination of IR4.0 technologies and ethical data analytics is revolutionising the world of Halal cosmetics, completely transforming the way consumers perceive and rely on these products. The captivating experiences provided by AR, VR, and MR, along with the valuable insights obtained from data analytics, are not just satisfying the needs of tech-savvy consumers, but also guaranteeing the adherence to Halal cosmetics standards in the constantly evolving beauty sector. With the rapid advancements in technology, the future of Halal cosmetics appears to be quite promising. This positive outlook is fuelled by a strong focus on innovation, transparency, and a sincere dedication to ethical practices.

This chapter emphasises the importance of halal cosmetics industry players embracing and utilising the advanced technologies provided by IR4.0. In order to reach a wider range of consumers and expand their market share, producers must respond to the increasing desire for natural and organic products. It is important for them to emphasise that halal certification is a universally recognised industry standard that is applicable to all consumers, regardless of their religious beliefs. In addition, small- and medium-sized enterprises (SMEs) can utilise blockchain technology in the context of Industry 4.0 to establish a trustworthy system for Halal traceability. This can help them cultivate strong relationships with customers and enhance their brand value.

Acknowledgement This research was funded by Universiti Sains Islam Malaysia under the Matching Grant Scheme (USIM/UNIDA/FEM/SEPADAN/72923).

References

Abdulkadir, A. (2022). Religiosity and consumer's attitude towards halal food and cosmetic products. *International Journal of Halal Research, 4*(22), 97–104.

Ali, A., Sherwani, M., Ali, A., Ali, Z., & Sherwani, M. (2021). Investigating the antecedents of halal brand product purchase intention: an empirical investigation. *Journal of Islamic Marketing, 12*(7), 1339–1362. https://doi.org/10.1108/JIMA-03-2019-0063

Ariffin, S. K., Azra, W. F., Wahid, N. A., & Nee, G. Y. (2019). Investigating the factors affecting purchase intention of Muslim women towards halal cosmetics. *Journal of Entrepreneurship, Business and Economics, 7*(2s), 78–105.

Azam, M. S. E., & Abdullah, M. A. (2020). Global halal industry: Realities and opportunities. *International Journal of Islamic Business Ethics, 5*(1), 47–59. https://doi.org/10.30659/ijibe.5.1.47-59

Belch, G. E., & Belch, M. A. (2004). Advertising and Promotion an Integrated Marketing Communications Perspective. New York.

Bhutto, Yaseen, M., et al. (2022). Adoption of halal cosmetics: Extending the theory of planned behavior with moderating role of halal literacy (evidence from Pakistan). *Journal of Islamic Marketing*. https://doi.org/10.1108/JIMA-09-2021-0295

Bilgen, E., & Kanata, C. (2016). New look at faith-based marketing: The global halal tourism marketing. *Business Horizons, 59*(3), 285–292.

Briliana, V., & Mursito, N. (2017). Exploring antecedents and consequences of Indonesian Muslim youths' attitude towards halal cosmetic products: A case study in Jakarta. *Asia Pacific Management Review, 22*, 176–184.

Cawood, S., & Fiala, M. (2008). Augmented reality: A practical guide. Pragmatic bookshelf. Retrieved August 24, 2011, from https://books.google.com.

Davis, F. D. (1989). Perceived usefulness, perceived ease of use, and user acceptance of information technology. *MIS Quarterly, 13*(3), 319–340. https://doi.org/10.2307/249008

Deloitte. (2021). Snap consumer AR global report 2021 Retrieved from Deloitte: https://www2.deloitte.com/content/dam/Deloitte/xe/Documents/AboutDeloitte/Snap%20Consumer%20AR_Global%20Report_2021.pdf

Dieck, M. C., Dieck, D. T., & Jung, T. H. (2014). A theoretical model of augmented reality acceptance.

Ertz, M. (2022). *Adoption of halal cosmetics: Extending the theory of planned behavior with moderating role of halal literacy (evidence from Pakistan), (April)*. https://doi.org/10.1108/JIMA-09-2021-0295

Firdaus, F. S., Ikhsan, R. B., & Fernando, Y. (2022). Predicting purchase behaviour of Indonesian and French Muslim consumers: Insights from a multi-group analysis. *Journal of Islamic Marketing*. https://doi.org/10.1108/JIMA-05-2021-0169

Forbes Technology Council. (2016, June 21). How consumers are impacting the adoption of new technology. Retrieved January 5, 2024, from https://www.forbes.com/sites/forbestechcouncil/2016/06/21/how-consumers-are-impacting-the-adoption-of-new-technology/?sh=692849517942.

Ghosh, S. (2016). Six things brands need to learn about Generation Z. Accessed from: https://www.mysalaam.com/en/story/six-things-brands-need-to-learn-about-generationz/SALAAM17102016061329

Gogoi, B. (2013). Study of antecedents of purchase intention and its effect on brand loyalty of private label brand of apparel. *International Journal of Sales & Marketing, 3*(2), 73–78.

Hashim, P., & Hashim, D. M. (2012). A review of cosmetic and personal care products: Halal perspective and detection of ingredient. *Pertanika*, 282–292.

Hashim, P., & Mat Hashim, D. (2013). A review of cosmetic and personal care products: Halal perspective and detection of ingredient. *Pertanika Journal of Science and Technology, 21*(2), 281–292.

Hays.com. (2017). Welcome Gen Z: Here's what employers want from you – and how far they'll adapt. Accessed from: https://www.hays.com.sg/press-releases/HAYS_1890617

Hong, M., Sun, S., Beg, A. B. M. R., & Zhou, Z. (2019). Determinants of halal purchasing behaviour: Evidence from China. *Journal of Islamic Marketing, 10*(2), 410–425.

Ishak, S., Che Omar, A.R., Khalid, K., Ab. Ghafar, I.S., Hussain, M.Y. (2019) "Cosmetics Purchase Behaviour of Educated Millennial Muslim Females" *Journal of Islamic Marketing 11*(5), pp. 1055–1071.

Javornik, A. (2016). Augmented reality: Research agenda for studying the impact of its media characteristics on consumer behavior. *Journal of Retailing and Consumer Services, 30*, 252–261.

Juaneda-Ayensa, E., Mosquera, A., & Murillo, Y. S. (2016). Omnichannel customer behavior: Key drivers of technology acceptance and use and their effects on purchase intention. *Frontiers in Psychology, 7*(1117), 1–11. https://doi.org/10.3389/fpsyg.2016.01117

Kamaljeet, K., Osman, S., Sukjeet, K., & Jaspal, S. (2018). Is halal a priority in purchasing cosmetics in Malaysia among muslim malaysian woman?. International Journal of Business, Economics and Law, 15(3), (April).

Leue, M. C., Jung, T., & Dieck, D. T. (2014). *A theoretical model of augmented reality acceptance, e-review of tourism research*. Manchester Metropolitan University.

Malinverni, L., Valero, C., Schaper, M. -M., & Pares, N. (2018). A conceptual framework to compare two paradigms of augmented and mixed reality experiences. In *Proceedings of the 17th ACM conference on interaction design and children—IDC '18*.

Masood, A. (2021). Factors influencing Halal cosmetics purchase behaviour of working adults and university students in Malaysia. *Jurnal Personalia Pelajar, 24*(1), 113–127.

Masood, A., Hijrah Hati, S. R., & Rahim, A. A. (2023). Halal cosmetics industry for sustainable development: A systematic literature review. *International Journal of Business and Society, 24*(1), 141–163. https://doi.org/10.33736/ijbs.5609.2023

Mohd Amir, R. I., Mohd, I. H., Saad, S., Abu Seman, S. A., & Tuan Besar, T. B. H. (2020). Perceived ease of use, perceived usefulness, and behavioral intention: The acceptance of crowdsourcing platform by using technology acceptance model (TAM). In N. Kaur & M. Ahmad (Eds.), *Charting a sustainable future of ASEAN in business and social sciences*. Springer. https://doi.org/10.1007/978-981-15-3859-9_34

Mohezar, S., Zailani, S., & Zainuddin, Z. (2016). Halal cosmetics adoption among young Muslim consumers in Malaysia: Religiosity concern. *Global Journal Al-Thaqafah, 6*(1), 47–59.

Naseri, R. N. N., & Raja Abdullah, R. N. (2022). Trends and growth of halal cosmetic industry in Malaysia. *International Journal of Academic Research in Business and Social Sciences, 12*(2), 487–495. https://doi.org/10.6007/IJARBSS/v12-i2/12892

Ngah, A. H., Ramayah, T., Ali, M. H., & Khan, M. I. (2020). Halal transportation adoption among pharmaceuticals and cosmetics manufacturers. *Journal of Islamic Marketing, 11*(6), 1619–1639. https://doi.org/10.1108/JIMA-10-2018-0193

Niu, Y., Ying, L., Yang, J., Bao, M., & Sivaparthipan, C. B. (2021). Organizational business intelligence and decision making using big data analytics. *Information Processing & Management, 58*(6), 102725.

Othman, Y., Awang, M. N., & Shah, H. A. R. (2023). Legal and regulatory challenges of halal certification: Insights from cosmetics manufacturers on halal built-in implementation. *International Journal of Academic Research in Business and Social Sciences, 13*(1), 322–335. https://doi.org/10.6007/IJARBSS/v13-i1/15785

Oztemel, E., & Gursev, S. (2020). Literature review of industry 4.0 and related technologies. *Journal of Intelligent Manufacturing, 31*(1), 127–182. https://doi.org/10.1007/s10845-018-1433-8

Panitapu, M. K. (2013). Models of consumer buying behavior. Available at SSRN: https://ssrn.com/abstract=2275391 or https://doi.org/10.2139/ssrn.2275391

Rezai, G., Mohamed, Z., & Shamsudin, M. N. (2015). Can halal be sustainable? Study on Malaysian consumers' perspective. *Journal of Food Products Marketing, 21*(6), 654–666. https://doi.org/10.1080/10454446.2014.883583

Rizkitysha, T. L., & Hananto, A. (2022). Do knowledge, perceived usefulness of halal label and religiosity affect attitude and intention to buy halal-labeled detergent? *Journal of Islamic Marketing, 13*(3), 649–670. https://doi.org/10.1108/JIMA-03-2020-0070

Shahid, S., Ahmad, F., & Hasan, U. (2018). A qualitative investigation into consumption of halal cosmetic products: The evidence from India. *Journal of Islamic Marketing, 9*(3), 484–503. https://doi.org/10.1108/JIMA-01-2017-0009

Shamsuddin, A., & Mohd Yusof, F. (2020). Exploring the issues and challenges in Malaysian cosmetic halal: A theoretical framework. *Nusantara Halal Journal, 1*(1), 10–14. https://doi.org/10.17977/um060.2020v1p010-014

Soltani, A., Johnston, L., & Longhurst, R. (2022). Fashioning hybrid Muslim women's veiled embodied geographies in Hamilton, Aotearoa New Zealand: #hijabi spaces. *Gender, Place and Culture, 29*(3), 393–418. https://doi.org/10.1080/0966369X.2021.1946487

State of Islamic Economic Report (2023) (SGIER 2023) pdf.

Sugibayashi, K., Yusuf, E., Todo, H., Dahlizar, S., Sakdiset, P., Arce, F. J., & See, G. L. (2019). Halal cosmetics: A review on ingredients, production, and testing methods. *Cosmetics, 6*(3), 1–17. https://doi.org/10.3390/cosmetics6030037

Thomas, D. (2019). Five tech trends shaping the beauty industry. *BBC News*. Retrieved from https://www.bbc.com/news/business-48369970

Vilkina, M. V., & Klimovets, O. V. (2019, May). Augmented reality as marketing strategy in the global competition. In *Institute of scientific communications conference* (pp. 54–60). Springer.

Vonkeman, C., Verhagen, T., & Van Dolen, W. (2017). Role of local presence in online impulse buying. *Information & Management, 54*(8), 1038–1048.

Widyanto, H. A., & Sitohang, I. A. T. (2022). Muslim millennial's purchase intention of halal-certified cosmetics and pharmaceutical products: The mediating effect of attitude. *Journal of Islamic Marketing, 13*(6), 1373–1394. https://doi.org/10.1108/JIMA-04-2020-0117

Wilson, J. A. J. (2019). A look at halal branding with the man who made it cool. Retrieved January 8, 2024, from https://halalfocus.net/a-look-at-halal-branding-with-the-man-who-made-it-cool/.

Wilson, J. A. J., & Liu, J. (2011). The challenges of islamic branding: Navigating emotions and halal. *Journal of Islamic Marketing* 2(1), 28-42. Emerald Group Publishing. Limited 1759–0833. Accessed from https://doi.org/10.1108/17590831111115222

Zaid, M. F. A. (2020). Propositions on the relationships between technology complexity, industry 4.0, and halal sustainability. *Journal of Engineering and Science Research, 4*(1), 52–58. https://doi.org/10.26666/rmp.jesr.2020.1.9

Adlin Masood is a Senior Fellow in the Faculty of Economics and Muamalat, Universiti Sains Islam Malaysia. Having proven dynamic 15 years track record in the hospitality industry with special focus in the field of marketing, public relations and communications has given Adlin a totally different perspective and outlook in the academic world. This has ultimately resulted towards a practitioner's approach in educating students, research, and consultancy. Preferring to look at the dynamics and framework of halal systems in various fields in research, from marketing extending to other functional business areas, even including education through experiential learning. Currently involved in simulation game innovation for halal training and consultancy in Taiwan and Zhengzhou, China and exploring other avenues with network in Korea and Uzbekistan. Other areas of interest include SDGs and ESG of Asia Pacific nations through halal economy and green sukuk as well as application of IR4.0 technologies.

Umi Hamidaton Mohd Soffian Lee is a Senior Lecturer in the Faculty of Economics & Muamalat at Universiti Sains Islam Malaysia (USIM) and possesses a bachelor's and master's degree in human science from IIUM, as well as a Ph.D. in Islamic Civilization from UKM. She is an accomplished author, having written a book on ethical leadership, multiple articles in scholarly publications, and book chapters, as well as presenting talks at seminars and conferences in her subject field. Appointed as the halal programme coordinator at the Graduate School of Muamalat (USIM) and Research Fellow at the Institute of Fatwa and Halal (IFFAH), Dr. Umi is also a journal and research grant reviewer, a guest lecturer both locally and abroad, and a primary investigator and member of numerous research programmes and initiatives. Her research interests are mostly in management, with an emphasis on organisational management and behaviour, the halal business, and Islamic philanthropy.

Nurul Adilah Hasbullah is a senior lecturer at Universiti Sains Islam Malaysia's (USIM) Faculty of Economics and Muamalat. She graduated from USIM in 2013 with a bachelor's degree and in 2016 with a master's degree. She then pursued a Doctor of Philosophy degree in Shariah and Economics at Universiti Malaya (UM), where she was awarded her degree in 2021. Her passion is Islamic social finance, particularly waqf. She has authored a number of publications in indexed journals in this discipline. The Routledge book 'Higher Education Finance and Islamic Endowments' was released by the author in 2024. She also dabbled in the halal industry at the same time, publishing a number of conference and journal publications in this area.

The Influence of Digital Transformation in Speeding Up the Empowerment of the Muslim Women's Economy to Improve the Performance of Micro, Small, and Medium Enterprises (MSMEs) in Ponorogo

Siti Nur Azizah , Suradi, Koeswinarno, M. Fakhruddin, and Irma Yuliani

Abstract This research aims to analyze the modeling of the impact of digital transformation in accelerating the empowerment of the Muslimah economy to enhance the performance of micro, small, and medium enterprises (MSMEs) in Ponorogo. This study used a mixed-method approach, namely quantitative and qualitative. This study involves 269 MSMEs in Ponorogo that women manage as members of Islamic organizations and have food, beverages, medicine (including herbal medicine), and cosmetic products. The model used is based on the APKM theory (Access, Participation, Control, and Benefits) in the gender approach to determine the effect of digital transformation on women's empowerment and the Resource-based View (RBV) theory to determine the improvement of MSME performance with a partial least square structural equation model as a data analysis tool and using a Phenomenological Approach to explore experiences by interviewing around 10% of respondents who represent all sample data. The study proved that digital transformation has a positive relationship with economic empowerment managed by Muslim women who are members of Islamic organizations. Overall, this signifies that the implementation of digital transformation positively affects women's economic empowerment and can improve the performance of micro, small, and medium enterprises (MSMEs). The practical implication of this research is that by analyzing

S. N. Azizah (✉)
Faculty of Islamic Economics and Business, Sunan Kalijaga State Islamic University, Yogyakarta, Indonesia
e-mail: siti.azizah@uin-suka.ac.id

Suradi · Koeswinarno · M. Fakhruddin
Research Centre of Public Policy, The National Research and Innovation Agency (BRIN), Jakarta, Indonesia

I. Yuliani
Faculty of Islamic Economics and Business, Ponorogo State Islamic Institute, Ponorogo, East Java, Indonesia

A. Rafiki et al. (eds.), *Strategic Islamic Business and Management*, Contributions to Management Science, https://doi.org/10.1007/978-3-031-61778-2_5

the modeling and influence of digital transformation on women's economic empowerment and the performance of micro, small, and medium enterprises (MSMEs), managers may realize the significant role of digital transformation in influencing women's economic empowerment and impacting operational and financial performance. It will attract more MSMEs to transform digitally to open up new and profitable business opportunities for MSMEs. It also empirically confirms that digital transformation can influence women's economic empowerment and improve the performance of MSMEs. Based on this research, we suggest advocating for the Ponorogo regional government to continue to review and control the use of digitization in MSMEs, as the implementation of tracking and reevaluation has so far been limited. Reevaluation so far is still limited. In addition, strict supervision should be accompanied by transparency, accountability, and engagement of various stakeholders, including Islamic organizations, business associations, and civil society.

1 Introduction

Women's empowerment is well-designed in the UN Sustainable Development Goals (SDGs). One way to achieve this empowerment is through female entrepreneurship (Crittenden et al., 2019). Female entrepreneurship has been of interest to academics for some time. It can drive economic growth and catalyze employment (Barney, 1991). The transformation of an agrarian society into an industrial society shows clear evidence of the importance of women's role in the economy (Sorgner, 2021).

Based on the 2022 population census, Indonesia's population is around 275,361.267 million people. 49.5% of this population is female, and 53.6% are of productive age (BPS, 2022). This data shows the great potential of women's resources to become an engine of economic growth. Referring to data published by the Coordinating Ministry for Human Development and Culture in 2020, the participation of women's Micro, Small, and Medium Enterprises (MSMEs) in gross domestic product (GDP) in Indonesia reached 9.1%. Meanwhile, more than 5% of women's MSME products in Indonesia contributed to the total export of domestic products. Even more than 60% of MSMEs in Indonesia are managed by women, including fashion, culinary, and crafts. These MSMEs continue to show good results. In addition, the COVID-19 event provided MSMEs with unique strengths, such as digital transformation in marketing and sales processes. According to data from the Coordinating Ministry for PMK, more than 18% of female MSMEs are involved in e-commerce, which will increase to 24% by 2022. This figure is higher when compared to the number of male MSMEs, which is only around 13.2% in e-commerce.

Undoubtedly, Indonesian women have a higher business ownership rate than the global average. Data from Google and Kantar in 2020 shows that currently, the response of women in Indonesia is increasingly positive toward entrepreneurship. The number of women in Indonesia who have become entrepreneurs is 49%, and the number of women who want to become entrepreneurs in the future is 45%. Unfortunately, women who use digital technology account for only about 35% of all online sales in Indonesia. One of the reasons for this is the unfinished gender gap in financial inclusion (Garz et al., 2020). Globally, the number of women who have undergone digital transformation, such as owning a financial access account and mastering e-commerce applications, is still about 65% lower than the number of men, which is 72% (Shofawati, 2019; Trivelli et al., 2018). This hurdle presents a challenge to MSMEs due to limited digital access, hindering progress toward financial inclusion by bringing individuals, households, and businesses into a more efficient system, particularly those women who face restricted access (Asian Development Bank, 2022).

Digital transformation offers substantial benefits for women engaged in SMEs, enhancing competitiveness and informed decision-making (Yang et al., 2022). Digital transformation is an ongoing process that changes the approach to conducting business through digital technology from personal lives (Ancillai et al., 2023; Nyagadza, 2022). It mainly involves the integration of novel technological innovations based on Internet services and modern information technology (Holzmann & Gregori, 2023). Ultimately, female involvement in MSMEs and their ability to embrace digital transformation and cultivate competitiveness (Summers et al., 2020) facilitate freedom and diminish discrimination (Hilbert, 2011), enhance gender equality, and empower women (GSMA, 2017), as well as foster direct customer communication (Papa et al., 2017), can tangibly augment the effectiveness of business organizations, specifically MSMEs (Khanna & Sharma, 2022). Small and medium enterprises to enhance their efficiency, productivity, and sustainability while maximizing their profits.

The high benefits of digital transformation should be an opportunity for government policies and financial institutions to design digital services to make women's economic empowerment more effective and indirectly improve the performance of MSMEs. A significant knowledge gap in studies on the effect of digital transformation on women's economic empowerment in the context of improving MSME performance has received attention from several researchers. Among the indicators used are mobile banking (Dorfleitner and Nguyen, 2022; Summers et al., 2020), digital financial inclusion (Gallego-losada et al., 2023; Pal et al., 2022; Soekarno et al., 2020), women and social media marketing (Aracil-Jord et al., 2023), gender equality and digital inclusion (Pawluczuk et al., 2021), financial literacy (Hasan & Ashfaq, 2022), and technology-empowered (Hilbert, 2011).

This research is different from previous research. This research focuses on the influence model of digital transformation in accelerating Muslimah economic empowerment to improve the performance of MSMEs in Ponorogo. The reason is that around 86.7% of Indonesia's population is Muslim, and around 53% are Muslim women. In addition, Ponorogo is one of the districts in East Java, Indonesia, with 51.9% female population. As well as having more than 65% of MSMEs are managed by women, with a continuous increase from year to year. In 2019, the total number of MSMEs in Ponorogo was 31,328 and rose to 35,025 in 2020, and the number of MSMEs rose again to 38,387 in 2021. Until this data is released, the number of MSMEs in Ponorogo Regency as of 2022 is 39,650. In addition, the variables used in this study differ from previous studies, including organizational culture, competitive pressure, organizational mindfulness, IT readiness, and strategic alignment, to determine how digital transformation affects the acceleration of women's economic empowerment and its impact on improving MSME performance. Despite these relevant studies, according to the authors' knowledge, there has yet to be a study that review the economic empowerment of Muslim women who are members of religious activities and as owners or managers of digital-based MSMEs to improve the performance of the MSMEs they manage.

This article is structured as follows. We will first discuss the role of digital transformation APKM on women's economic empowerment to analyze and identify gender-responsive policies, programs, or intervention activities. Next, we will elaborate on the specific model of the antecedence of digital transformation to Muslim women's economic empowerment. The next discussion elaborates on the effect of Muslim women's economic empowerment on improving the performance of MSMEs. Then, we will explain the methodology applied using the quantitative and qualitative methods. Subsequently, we will highlight the discussion points based on the study context and present the theoretical and practical implications. The article will be closed with a conclusion and suggestions for local government policy and future work.

2 Literature Review (and Hypotheses Development)

2.1 Digital Transformation and Muslimah Economic Empowerment

Digital transformation is a multidisciplinary topic, analyzed from diverse perspectives, including business strategy and innovation, information technology, human resources, and marketing (Llopis-Albert et al., 2021; Nwankpa & Roumani, 2016; Verhoef et al., 2021). Subjectively, digitalization and digital transformation differ from each other. Digitization involves converting information into digital format to enable computers to store, process, and transmit it effectively (Gong & Ribiere, 2021; Verhoef et al., 2021). Furthermore, digital transformation employs technology to foster business innovation that can lead to increased revenues, value, and financial performance. It also involves a significant alteration in the makeup and functionality of a system to create new value. Ahmed et al. (2022) show that digital transformation enhances women's resilience and aids them in overcoming financial difficulties. All three studies confirm the positive impact of digital transformation on women's lives and businesses. Suwana and Lily (2017) also found that digital transformation advances the quality of life for women. Additionally, Crittenden and Crittenden research suggests that digital transformation enhances women's self-efficacy and social capital ties in MSMEs, ultimately leading to greater economic empowerment. According to Nord et al. (2016), technology plays a significant role in the economic empowerment of women in both developing and developed countries. It has the potential to enhance national economies. This study's hypotheses consist of

H1: There is a positive relation between digital transformation and the economic empowerment of Muslimah.

Analyzing the effect of digital transformation on women's economic empowerment, this study involves the Gender and Development (GaD) approach, which is an approach to development policies and programs based on social construction, gender roles, and relations between men and women (ADB, 2022). The Gender and Development approach ensures that the digital transformation of MSME groups in Ponorogo has changed inequalities, especially those covering access, participation, control, and benefits (APKM) for women. So that women's rights, roles, and opportunities can be achieved in the context of economic empowerment. GaD is influenced by the socialist feminist approach and is known as women's empowerment, often called gender mainstreaming.

These four aspects become a reference to determine gender gaps and problems between men and women. In the data analysis process, this article uses the Gender Analysis Pathway (GAP) analysis technique. GAP is an analytical tool for identifying gender-responsive policies, programs, or intervention activities (Andersen, 1988), (Fig. 1). There are five steps in the GAP model analysis, including (1) identifying the goals and objectives of the activity program, (2) presenting quantitative or qualitative data selected based on gender, (3) analyzing the factors that cause gender gaps, (4) identifying gender gap problems based on the factors that cause gaps by answering the 5W+1H questions, and (5) identifying where gender gaps occur.

2.2 *Antecedent Digital Transformation*

Indicate that digital transformation can change business models. Soluk et al. (2023) argue that for businesses to operate efficiently, they must adapt to new technologies. Technology-driven digital transformation can be highly competitive and innovative and transform corporate strategies (Vial, 2019). It is due to businesses catering to customer demands and satisfaction levels. Digital transformation maximizes profits, reduces cost inefficiencies, and enhances production effectiveness (Khanna & Sharma, 2022). Accordingly, the digital revolution offers opportunities to leverage corporate assets, including Muslim women or women involved in MSMEs, for the performance improvement of MSMEs. The following are the preconditions required for developing the digital transformation model:

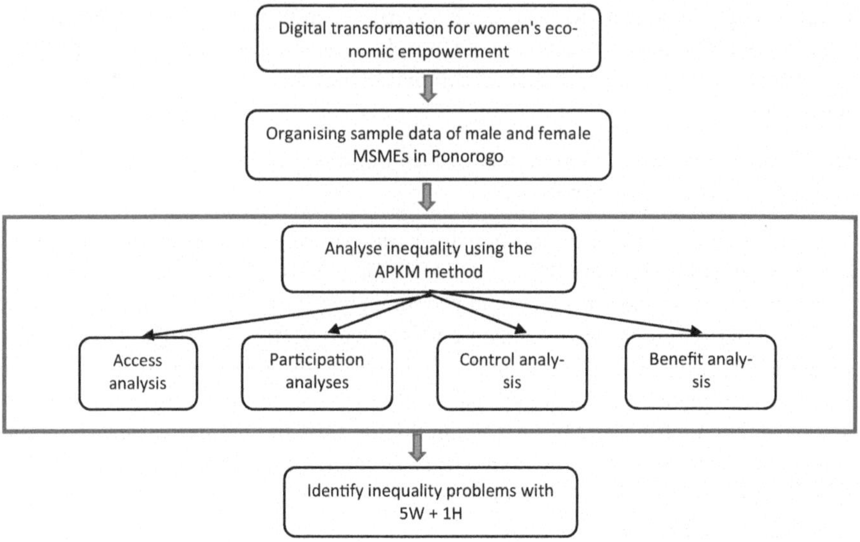

Fig. 1 APKM of digital transformation on women's economic empowerment

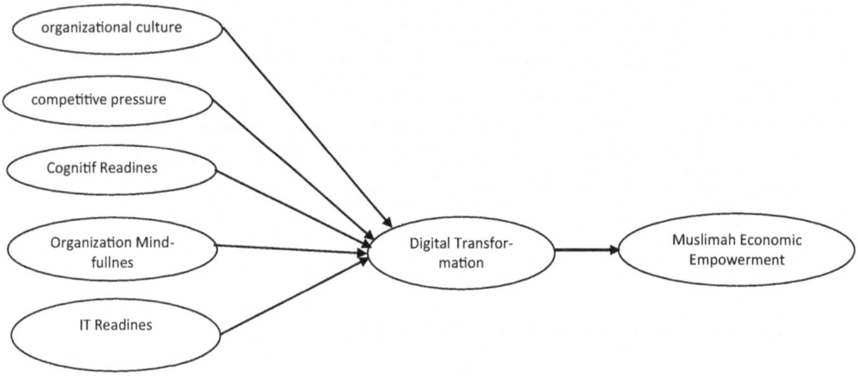

Fig. 2 Antecedence model of digital transformation on Muslimah economic empowerment

2.2.1 Organizational Culture

The organizational culture comprises company-specific attributes, adopted norms, and guiding beliefs integral to the operation (Srivastava et al., 2020), (Fig. 2). It relates to the shared beliefs and understandings concerning the organization's functions (Moore, 2020). This is relevant in incorporating new personnel and current leadership during external integration. Thus, resistance to change is a facet of organizational culture that may impact digital transformation (Martínez-Caro et al., 2020; Usai et al., 2021). Consequently, the hypothesis that can be derived from this explanation is

H1a: There is a positive relation between *organizational culture and the digital transformation process.*

Moreover, the economic empowerment of women is influenced by organizational culture (Hibbs, 2022). There is a debate concerning the sustainability of women's competitive advantage (Eger et al., 2022). It is based on the idea that a company's performance can be significantly impacted by a positive organizational culture, facilitating technological transformation.

H1b: There is a positive relation between organizational culture and women's economic empowerment.

2.2.2 Competitive Pressure

Fierce competition compels companies to adjust and embrace novel technologies for enduring competitiveness persistently (Hsu et al., 2014). Digital technology compels companies to meet all contest modes (Adjei et al., 2021). Heightened competitive pressure obliges companies to undertake digital transformation (Raut et al., 2017). Digital transformation involves introducing novel technologies, managing change within the organization, optimizing business processes, and redefining business models (Dinçer et al., 2017). Companies that invest in digital transformation will emerge as significant players in their respective industries. Ultimately, competitive pressure becomes the norm in digital transformation.

H2a: There is a positive relation between competitive pressure and digital transformation

Competitive pressure makes women involved in business, especially MSMEs, more innovative. It is because competitive pressure leads to the exploration of individual innovation so that organizations can face stronger competitors in the market (Holzmann & Gregori, 2023). So, women involved in MSMEs will tend to experiment with ideas and innovate their products and services (Meyer et al., 2021). Thus, MSMEs managed by women can improve product standards and prevent barriers in the market (Hilbert, 2011). Therefore, the researcher hypothesized:

H2b: There is a positive relation between competitive pressure and Women's Economic Empowerment

2.2.3 Cognitive Readiness

Cognitive readiness is considered a knowledge-based force in organizations and facilitates digital innovation (Hussain & Papastathopoulos, 2022). Cognitive readiness involves skills, adaptability, and knowledge in adopting technological transformation (Halpern et al., 2021). Cognitive readiness is essential in adapting to an unpredictable industrial environment (Seong & Hong, 2022). It is also one of the factors for adapting to technology. Therefore, the hypothesis in this study is as follows:

H3a: There is a positive relation between cognitive readiness and digital transformation

Furthermore, cognitive readiness is considered a strategic business resource (Halpern et al., 2021). Cognitive readiness is considered a valuable, rare, irreplaceable, and inimitable resource, but it can be learned (Hussain & Papastathopoulos, 2022). Women with cognitive readiness are considered to have a high level of resilience (Papa et al., 2017). Women with cognitive readiness will be able to adapt to technology (Tay et al., 2022). Therefore, women will be far from financial stress, income concerns, and gender violence (Hilbert, 2011). In addition, indirectly, women will manage MSMEs and be ready to compete (Yang et al., 2022). From previous research, this study has the following hypothesis:

H3b: There is a positive relation between cognitive readiness and Women's Economic Empowerment

2.2.4 Organization Mindfulness

Organizational mindfulness encourages selective companies to map threats and opportunities quickly and precisely in the digital transformation era (Verma et al., 2022). Periodic evaluation and monitoring processes make it easier for companies to map opportunities and threats (Pagani & Pardo, 2017). This action can encourage innovation and high efficiency in the company's operating system (Soluk et al., 2023). Companies aware of the rapid changes in the digital transformation era will actively utilize information technology and intelligence as a form of awareness of technological changes. The selection of the type of digital platform, for example, in the sales process, is always in line with the company's marketing strategy. If company managers do not pay attention to the development of digital transformation, then decision-making in business will only be based on past knowledge and experience (Eger et al., 2022). Organizational mindfulness allows companies to set realistic expectations and is the key to success in digital transformation (Li et al. 2021). Organizational mindfulness helps businesses enrich current knowledge to adjust, positively impacting digital transformation (Poláková-Kersten et al., 2023). Therefore, this study develops hypotheses:

H4a: There is a positive relation between organization mindfulness and digital transformation

In the context of women's empowerment, organizational mindfulness of changes in the digital transformation era is needed to achieve a competitive advantage (Oeij et al., 2022). This step is needed by women entrepreneurs so that they can easily make business decisions (Papa et al., 2017). With organizational mindfulness of digital transformation, companies can gain much current information in their business processes (Tay et al., 2022). Organizational mindfulness is also useful in detecting factors that drive changes in the business environment and informing future competitor conditions. These strategic actions encourage companies to perform well (Aracil-Jord et al., 2023). Therefore, we develop a hypothesis:

H4b: There is a positive relation between organizational awareness and Women's Economic Empowerment

2.2.5 IT Readiness

Information technology (IT) readiness is important for companies to achieve high innovation in the digital era (Nwankpa & Roumani, 2016). The emergence of information technology encourages companies to respond more quickly to market needs and conduct business processes. If companies are equipped with IT in their business processes, it can reduce the conventional challenges they generally face (GSMA, 2017). IT readiness generally encompasses relevant processes and operational units of procedure required to select the right IT for the company (Barney, 1991). Companies with IT resources can apply digital technology as a basis for big data analysis, especially social media data, as a marketing channel. Companies with the latest IT infrastructure will also benefit more because they can design business models as business solutions in the digital era. Therefore, we develop hypotheses:

H5a: IT readiness has a significant positive effect on digital transformation.

With increasing competition in the digital era, women, as part of business actors, need to consider IT as their business capital (Pawluczuk et al., 2021). It plays an important role in creating market opportunities and delivering product differentiation quickly and precisely (Hasan & Ashfaq, 2022). Companies with high IT readiness can achieve a competitive advantage (Yang et al., 2022). Superior IT infrastructure will also encourage opportunities to collaborate with other business partners. So, new businesses emerge that impact women's economic empowerment in the era of digital transformation (Barney, 1991). Therefore, we develop a hypothesis:

H5b: IT readiness has a significant positive effect on women's economic empowerment.

2.3 Muslimah Economic Empowerment in Improving MSME Performance

In assessing the performance of micro, small, and medium enterprises, this study uses the Resource-based View (RBV) theory. RBV theory has been believed to be one of the established theories for assessing business performance. Barney says the

View Theory is a theory that describes how a company can achieve a competitive advantage by relying on resources so that it can direct the company to be sustainable continuously. Barney (1991) explains that all resources available to the company if used effectively and efficiently can produce profitable business results. The foundation of RBV theory is based on the assumption that a firm's tangible or intangible attributes if properly executed may be able to create inimitable, rare, irreplaceable, and valuable holdings. Owning such resources can benefit the company with a sustainable competitive advantage.

Based on the RBV concept (Fig. 3), this research argues that empowered women are the resources micro, small, and medium enterprises use to achieve sustainable competitive advantage. This research argues that MSMEs are aggressive in producing and selling homogeneous products but must face intense, competitive, saturated, and highly volatile market competition. So, MSMEs are competing to create creative products and always try to find ways to survive and even lead the market by offering products that meet consumer expectations. Therefore, MSMEs rely on women's abilities to convince consumers of product excellence.

To support this argument, evidence from previous research shows that companies, especially MSMEs, use women's resources as instruments to achieve the desired business results. Research by Nurazira et al. (2023) found a positive relationship between empowered women, technological literacy, and operational performance, especially financial performance. Research by Malodia et al. (2023) found performance differences between MSMEs managed by women and men, especially in the fashion and food sectors. Furthermore, Khanna and Sharma (2022) found that empowered women have high resilience and influence the operational performance of MSMEs. Apart from being a Business manager in MSMEs, MSME players use the influence of women's communication prowess as a mechanism to improve organizational performance. Therefore, this study proposes the following hypothesis:

H1. Women empowerment has a positive effect on the performance of MSMEs
H2. Women's empowerment has a positive effect on financial performance
H3. Operational performance has a positive influence on financial performance
H4. Operational performance positively mediates the relationship between MSMEs and financial performance

Based on the review of relevant literature and using the RBV theory as the foundation of the theoretical framework

3 Research Methodology

This research uses quantitative and qualitative methods. The complexity of many factors affecting the justification of research results supports the use of mixed approaches in research. In addition, interpretation, clarification, description, and validation of quantitative results, as well as grounding and modification of

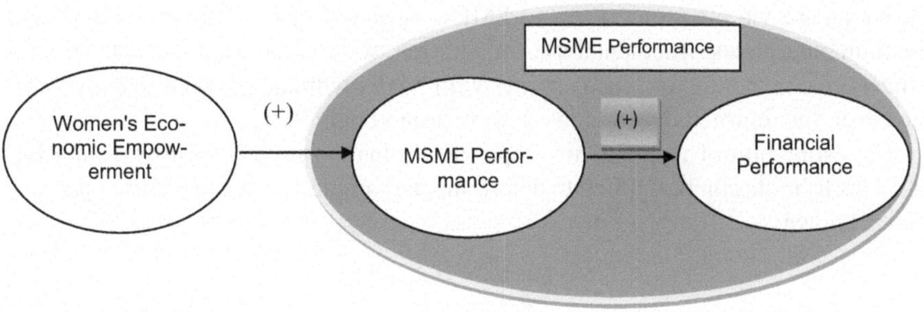

Fig. 3 Proposed theoretical model

theoretical perspectives, can all be aided by using qualitative data. Thus, this study used mixed methods to strengthen the results. This is because conclusions can be strengthened through cross-validation triangulation obtained when different types and sources of data converge consistently. At the data collection level, Morgan proposed a matrix approach to bridge qualitative and quantitative research, where classification is based on two types of decisions: priority and order. Based on this matrix, the primary methodology of this research is quantitative, and the secondary method, small-scale qualitative methods, is used to assess and understand the findings.

3.1 Study 1 (Quantitative Approach)

As this study aims to look at modeling the effect of digital transformation on women's economic empowerment to improve the performance of MSMEs, it applies a quantitative research design and uses a survey methodology. Data was collected using a self-administered questionnaire. The questionnaire was designed based on a review of past literature and sourced from several published instruments similar to this study. The questionnaire was divided into four sections (Sections A, B, C, and D). Section A involves five items designed to collect demographic information about the respondents. At the same time, section B consists of 5 items relating to the components of digital transformation. Section C has 12 items relating to women's economic empowerment, and section D has 12 items on operational (eight items) and financial (four items) performance. Section A's nominal scale was used to collect personal and organizational background information. Sections B, C, and D used a five-point Likert scale ranging from 1 (strongly disagree) to 5 (strongly agree). Before data collection, the questionnaire was pretested to check for wording, sequence, clarity, relevance, adequacy, and overall structure.

We sent the questionnaire to MSMEs managed by Muslim women, with about 269 respondents meeting the criteria of belonging to an Islamic organization and having food, beverage, medicine (including herbal), and cosmetic products in Ponorogo. We used both electronic and conventional media. We sent the questionnaire via electronic form to SMEs registered in the Ponorogo small and medium enterprises office database. In addition, we also enlisted the help of field officers to conduct face-to-face surveys with SMEs without any recorded contacts. However, the returned questionnaires were around only 269.

The estimation of respondent sample determination used Slovin's formula Eq. (1). This formula can be applied to determine the sample size when the total population is known:

$$n = \frac{N}{1 + Ne^2} \tag{1}$$

where n is the sample size, N is the population, and e is the error term (0.05). Thus, by entering the variables into Eq. (1), the minimum sample size (n) is 269. However, for data validity reasons, only 268 data could be processed by the author. The demographic data of the respondents is presented in Table 1.

3.2 *Study 2 (Qualitative Method)*

Qualitative methods were used to confirm the results of the first study. Qualitative research's main purpose is to find out the relevance of theoretical conceptualization to existing field conditions to get results and an in-depth understanding of an event. In addition, qualitative methods are also used in this research because the data obtained represents what happened and does not experience the impact of data reduction into numbers, such as data from quantitative research. Qualitative data is believed to be rich and holistic. Thus, the complexity of the effects of digital transformation on women's economic empowerment to improve the performance of MSMEs will be more likely to be revealed than the results of quantitative data. This research combines two approaches, namely phenomenology and literature. Because researchers need the two data to complement each other. Our phenomenological approach tries to explore experiences by interviewing around 10% of respondents who represent all sample data used, namely around 26 respondents, to find phenomena that occur in the field, including the criteria of belonging to an Islamic organization and having food, beverage, medicine (including herbs), and cosmetics products located in Ponorogo (Fig. 4).

4 Analysis Method

Once the data was collected, the correlational relationship between digital transformation and Muslimah economic empowerment in order to improve MSMEs performance was analyzed using the partial least squares structural equation modeling (PLS-SEM) technique. The PLS-SEM technique was chosen for data analysis as

Table 1 Respondent description

Characteristics	Description	Frequency	Percentage (%)
Status	Married	220	0.78
	Not married	49	0.22
Age	<25 years old	54	0.20
	26–35 years old	89	0.33
	36–45 years old	126	0.47
Education	Elementary school	1	0.00
	Junior high school	35	0.13
	Senior high school	126	0.47
	Bachelor's degree	99	0.37
	Master's degree	8	0.03
Business scale	Large enterprises	24	0.09
	Small enterprises	40	0.15
	Medium enterprises	33	0.12
	Micro enterprises	172	0.64
Types of MSMEs	Food and beverage	110	0.41
	Fashion	57	0.21
	Manufacturing	21	0.08
	Retail	38	0.14
	Services	24	0.09
	Travel	19	0.07
Type of Islamic organizations followed	Fatayat Cadre	116	0.43
	Muslimatan	91	0.34
	Aisyah	62	0.23
Types of technology used	E-commerce	84	0.31
	Mobile banking	51	0.19
	Qris	16	0.06
	Situs website	29	0.11
	Social media	89	0.33
Duration of MSME establishment	<1 Years	54	0.20
	1–3 Years	38	0.14
	4–6 Years	148	0.55
	6 Years	29	0.11
MSME Omzet per month	IDR. 5,000,000–10,000,000	118	0.44
	IDR. 10,100,000–Rp. 20,000,000	65	0.24
	IDR. 20,000,000–Rp. 50,000,000	67	0.25
	>Rp. 50,000,000	19	0.07

(continued)

Table 1 (continued)

Characteristics	Description	Frequency	Percentage (%)
Monthly income	IDR. 2,000,000–Rp. 5,000,000	43	0.16
	IDR. 5,000,000–Rp. 10,000,000	175	0.65
	>Rp. 100,000,000	51	0.19

Note: US$ 1 on average equals IDR 15,000
Source: Primary data from the survey

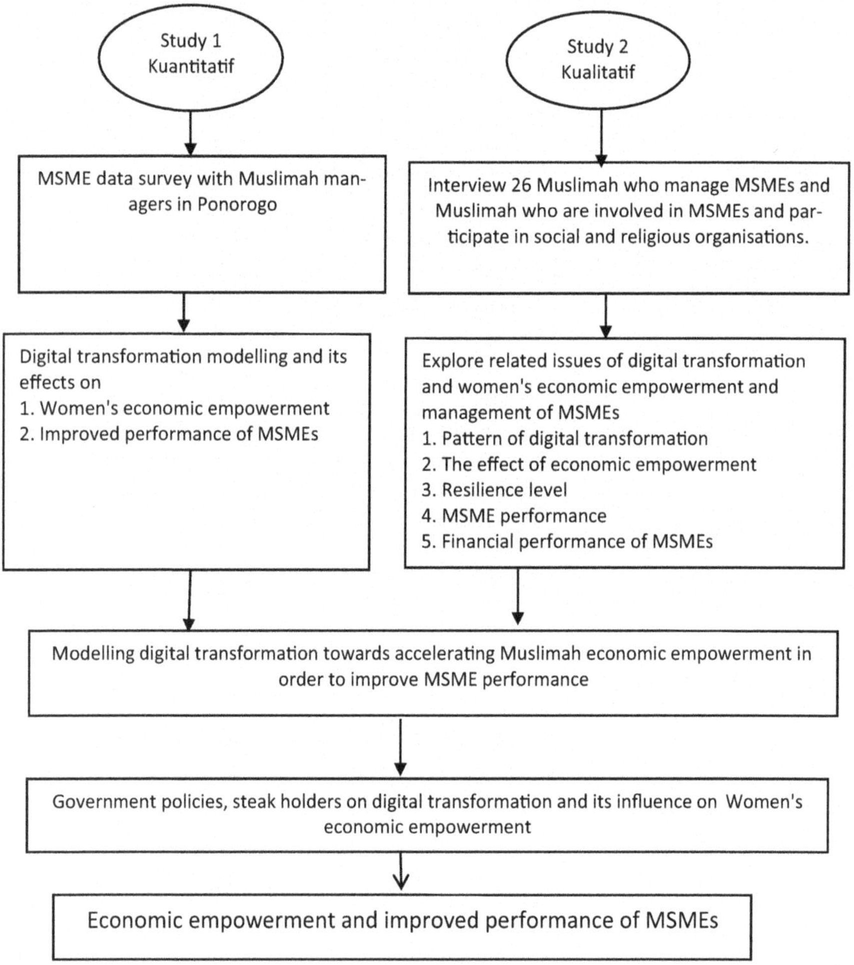

Fig. 4 The research flow. Source: Authors' construct

this approach is suitable for exploratory research where understanding and substantial information are somewhat limited. In addition, PLS-SEM is a latent variable modeling technique that simultaneously observes multiple variables and recognizes measurement errors. In addition, PLS-SEM is widely applied because it has relatively high statistical power to analyze complex structural equation models with multiple constructs. The SmartPLS 2.0 application was used for data analysis.

Meanwhile, the results of study 2 research were analyzed using a phenomenological approach. Phenomenology is the science of phenomena. Phenomenology is employed to gain an understanding of various social symptoms or phenomena within society, as well as a medium for examining and analyzing the inner life of individuals in the form of experiences regarding phenomena or appearances as they are, which are commonly called streams of consciousness. Craib says that the phenomenological approach connects scientific knowledge with everyday experience, while everyday activities and experiences are the source and root of scientific knowledge. Thus, in this study, the phenomenological approach, by quoting Creswell (2017), seeks to explain the meaning of the lived experience of several people (respondents) about a concept or symptom, including their self-concept or outlook on life. They are related to the influence of digital transformation on Muslim women's economic empowerment and improvement in the performance of MSMEs.

This study uses interpretative phenomenological analysis (AFI) or Interpretative Phenomenology Analysis (IPA). According to Smith and Osborn, IPA aims to reveal in detail how participants interpret their personal and social worlds. The main target is the meaning of various experiences, events, and statuses owned by participants. Through the IPA method, researchers explore personal experiences and emphasize an individual's perception or opinion about objects or events.

IPA seeks to understand "what it is like" from the participant's point of view to be able to stand in their shoes. "Understanding," in this case, has two meanings, namely understanding-interpretation in the sense of identifying or empathizing and the second meaning of understanding in the sense of trying to make meaning. IPA emphasizes meaning-making on the part of both participants and researchers. As cognition becomes the central analysis, there is an interesting theoretical alliance with the cognitive paradigm often used in contemporary psychology that addresses mental processes.

The stages of interpretative phenomenological analysis were carried out as follows: (1) Reading and re-reading; (2) Initial-noting; (3) Developing emergent themes; (4) Searching for connections across emergent themes; (5) Moving the next cases; and (6) Looking for patterns across cases.

5 Results and Discussion

5.1 *Analysis of Model 1*

5.1.1 Measurement Model Assessment (Outer Model)

The measurement model analysis examined many components within the assessment context, such as indicator loadings, average variance, and composite reliability. The assessment also included concerns about discriminant validity, which involved examining factor loadings and the connection between latent variables. The main objective of the inquiry on convergent and discriminant validity was to ascertain content validity by verifying item factor loadings; as described by Hair et al. (2019), the factor loading of the 61 questions in the instrument was determined to be 0.70, suggesting a substantial impact on clarifying underlying structures. The assessment procedure for determining the convergent and discriminant validity of the model constructs adhered to a reflective measuring technique, which involved multiple stages. The discriminant validity of this study refers to the ability of individual latent variables to distinguish themselves from each other (Hair et al., 2019). This concept underscores the idea that the correlation between indicators belonging to different latent variables should be substantial but weak concerning other variables. The evaluation of the measurement model's discriminant validity involved examining cross-loading values. The square root of the Average Variance Extracted (AVE) score for each construct in every diagonal section is greater than the correlation with other constructs. It indicates that each construct has passed the test of discriminant validity (Hoyle, 2012), as indicators of different constructs do not exhibit high correlations. Measurement model indicators are shown in Table 2.

Based on table above, the outer loading value shows one invalid item with the code RC3. This is because the outer loading value of item RC3 is less than 0.7 (Shmueli et al., 2019). However, according to Sarwono and Narinawati, an outer loading value of more than 0.5 is qualified and considered significant. According to Brown, a factor loading of more than 0.4 is still valid in confirming a factor construct or variable. As a criterion for selecting items based on factor analysis, the Kaiser–Meyer–Olkin Measure of Sampling Adequacy (KMO) value limit is > 0.50. If the KMO value is above 0.50 and the loading factor is> 0.4, the respondent's answer to the question item is considered valid enough to be analyzed further. So, RC3 items are still included in the measurement because there is still a possibility that the reliability value will improve with the increase in the number of samples.

Furthermore, model measurements to see the reliability of research constructs are reviewed from Cronbach's alpha (α) and composite reliability (CR) (Byrne, 2013; Hair et al., 2019; Hoyle, 2012). Hair et al. (2019) state that the safe limit of α and CR values ranges from 0.70 to 0.95. In this research, the α value is obtained in the range of 0.817–0.897, while CR is in the range of 0.873–0.948. These results indicate that all constructs are consistent, which is reliable (Aji et al., 2020). In

Table 2 Reflective measurement model

Construct	Item	Loading	Cronbach alpha	rho_A	Composite reliability	AVE
Organizational culture	BO1	0.725	0.82	0.836	0.873	0.579
	BO2	0.770				
	BO3	0.734				
	BO4	0.821				
	BO5	0.751				
Competitive pressure	TKI	0.709	0.798	0.81	0.868	0.623
	TK2	0.772				
	TK3	0.855				
	TK4	0.813				
IT readiness	IR1	0.839	0.869	0.876	0.911	0.718
	IR2	0.881				
	IR3	0.863				
	IR4	0.806				
Organization mindfulness	OM1	0.850	0.892	0.925	0.919	0.695
	OM2	0.773				
	OM3	0.787				
	OM4	0.873				
	OM5	0.878				
Cognitive readiness	RC1	0.784	0.874	0.885	0.905	0.615
	RC2	0.740				
	RC3	0.686				
	RC4	0.835				
	RC5	0.849				
	RC6	0.798				
Digital transformation	DT1	0.812	0.912	0.913	0.932	0.695
	DT2	0.823				
	DT3	0.822				
	DT4	0.866				
	DT5	0.826				
	DT6	0.851				
Muslimah economic empowerment	MEE1	0.835	0.937	0.940	0.948	0.694
	MEE2	0.764				
	MEE3	0.795				
	MEE4	0.823				
	MEE5	0.890				
	MEE6	0.889				
	MEE7	0.867				
	MEE8	0.792				

Source: Authors' calculation

addition to the construct reliability test, a convergent and discriminant validity test is carried out in terms of outer loadings and Average Variance Extracted (AVE). The AVE value shows that each variable has a value of more than 0.5. Therefore, all variables in this study passed the convergent validity test. All variables in this study are reliable because they have composite reliability and Cronbach alpha values that meet the requirements of more than 0.7. Moreover, the overall correlation between reflective constructs does not exceed the square root of the AVE value (Hoyle, 2012). This means that all indicators used have good accuracy, consistency, and accuracy in measuring each construct in this study.

5.2 *Structural Model Assessment (Inner Model)*

Model suitability can be seen from the standardized root mean square residual (SRMR). Hair et al. (2014) state that the SRMR value threshold is <0.080. The SRMR results of this study are by the specified criteria, namely 0.058 for the saturated and estimated models. The variance inflation factor (VIF) value is reviewed to see multicollinearity in the model. According to Hair et al. (2019), the threshold for the VIF value is <3. The VIF value in this study ranges from 1000 to 2895. A VIF value of smaller than 10 indicates that the model has no multicollinearity problem, so the model used is free from multicollinearity disorders.

Based on Table 3, the value of *R* squared can be concluded that the weight of the effect of the DT determinant on MT of 0.620 is in the fit and moderate category because it is in the interval 0.33–0.67. The model that connects exogenous and intervening variables with endogenous variables is in a high fit condition with an R^2 value of 0.719 or 71.9%. Thus, the explanatory variables in the model can explain variations in endogenous variables by 71.9%.

As shown in Table 4, the bootstrap procedure results revealed significant structural relationships between the constructs. In general, the direct effect of an exogenous construct on an endogenous one is considered significant when the *t*-value (*t*-statistic) exceeds the critical value from the t-table at the 5% (0.05) significance level, while a *p*-value less than 0.05 leads to the rejection of the null hypothesis (Ho). Since the significance level indicates that the *t*-count value (*t*-statistic) > *t*-table at a 5% (0.05) significance level, the result is declared statistically significant. According to the analysis, BO does not influence DT (β = –0.021, *t*-stat = 0.290, $p < 0.00$). IR does not influence DT (β = –0.027, *t*-stat = 0.640, $p < 0.05$). OM does not affect DT (β = 0.092, *t*-stat = 1.215, $p < 0.05$). RC significantly positively affects DT (β = 0.453, *t*-stat = 5.233, $p < 0.00$). TK significantly influences DT (β = 0.341, *t*-stat = 5.060, $p < 0.00$). Digital transformation is insignificant in moderating the relationship between organizational culture and Muslimah economic empowerment (β = –0.018, *t*-stat = 0.290, $p < 0.00$). Digital transformation is insignificant in

Table 3 *R* squared value

	R squared	*R* squared adjusted
Digital Transformation_	0.620	0.613
Muslimah Economic Empowerment_	0.720	0.719

Source: Authors' calculation

Table 4 Path coefficient

	β	*t* statistic	*p* values	Conclusion
BO → DT	−0.021	0.290	**0.772**	Unsupported
IR → DT	−0.027	0.640	**0.522**	Unsupported
OM → DT	0.092	1.215	**0.225**	Unsupported
RC → DT	0.453	5.233	**0.000**	Supported
TK → DT	0.341	5.060	**0.000**	Supported
BO → DT → MEM	−0.018	0.290	0.772	Unsupported
IR → DT → MEM	−0.023	0.637	0.525	Unsupported
OM → DT → MEM	0.078	1.215	0.225	Unsupported
RC → DT → MEM	0.385	5.032	0.000	Supported
TK → DT → MEM	0.289	5.037	0.000	Supported

Supported as per significance value

moderating the relationship between IT readiness and Muslimah economic empowerment ($\beta = -0.023$, *t*-stat 0.637, $p < 0.00$). Digital transformation is insignificant in moderating the relationship between organizational mindfulness and Muslimah economic empowerment ($\beta = 0.078$, *t*-stat 1.215, $p < 0.05$). Digital transformation significantly moderates the relationship between cognitive readiness and Muslimah economic empowerment ($\beta = 0.385$, *t*-stat 5.032, $p < 0.00$). Digital transformation significantly moderates the relationship between competitive pressure and Muslimah economic empowerment ($\beta = 0.289$, *t*-stat 5.037, $p < 0.00$) (Fig. 5).

5.3 *Model 2 Analysis*

5.3.1 Assessment of the Measurement Model (Outer Model)

The measurement model to see the reliability of the research constructs is checked by Cronbach's alpha (α) and composite reliability (CR) (Byrne, 2013; Hair et al., 2019; Hoyle, 2012). Hair et al. (2019) state that the safe limits of α and CR values range from 0.70 to 0.95. In this study, α values ranged from 0.892 to 0.937, and CR values ranged from 0.916 to 0.948. These results indicate that all constructs are consistent, which is reliable (Aji et al., 2020). In addition to the construct reliability test, a convergent and discriminant validity test is carried out regarding external loadings and Average Variance Extracted (AVE). The discriminant validity test is based on the Fornell–Larcker validity criteria (Hair et al., 2017). The output shows that the overall loadings are higher than 0.70. Therefore, all items can be included to continue with the next measurement process. The AVE value is in the range of 0.579 to 0.694 (> 0.50), and the overall correlation between the reflective constructs does not exceed the square root of the AVE value (Hoyle, 2012). This means that all the indicators used have good accuracy, consistency, and precision in measuring each construct in this study. Finally, discriminant and convergent validity were met in this study, as can be seen in Table 5 (Fig. 6).

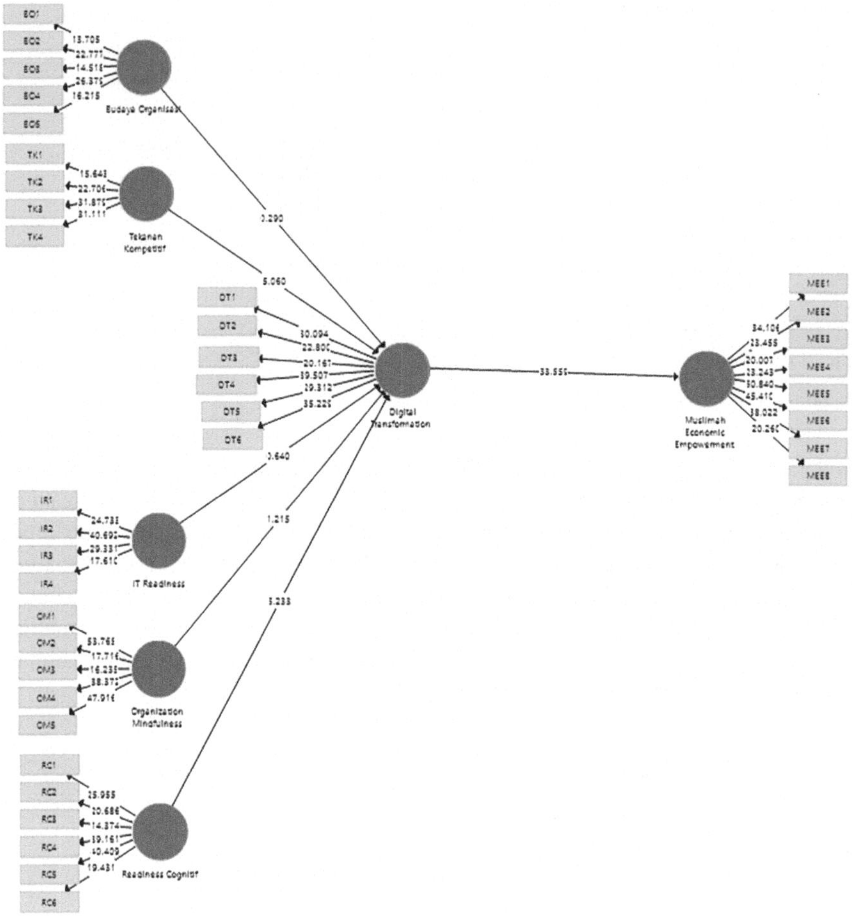

Fig. 5 Inner model

5.4 Structural Model Assessment (Inner Model)

The Standardized Root Mean Square Residual (SRMR) can be used to assess model fit. According to Hair et al. (2014), the SRMR value should be less than 0.080. With values of 0.058 for both the saturated and estimated models, the SRMR results of this study meet the specified criteria. The variance inflation factor (VIF) is checked to check for multicollinearity in the model. Hair et al. (2019) suggest a threshold of less than 3 for the VIF value. In this study, the VIF value ranges from 1.000 to 2.895. A VIF value below 10 indicates the absence of multicollinearity in the model, meaning it is free from such disorders. In order to test the hypothesis, this study used a bootstrap approach of 5000 resamplings with bias-corrected confidence intervals and a two-sided significance level of $p < 0.05$ (*), $p < 0.01$ (**), and $p < 0.001$ (***).

Table 5 Reflective measurement model

Construct	Item	Loading	Cronbach alpha	rho_A	Composite reliability	AVE
Muslimah economic empowerment	MEE1	0.832	0.937	0.940	0.948	0.694
	MEE2	0.759				
	MEE3	0.795				
	MEE4	0.825				
	MEE5	0.891				
	MEE6	0.889				
	MEE7	0.868				
	MEE8	0.797				
Operational performance	OP1	0.756	0.909	0.910	0.925	0.579
	OP 2	0.787				
	OP 3	0.705				
	OP 4	0.715				
	OP5	0.792				
	OP6	0.728				
	OP7	0.768				
	OP8	0.786				
	OP9	0.805				
Financial performance	FP1	0.874	0.885	0.898	0.910	0.593
	FP 2	0.716				
	FP 3	0.803				
	FP 4	0.799				
	FP 5	0.738				
	FP6	0.687				
	FP7	0.756				

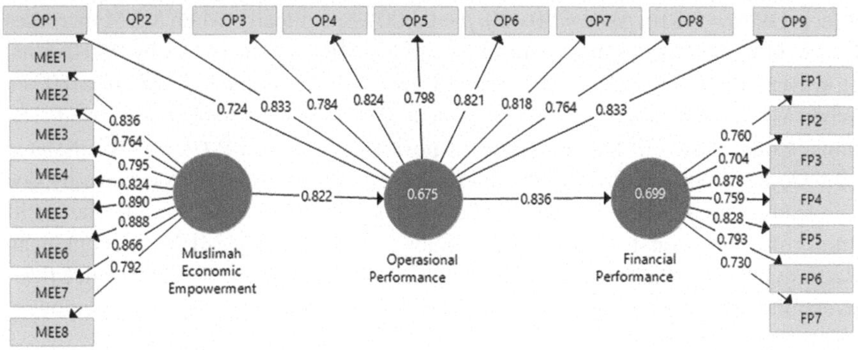

Fig. 6 Outer model

Table 6 shows that the bootstrap procedure has resulted in significant structural relationships between the constructs. The assumption is that the direct influence of an exogenous construct on an endogenous construct is significant when the *t*-value

Table 6 Structural model

	β	t Statistic	p values	Supported
MEM → OP	0.816	30.480	**0.000**	Yes
MEM → FP	0.152	2.166	**0.031**	Yes
OP → FP	1.012	19.932	**0.000**	Yes
MEM → OP → FP	0.687	20.726	**0.000**	Yes

Supported as per significance value

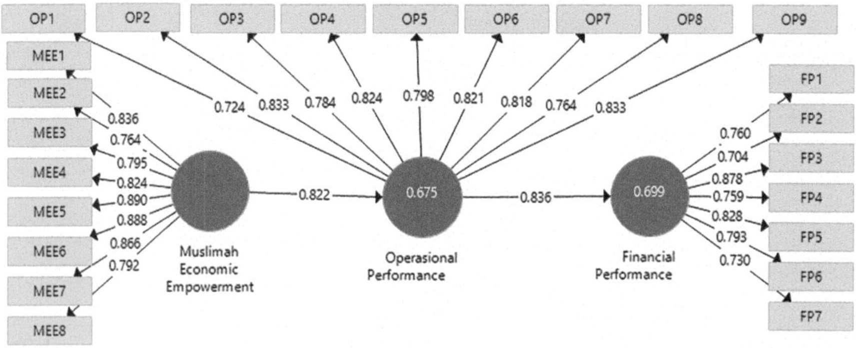

Fig. 7 Inner model

(t-statistic) is greater than the critical value from the t-table at a 5% (0.05) significance level, whereas a p-value less than 0.05 leads to the rejection of the null hypothesis (Ho). The statistical analysis shows that the t-count (t-statistic) value is greater than the t-table at the 5% (0.05) significance level, indicating that the results are statistically significant. The results indicate that MEM significantly positively affects OP ($\beta = 0.816$, t-stat = 30.480, $p < 0.000$). Specifically, when MEM increases by one unit, the Operational Performance of MSMEs also increases by 0.816 units. However, MEM negatively influences FP ($\beta = -0.152$, t-stat = 2.166, $p < 0.031$). When MEM decreases by one unit, Financial Performance also decreases by –0.152 units. Conversely, OP has a significant positive effect on FP ($\beta = 1.012$, t-stat = 19.932, $p < 0.000$), indicating that an increase in Operational Performance by one unit leads to an increase in Financial Performance by 1.012 units. Finally, the mediation of Operational Performance yields an impact of MEM on Financial Performance ($\beta = 0.687$, t-stat = 20.726, $p < 0.000$). Operational Performance positively and significantly mediates the effect of MEM/women's economic empowerment (Fig. 7).

5.4.1 Robustness Check

Based on Table 7, the value of R-squared can be determined. The weight of the influence of the variables that affect OP on FP is 0.675, which is in the high category as it lies in the interval 0.67–100. The model that links exogenous and intervening

Table 7 *R* squared value

	R squared	*R* squared adjusted
Operational Performance	0.675	0.674
Financial Performance _	0.699	0.698

Table 8 Coefficient determination and predictive relevance

Construct	SSO	SSE	Q^2	R^2	R^2 adjusted
Operational performance (OP)	1.876	1,024,791	0.454	0.797	0.795
Financial performance (FP)	2.412	1,505,526	0.376	0.666	0.664

variables with endogenous variables is in good condition with an R^2 value of 0.698 or 69.8%. Therefore, the explanatory variables in the model can account for 69.8% of the variations in endogenous variables. Hair et al. (2019) categorize the R^2 value at 0.75, 0.50, and 0.25. However, it is important to note that the R^2 value only captures the explanatory power of the sample used and does not predict performance outside of the sample. Therefore, this study utilizes the PLS-predict approach, focusing on the MSME performance construct as the main target. The study demonstrates that Q^2 has an overall predictive value greater than 0. Additionally, the PLS-SEM model's root mean squared error (RMSE) and mean absolute error (MAE) indicators have lower values than the linear model's. Hence, it can be concluded that the model in this study has high predictive power (Shmueli et al., 2019). Furthermore, the predictive relevance analysis employs Stone–Geisser's Q^2. The Q^2 value, as defined by Hair et al. (2019), can serve as a reference for the predictive relevance of independent and dependent variables. In this case, the Q^2 value of the variable switching between operational and financial performance is between 0.164 and 0.502 (> 0). 0.164–0.502, indicating a high level of predictive accuracy (Tables 8 and 9).

6 Discussion and Findings

In overall terms, this study proved two things empirically. Firstly, the Digital Transformation Antecedence positively influences Muslimah's Economic Empowerment. Digital transformation awareness is measured through organizational culture, competitive pressure, IT readiness, organizational mindfulness, and cognitive readiness. Secondly, it establishes the impact of women's economic empowerment on the operational and financial performance of MSMEs led by women.

Table 9 PLS predict

	PLS			LM		
Indicator	RMSE	MAE	Q^2_predict	RMSE	MAE	Q^2_predict
FP1	0.446	0.336	0.408	0.466	0.331	0.355
FP2	0.559	0.455	0.080	0.569	0.475	0.050
FP3	0.471	0.356	0.409	0.484	0.334	0.375
FP4	0.530	0.405	0.202	0.556	0.434	0.123
FP5	0.516	0.374	0.226	0.539	0.384	0.156
FP6	0.528	0.434	0.164	0.532	0.430	0.151
FP7	0.511	0.400	0.276	0.513	0.375	0.271
OP1	0.424	0.306	0.480	0.430	0.282	0.464
OP2	0.422	0.291	0.502	0.424	0.271	0.495
OP3	0.532	0.398	0.197	0.556	0.434	0.123
OP4	0.539	0.384	0.282	0.540	0.379	0.280
OP5	0.489	0.357	0.325	0.503	0.384	0.287
OP6	0.507	0.329	0.386	0.521	0.325	0.353
OP7	0.472	0.346	0.300	0.492	0.372	0.240
OP8	0.413	0.281	0.513	0.416	0.271	0.506
OP9	0.465	0.322	0.423	0.484	0.334	0.375

6.1 Discussion of First Empirical Results

6.1.1 Antecedence Analysis

The results of the analysis show that Organizational culture (BO) does not affect DT ($\beta = -0.021$, t-stat = 0.290, $p < 0.00$) and does not successfully mediate MEM ($\beta = -0.018$, t-stat = 0.290, $p < 0.00$). These results are consistent with previous studies (Martínez-Caro et al., 2020; Usai et al., 2021), which found that digital transformation requires adapting employees and all existing management in the context of external integration. Organizational culture can impact digital transformation, particularly regarding resistance to change. This is because organizational culture is shaped by systems, procedures, and structures that must align with the values of individuals within the organization. The wider the organizational culture adopted and the greater the agreement among members regarding its values, the more challenging it will be to change the culture.

Furthermore, digital transformation is causing changes in business at a rapid pace and in a comprehensive manner. Digital technologies speed up innovation, disruption, and competition in the business environment. In order to survive in a disrupted environment, organizations must transform, change, and restructure themselves due to rapidly changing environmental conditions and the acceleration of these changes caused by technological innovations. If a value construct embraced by the organizational culture is not supportive and is difficult, it will impede business transformation. Stereotypes can hurt interest, dedication, and performance in technological fields, particularly in women-led business organizations. No new content has been added beyond the original text. The language is clear, concise, and value-neutral, with a formal register and precise word choice. The text adheres to conventional structure and formatting features, with consistent citation and footnote style.

The logical progression of information is maintained through causal connections between statements, and the text is free from grammatical, spelling, and punctuation errors. It has also been identified that women may find it more challenging to identify with technology. To ensure objectivity, subjective evaluations have been excluded. According to the United Nations, women are often perceived as more proficient in grammar and reading, while men are perceived as more proficient in maths, science, and technology. Parents and teachers appear to reinforce this belief, which affects women's outcomes in education in these areas. This obstacle hinders digital transformation for women's economic empowerment. Some of the respondents' comments regarding organizational culture include:

> Actually, we know technology helps us in terms of marketing faster through the use of chatting, Facebook, and Instagram. It makes it easier for us to complete orders on time (Al. BB1)

> The use of digital creates a progressive and technology-friendly organisational culture that encourages creativity and innovation, which, in turn, increases our business revenue. However, we have limitations in terms of knowledge, skills, and training (F1).

> Digital technology has greatly improved the effectiveness and efficiency of our business, particularly in the distribution of goods. However, our main challenge is that we still need more human resources. J1

> We have been able to take on more orders since implementing sophisticated tools in our production process, which has increased our output and streamlined our stock ordering process. F&B1

The second antecedent is IT readiness. The study results indicate that IT readiness does not influence DT ($\beta = -0.027$, t-stat = 0.640, $p < 0.05$). DT did not successfully mediate IT readiness on MEM ($\beta = -0.023$, t-stat = 0.637, $p < 0.00$). These findings are consistent with the research conducted by Parasuraman and Colby (2015), which suggests that adopting information and communication technology (ICT) requires an adequate level of readiness, as there are many things to learn and adjustments to be made. It is a fact that implementing new technology requires a high level of preparedness. Individual factors, including security and comfort, influence technology acceptance in a business organization. Discomfort and insecurity hurt digital usage (Liljander et al., 2006). According to Erdogmus and Esen (2011), the successful adoption of new technology, particularly ICT in organizations, is largely determined by the readiness of human resources. In a business organization, there is a perception that women are less likely to adopt digital technology than men. This perception may be due to women's tendency to focus on the difficulty of tasks and the process of achieving their goals. While men are typically more task-focused and willing to accept change, including adopting technology, women may take longer to adapt. However, the success of adopting new technology, particularly ICT in organizations, is largely determined by the readiness of women leaders in MSMEs, affecting their economic empowerment. The results of interviews related to the use of technology are facing competitive pressures, while the respondents' comments include:

> The use of technology is due to fulfilling orders from our TKW friends who work in Hong Kong, Taiwan, and Malaysia. They become our resellers abroad; orders are only made through the TKW group because we are former TKWs. B&F. 1.

> We use technology because we want our products to be well-known by the wider community. This is because there are many products like the ones we sell in Ponorogo. INK 1.

> We initially only used technology for tiktokan, FB, and Instagram, and looked at YouTube so that we had additional innovations for our products, but we ended up using it for marketing media, but we found it difficult that our resources lacked support to receive many orders. C1.

Additionally, examining the effect of organizational mindfulness/OM revealed no effect on DT ($\beta = 0.092$, t-stat = 1.215, $p < 0.05$). Furthermore, DT does not play a significant role in mediating Muslimah Economic Empowerment (MEM) ($\beta = 0.078$, t-stat = 1.215, $p < 0.05$). Mindfulness refers to a state of mind or mode of practice that enables questioning of expectations, knowledge, and the adequacy of routines in complex and unpredictable social, technological, and physical settings. Therefore, mindfulness is a valuable attribute that helps individuals respond better to their environment, including using technology. This research aligns with Li et al. (2021) in that organizations that recognize the importance of digital transformation are better equipped to establish effective digital technology infrastructure, utilize digital technology to support internal and external relationships, and achieve digital business-technology strategic alignment. This, in turn, enhances their ability to respond promptly to environmental changes in the marketplace. However, despite increased awareness of digital transformation, it has not led to greater economic empowerment for women. This is because women need to improve their cognitive and learning skills to fully benefit from technology adoption, which can facilitate their engagement in the digital economy and online education (Sak et al., 2023).

The response from fashion Muslim owners indicates that:

> We focus on developing our business as it has been our primary occupation and a family-owned enterprise since our grandmother's time. INK 1.

The next antecedent is Cognitive Readiness (CR). Testing results demonstrate a significant positive influence on Digital Transformation (DT) ($\beta = 0.453$, t-stat = 5.233, $p < 0.00$). CR plays a positive and significant role in mediating the relationship between Cognitive Readiness and MEM. The impact of cognitive readiness on digital transformation reflects the extent to which an individual's or organization's level of cognitive preparedness can influence their ability to adopt and adapt to changes involving digital technology. Cognitive readiness encompasses the ability to understand, process, and utilize information quickly and effectively. This study's findings align with those of Neukirchen and Klumpp (2019), suggesting that the capacity to understand, process, and internal motivation are opportunities for leveraging digitalization in business. Human resources with high cognitive readiness consistently think about productivity effectively and efficiently. Acceptance and adaptation to digital transformation render the managed business more effective and efficient, yet with high productivity (Kaya & Bozbura, 2023). This also enhances knowledge, expands networks, and increases the innovation of the managed business (Giang et al., 2021). Indirectly, digital usage readiness promotes the resilience of the business performance they build, which in turn affects the economic empowerment of the business owners (Gurumurthy, 2004). Women are perceived to have a greater ability to read and learn actively compared to men. Thus, in business, women are considered to have higher cognitive readiness than men (Tsourela & Roumeliotis, 2015). Women with high cognitive capabilities drive effectiveness in the management of their businesses. The cognitive dimension for women refers to their understanding of their conditions and the causes of their subordination at both the micro and macro levels. It enables women to have access and control over productive resources, ensuring financial autonomy (Sorgner & Boden, 2018), ultimately leading to Economic Empowerment (Arrawatia & Meel, 2012). It helps in increasing monthly income, providing jobs, and opportunities to join major industries. Empowering women economically is a source of all other types of women's empowerment (Kaya & Bozbura, 2023). Improved women's access to the labor market, the growth of entrepreneurship supported by technological assistance, and the rise in average household income. Digital transformation can accelerate the empowerment process for women with mental readiness (cognitive readiness) by providing them with education, information, knowledge, training, employment opportunities, market information, financial services, information about relevant government programs, information about health and labor institutions, feedback mechanisms, support systems, and networks (Anjum & Tiwari, 2012).

The response from retail business owners indicates that:

> The use of digital platforms is confined solely to marketing aspects, and therefore, they are not employed comprehensively. BB2.

TK significantly influences DT ($\beta = 0.341$, t-stat $= 5.060$, $p < 0.00$). Digital transformation significantly moderates the relationship between competitive pressure and Muslim women's economic empowerment ($\beta = 0.289$, t-stat 5.037, $p < 0.00$). This study's results align with Mihu et al. (2023), who state that companies face increasing competitive pressures, which drive them to achieve better performance with the help of digital transformation. In today's dynamic environment, digital transformation is essential for modern businesses to remain competitive. Digital transformation is important in manufacturing and helping businesses achieve competitive advantage (Shehadeh et al., 2023). However, this is not an easy task; it is a long journey involving adopting digital solutions, rethinking and reshaping business strategies, and ensuring business continuity. The motivation for digital transformation stems from the desire to remain competitive, enhance performance, and ensure business continuity. (Metzler & Muntermann, 2021). This research also supports several other studies because it is undeniable that the current business landscape is characterized by tight competition (Elizabeth, 2017; Kathuria et al., 2010; Knowles, 2017), which puts pressure on companies to acquire customers, increase market share, and survive. Organizations are encouraged by this competitive pressure to pursue creativity, embrace change, and develop innovative competencies to outperform market competitors (Kathuria et al., 2010).

Digital transformation is increasingly driven by competitive pressure worldwide, with online businesses competing in new areas and opening markets wide for increased online competition and increasing competition. Thus, businesses must be truly bold in innovating and taking risks. This includes MSMEs led by women. Female entrepreneurs have an innovative mindset in many ways. Women are smarter in responding to competitive pressure and more diligent in new projects and opportunities than men. The desire to be successful makes women more daring to take risks. Valuing personal independence, female entrepreneurs strive to achieve economic independence. Digital transformation is considered a strategy for facing business competitive pressures in the era of globalization by women. Thus, female entrepreneurs have an innovative perspective, build businesses, and manage processes. They show a confident attitude by taking part in working life in a society dominated by men. Female entrepreneurs who value their independence strive to achieve economic independence. At this point, they act with a perspective to balance family life, personal life, and other responsibilities.

The interview results related to the use of technology in facing competitive pressure included comments from respondents as follows:

> We utilize technology to fulfill orders from friends working as migrant workers in Hong Kong, Taiwan, and Malaysia. They have become our resellers abroad. Orders are placed exclusively through migrant worker groups, as we ourselves are former migrant workers. B&F. 1.

> We use technology because we want our products to be well-known among the wider community. This is due to the abundance of similar products to ours in Ponorog. INK 1.

> Initially, we casually used technology for TikTok, Facebook, Instagram, and browsing YouTube to gather additional innovative ideas for our products. But eventually, we started using it for marketing purposes. Nonetheless, handling numerous orders is challenging due to our limited resources. C1.

6.1.2 Analysis of SME Performance

SMEs face various challenges, including limited access to financing, a lack of technical knowledge, and difficulties accessing markets. Consequently, this impacts the operational performance of SMEs. A company's operational performance directly affects the financial performance of SMEs and the overall continuity of operations. Women's economic empowerment should provide opportunities to overcome these challenges and enhance the company's sustainability and growth as a whole. This also proves the influence of women's economic empowerment on women-led SMEs' operational and financial performance.

MEM significantly influences the Operational Performance of SMEs in Ponorogo ($\beta = 0.816$, t-stat = 30.480, $p < 0.00$). This study offers a new perspective by showing that women, in this research context, are active in social and religious activities like fatayer, Muslim, and Aaliyah, possess the ability and independence, and have strategic concepts for accessing different resources, including technology adoption. These findings are consistent with the idea that women's empowerment in business can include improved access to resources, skill training, financial support, and active participation in decision-making. All these can contribute to increased productivity and efficiency of SMEs. Thus, women are capable of making economic decision choices. Empowered women can make effective business or entrepreneurial decisions to reduce social and economic barriers. Business decisions are closely linked with the targeted business outputs. This finding aligns with previous research indicating that a high level of women's economic empowerment influences the high productivity level of SMEs and enhances work efficiency. Empowerment can give women more control over themselves and their environment and make the necessary changes to work effectively. As a result, it can improve the performance of SMEs and cost-effectiveness. Women-led, more empowered SMEs have higher company revenues than those led by less empowered women.

The interview results are as follows:

> By transitioning to online sales, [they] can discover new ways to expand and experience growth. J1.

MEM significantly influences the Operational Performance of SMEs in Ponorogo ($\beta = 1.52$, t-stat = 2.166, $p < 0.05$). Women's spirit, creativity, innovation, and capabilities make SMEs robust and resilient. This study supports findings from previous research, such as Febriani and Indiworo, which suggest that women's success in entrepreneurship is also supported by their strengths, including attention to detail, perseverance, seriousness, patience, determination, honesty, strong will, high dedication, and discipline in administration and financial management. This is in line with studies by Pathak and Emah, Welsh et al., Sumantri et al., and Sumarwati and Rachman, which show that empowered women can improve the financial performance of SMEs. Furthermore, empowered women with the advantages above have innovative marketing strategies, enabling marketers to differentiate their offerings from competitors in the current market. Effective marketing drives productivity improvements. Increased productivity in production processes or service provision can, in turn, contribute to financial performance.

Women with high levels of economic empowerment initially have meticulousness, keenness, and perseverance, enabling them to diversify their businesses. Business diversification can encourage the diversification of products or services offered by SMEs, potentially expanding market share and revenue. It allows SMEs to retain customers, expand market share, and reduce marketing costs. As a result, SMEs no longer need to allocate a large budget for advertising and promotions, the results of which are still being determined. Indirectly, the capability to diversify products potentially impacts SMEs by reducing operational costs, facilitating rapid capital turnover, and increasing profits. Women are also considered to have better financial management. With training and support in financial management, women in SMEs can help improve efficiency and financial transparency. Ultimately, economically empowered women can become a resource for SMEs' financial access. Women's empowerment can help improve SMEs' access to capital and financial support, facilitate rapid capital turnover, and increase profits, positively impacting the company's financial health. Ultimately, this contributes to the longevity and growth of the business over time.

This analysis aligns with the condition of women-led SMEs in Ponorogo:

> Adopting digital platforms has increased sales, aided by implementing COD (Cash On Delivery) services in less than a year. INK. 2.

> Transitioning to online sales enables [them] to discover new avenues for expansion and growth. J1.

> Digital transformation has a positive impact, and it's recognized that the prudent utilization of technology can produce extraordinary outcomes. J2.

The next hypothesis analyzes the relationship between operational performance and the positive mediating influence between MEM and financial performance metrics. MEM successfully mediates the operational performance of SMEs in Ponorogo toward financial performance ($\beta = 0.687$, t-stat = 20.726, $p < 0.000$). As suggested by the Resource-Based View (RBV), it explains the relationship between women's economic empowerment and performance, where women's economic empowerment is considered a resource that partially helps achieve a company's sustainable competitive advantage. Previously influenced by digital transformation, women's economic empowerment indicates that women entrepreneurs or managers can enhance performance if supported by information technology, especially in the current digital era. The development of information technology, both software like Internet networks and applications, and hardware such as smartphones and laptops, is highly beneficial for women to enhance their performance on the go. Especially now, with many social media applications like WhatsApp, Instagram, Facebook, and others, and marketplaces like Shopee, Bukalapak, Tokopedia, and others, women can utilize these to promote and sell SME products.

Additionally, SMEs can use applications to record business transactions or create financial reports. Thus, SMEs can expand their market share and reduce marketing costs. As a result, there is a moderate mediating effect between operational, financial, and market performance. The partial mediation effect reveals that operational performance leads to good market and financial performance in Ponorogo SMEs.

The interview confirmations are as follows:

> Wise utilization of digital tools brings business profits. C1.

> Since using the online platform, we receive more orders, enabling us to recruit neighbors as daily workers, thus increasing our profits. B&F 2.

> The high volume of online orders from customers has led us to use large-scale delivery services, thereby creating job opportunities. INK 1.

7 Conclusions

The driving factors and their impact on business sustainability through digital transformation and Muslim women's economic empowerment is an intriguing topic in digital and female entrepreneurship research. Unlike previous studies, this research integrates the antecedents of digital transformation and its impact on women's economic empowerment. It also delves deeper into analyzing the influence of women's economic empowerment on SME performance, which can mediate their financial performance. Previous research on transformation and women's economic empowerment has yet to explore this relationship. This model has been empirically validated using the PLS-SEM method, utilizing data from 269 SMEs across various sectors, including Food and Beverage, Fashion, Manufacturing, Retail, Services, and Travel, in Ponorogo. Digital transformation impacts women's economic empowerment, particularly through cognitive readiness antecedents.

Other antecedents, like organizational culture, do not influence DT and fail to mediate MEM. Meanwhile, IT readiness does not impact DT, and DT does not successfully mediate IT Readiness to MEM. Organizational mindfulness (OM) is found to have no impact on DT, and DT is not significant in mediating MEM. Cognitive Readiness (RC) testing results show a significant positive influence on DT. DT is positive and significant in mediating RC to MEM. It is because, in business, women are considered to have higher cognitive readiness than men. TK has a significant positive influence on DT. Digital transformation significantly influences moderating the relationship between competitive pressure and Muslim women's economic empowerment. Female entrepreneurs have an innovative mindset in many aspects. Women respond more intelligently to competitive pressure and are more active in new projects and opportunities than men. The desire to succeed makes women more willing to take risks.

Women's economic empowerment, analyzed through RBV theory, impacts SME performance. Furthermore, this research shows that SME performance fully mediates the relationship between women's economic empowerment and SME financial performance. The findings of this study confirm the hypotheses proposed in this research and several previous studies showing that digital transformation positively impacts women's economic empowerment. Subsequently, this economic empowerment affects SMEs' operational and financial performance. Additionally, operational performance fully mediates the relationship between women's economic empowerment and financial performance. This empirical evidence is expected to provide additional confidence and motivation for SME managers, especially those managed by women, not to hesitate to adopt digital technologies, strengthening the antecedents of digital transformation so that SME managers become economically empowered women capable of enhancing the performance of these SMEs.

A further practical implication of this research is advocating for the government, particularly the local government of Ponorogo, to continuously review and monitor the use of digitalization in SMEs, as implementation tracking and reevaluation have been limited so far. Moreover, strict oversight should be accompanied by transparency, accountability, and involvement of various stakeholders, including Islamic organizations, business associations, and civil society. Active involvement from all parties can help avoid conflicts and ensure that using digital tools benefits SMEs and all stakeholders in empowering SMEs within the digital ecosystem.

Furthermore, this study has limitations, including the research location being focused only on one area. Therefore, it needs more diversity in results. The study also focuses on two major Islamic organizations in Ponorogo, Muhammadiyah and NU. Future research is expected to delve deeper and more broadly, encompassing women in Islamic organizations and those from various regions and ethnic groups. This would help to understand the extent of women's roles in developing SMEs through digital transformation.

Author Contributions Conceptualization, S.N.A., S., IRY and F.A; methodology, IRY, K.W. and S..; validation, S.., F.A and S.N.A.; formal analysis, S.N.A.; IRY investigation, S.N.A. and K.W, resources, S. and K.W., data curation, K.W. and S.N.A.; writing, S.N.A.; IRY; preparation, S.. and F.A; writing—review and editing, K.W and S.N.A.; IRY; supervision, S..; project administration, S.N.A.; funding acquisition, S.N.A. All authors have read and agreed to the published version of the manuscript.

Funding: This research received funding from BRIN. The authors would like to express their gratitude to BRIN.

Institutional Review Board Statement: Not applicable.

Informed Consent Statement: Informed consent was obtained from all subjects involved in the study.

Data Availability Statement: Data supporting reported results can be found by asking directly of the first author.

Conflicts of Interest: The authors declare no conflict of interest.

References

Adjei, J. K., Adams, S., & Mamattah, L. (2021). Cloud computing adoption in Ghana; accounting for institutional factors. *Technology in Society, 65*, 101583. Elsevier.

Ahmed, N., Tasmin, M., & Nasim Ibrahim, S. M. (2022, December). Technology for empowerment: context of urban Afghan women. *Technology in Society, 70*, 102058.

Aji, H. M., Berakon, I., & Md Husin, M. (2020). COVID-19 and e-wallet usage intention: A multigroup analysis between Indonesia and Malaysia. *Cogent Business and Management, 7*, 1–16. Available at: https://doi.org/10.1080/23311975.2020.1804181

Ancillai, C., Sabatini, A., Gatti, M., & Perna, A. (2023). Technological forecasting & social change digital technology and business model innovation: A systematic literature review and future research agenda. *Technological Forecasting & Social Change, 188*, 122307. Elsevier.

Andersen, ML. (1988). *Thinking about women: Sociological perspective on sex and gender*. Macmillan Publishing Company.

Anjum, B., & Tiwari, R. (2012). Role of information technology in women empowerment. *International Journal of Multidisciplinary Management Studies, 2*, 226–223.

Aracil-Jord, J., Clemente-Almendros, J.-A., Jiménez-Zarco, A.-I., & Gonzalez-Gonzalez, I. (2023, March). Improving the social performance of women-led microenterprises: The role of social media marketing actions. *Technological Forecasting & Social Change, 191*. https://doi.org/10.1016/j.techfore.2023.122484

Arrawatia, M. A., & Meel, P. (2012, October). Information and communication technologies & woman empowerment in India. *International Journal of Advanced Research in Computer Engineering & Technology (IJARCET), 1*, 34–52.

Asian Development Bank. (2022). Financial instruments to strengthen women's economic resilience to climate change and disaster risks.

Barney, J. (1991). Firm resources and sustained competitive advantage. *Journal of Management, 17*, 99–120.

BPS. (2022). Statistical year book of Indoneisa. Available at: https://www.bps.go.id/id/publication/2022/02/25/0a2afea4fab72a5d052cb315/statistik-indonesia-2022.html

Byrne, B. M. (2013). *Structural equation modeling with AMOS: Basic concepts, applications, and programm* (2nd ed., pp. 17–49). Routledge, Taylor and Francis. https://doi.org/10.4324/9781410600219

Creswell, J. W. (2017). Research design: Qualitative, quantitative, and mixed methods, Third Edition (Terjemah). Pustaka Pelajar.

Crittenden, V. L., Crittenden, W. F., & Ajjan, H. (2019). Empowering women micro-entrepreneurs in emerging economies : The role of information communications technology. *Journal of Business Research, 98*, 191–203.

Dinçer, H., Hacıoğlu, Ü., & Yüksel, S. (2017). Balanced scorecard based performance measurement of European airlines using a hybrid multicriteria decision making approach under the fuzzy environment. *Journal of Air Transport Management, 63*, 17–33.

Dorfleitner, G., & Nguyen, Q. A. (2022). Mobile money for women's economic empowerment : the mediating role of financial management practices. *Review of Managerial Science*. https://doi.org/10.1007/s11846-022-00564-2. Springer.

Eger, C., Fetzer, T., Peck, J., & Alodayni, S. (2022). Organizational, economic or cultural? Firm-side barriers to employing women in Saudi Arabia. *World Development, 160*, 106058.

Elizabeth, R. (2017). Increasing competitive threats are driving digital transformation. *Technology Record*, October, 10–30. https://www.technologyrecord.com/Article/increasing-competitive-threats-are-driving-digital-transformation-60443.

Erdogmus, N., & Esen, M. (2011). An investigation of the effects of technology readiness on technology acceptance in e-HRM. *Procedia Social and Behavioral Sciences, 24*, 487–495.

Gallego-losada, M., Montero-navarro, A., García-abajo, E., & Gallego-losada, R. (2023). Digital financial inclusion: Visualizing the academic literature. *Research in International Business and Finance, 64*, 101862.

Garz, S., Heath, R., Kipchumba, E., & Sulaiman, M. (2020). *Evidence of digital financial services impacting women's economic empowerment* (WEE-DiFine White Paper, No. October). https://doi.org/10.13140/RG.2.2.10430.82242.

Giang, N. T. H., Hai, P. T. T., Tu, N. T. T., & Tan, P. X. (2021). Exploring the readiness for digital transformation in a higher education institution towards industrial revolution 4.0. *International Journal of Engineering Pedagogy, 11*, 4–24.

Gong, C., & Ribiere, V. (2021, December). Developing a unified definition of digital transformation. *Technovation, 102*, 102217.

GSMA. (2017). *Mobile access and usage in low—Table of contents* (pp. 35–45).

Gurumurthy, A. (2004). *Bridging the digital gender divide: Issues and insights on ICT for women's economic empowerment*. UNIFEM. http://www.itforchange.net/resources/Pat.htm

Hair, F. J., Sarstedt, M., Hopkins, L. G., & Kuppelwieser, V. (2014). Partial least squares structural equation modeling (PLS-SEM). *European Business Review, 26*, 106–121.

Hair, J. F., Hult, G. T. M., Ringle, C. M., & Sarstedt, M. (2017). *A primer on partial least squares structural equation modeling (PLS-SEM)* (pp. 40–57). Sage.

Hair, J. F., Risher, J. J., Sarstedt, M., & Ringle, C. M. (2019). When to use and how to report the results of PLS-SEM. *European Business Review, 31*, 2–24.

Halpern, N., Mwesiumo, D., Suau-Sanchez, P., Budd, T., & Bråthen, S. (2021). Ready for digital transformation? The effect of organisational readiness, innovation, airport size and ownership on digital change at airports. *Journal of Air Transport Management, 90*, 101949.

Hasan, R., & Ashfaq, M. (2022). Financial inclusion–does digital financial literacy matter for women entrepreneurs ? entrepreneurs. *International Journal of Social Economics*. https://doi.org/10.1108/IJSE-04-2022-0277

Hibbs, L. (2022). I could do that!'—The role of a women's non-governmental organisation in increasing women's psychological empowerment and civic participation in Wales. *Women's Studies International Forum, 90*, 102557.

Hilbert, M. (2011). Digital gender divide or technologically empowered women in developing countries ? A typical case of lies , damned lies , and statistics. *Women's Studies International Forum, 34*, 479–489.

Holzmann, P., & Gregori, P. (2023). The promise of digital technologies for sustainable entrepreneurship : A systematic literature review and research agenda. *International Journal of Information Management, 68*, 102593.

Hoyle, R. H. (2012). *Handbook of Structural Equation Modeling* (pp. 29–45). Guilford Press.

Hsu, P. F., Ray, S., & Li-Hsieh, Y. Y. (2014). Examining cloud computing adoption intention, pricing mechanism, and deployment model. *International Journal of Information Management, 34*, 474–488.

Hussain, M., & Papastathopoulos, A. (2022, October). Organizational readiness for digital financial innovation and financial resilience. *International Journal of Production Economics, 243*, 108326.

Kathuria, R., Partovi, F., & Greenhaus, J. (2010). Leadership practices, competitive priorities, and manufacturing group performance. *International Journal of Operations & Production Management, 30*, 1080–1105. https://doi.org/10.1108/01443571011082535

Kaya, Y., & Bozbura, F. T. (2023). Digital transformation: a cognitive study for organizations to shape their journeys. *International Journal of Professional Business Review, 8*, 01825.

Khanna, R., & Sharma, C. (2022). Impact of information technology on firm performance : new evidence from Indian manufacturing. *Information Economics and Policy, 60*, 100986.

Knowles, H. (2017, October 9). Digital transformation driven by increasing competitive threats rather than innovation. *Compare the Cloud*, 51–78. https://www.comparethecloud.net/news/digital-

Li, H., Wu, Y., Cao, D., & Wang, Y. (2021). Organizational mindfulness towards digital transformation as a prerequisite of information processing capability to achieve market agility. *Journal of Business Research, 122*, 700–712.

Llopis-Albert, C., Rubio, F., & Valero, F. (2021). Impact of digital transformation on the automotive industry. *Technological Forecasting and Social Change, 162*, 120343.

Malodia, S., Mishra, M., Fait, M., Papa, A., & Dezi, L. (2023). To digit or to head? Designing digital transformation journey of SMEs among digital self-efficacy and professional leadership. *Journal of Business Research, 157*, 113547. Elsevier.

Martínez-Caro, E., Cegarra-Navarro, J. G., & Alfonso-Ruiz, F. J. (2020, February). Digital technologies and firm performance: the role of digital organisational culture. *Technological Forecasting and Social Change, 154*, 119962.

Metzler, D. R., & Muntermann, J. (2021, December). The impact of digital transformation on incumbent firms: An analysis of changes, challenges, and responses at the business model level. In *International conference on information systems, ICIS, making digital inclusive: Blending the local and the global* (pp. 23–40).

Meyer, J., Toldos, M. P., & Gonz, E. M. (2021, November). What women want ? How contextual product displays influence women's online shopping behavior. *Journal of Business Research, 123*, 625–641.

Mihu, C., Pitic, A. G., & Bayraktar, D. (2023). Drivers of Digital transformation and their impact on organizational management. *Studies in Business and Economics, 18*, 149–170.

Moore, F. (2020). Multiple interpretations of 'national culture' and the implications for International business: The case of Taiwan. *Journal of World Business, 55*, 101128.

Neukirchen, T., & Klumpp, M. (2019). Digital logistics, strategic cognitive readiness and employee training. *International Conference of Logistics (HICL), 27*, 117–150. https://doi.org/10.15480/882.2467

Nord, J. H., Lee, T. R., Jiun, S., Çetin, F., Atay, Ö., & Paliszkiewicz, J. (2016). Examining the impact of social technologies on empowerment and economic development. *International Journal of Information Management, 36*, 1101–1110.

Nurazira, S., Daud, M., & Ahmad, A. H. (2023). Financial inclusion, economic growth and the role of digital technology. *Finance Research Letters, 53*, 103602.

Nwankpa, J. K., & Roumani, Y. (2016). IT capability and digital transformation: A firm performance perspective *2016 International Conference on Information Systems. ICIS, 2016*, 1–16.

Nyagadza, B. (2022). Sustainable digital transformation for ambidextrous digital firms: systematic literature review, meta-analysis and agenda for future research. *Sustainable Technology and Entrepreneurship, 1*, 1–13.

Oeij, P. R. A., Hulsegge, G., & Preenen, P. T. Y. (2022, May). Organisational mindfulness as a sustainable driver of employee innovation adoption: Individual and organisational factors. *Safety Science, 154*, 105841.

Pagani, M., & Pardo, C. (2017, May). The impact of digital technology on relationships in a business network. *Industrial Marketing Management, 67*, 185–192.

Pal, M., Gupta, H., & Joshi, Y. C. (2022). Social and economic empowerment of women through financial inclusion: empirical evidence from India. *Equality, Diversity and Inclusion: An International Journal, 41*, 294–305. https://doi.org/10.1108/EDI-04-2021-0113

Papa, A., Dezi, L., Gregori, G. L., Mueller, J., & Miglietta, N. (2017). Improving innovation performance through knowledge acquisition : the moderating role of employee retention and human resource management practices. *Journal of Knowledge Management, 24*, 589–605.

Parasuraman, A., & Colby, C. L. (2015). An updated and streamlined technology readiness index: TRI 2. 0. *Journal of Service Research, 18*, 59–74.

Pawluczuk, A., Lee, J., & Gamundani, A. M. (2021). Bridging the gender digital divide : an analysis of existing guidance for gender digital inclusion programmes' evaluations. *Digital Policy, Regulation and Government, 23*, 287–299.

Poláková-Kersten, M., Khanagha, S., van den Hooff, B., & Khapova, S. N. (2023). Digital transformation in high-reliability organizations: A longitudinal study of the micro-foundations of failure. *The Journal of Strategic Information Systems, 32*, 101756.

Raut, R. D., Gardas, B. B., Jha, M. K., & Priyadarshinee, P. (2017). Examining the critical success factors of cloud computing adoption in the MSMEs by using ISM model. *Journal of High Technology Management Research, 28*, 125–141.

Sak, M., Alnoor, A., Valeri, M., & Bayram, G. E. (2023). The role of digital transformation on women empowerment for rural areas: The case of Turkey. *Tourism Innovation in the Digital Era: Big Data, AI and Technological Transformation, 2*, 91–105.

Seong, B. H., & Hong, C. Y. (2022). Corroborating the effect of positive technology readiness on the intention to use the virtual reality sports game 'Screen Golf': Focusing on the technology readiness and acceptance model. *Information Processing and Management, 59*, 102994.

Shehadeh, M., Almohtaseb, A., Aldehayyat, J., & Abu-AlSondos, I. A. (2023). Digital transformation and competitive advantage in the service sector: a moderated-mediation model. *Sustainability, 15*, 2077. https://doi.org/10.3390/su15032077

Shmueli, G., Sarstedt, M., Hair, J. F., Cheah, J.-H., Ting, H., Vaithilingam, S., & Ringle, C. M. (2019). Predictive model assessment in PLS-SEM: Guidelines for using PLS predict. *European Journal of Marketing, 53*, 2322–2347.

Shofawati, A. (2019). The role of digital finance to strengthen financial inclusion and the growth of SME in Indonesia. In *2nd ICIEBP* (Vol. 5, pp. 389–407).

Soekarno, S., Mambea, I. Y., & Setiawati, M. (2020). Women and digital financial inclusion in Indonesia as emerging market. *International Review of Management and Marketing, 10*, 46–49.

Soluk, J., Decker-Lange, C., & Hack, A. (2023). Small steps for the big hit: A dynamic capabilities perspective on business networks and non-disruptive digital technologies in SMEs. *Technological Forecasting and Social Change, 191*, 122490.

Sorgner, A. (2021). Gender and industrialization: developments and trends in the context of developing countries. *SSRN Electronic Journal*. https://doi.org/10.2139/ssrn.3798913

Sorgner, A., & Boden, C. K. (2018). Empowering women in the digital age: where do we stand?. *G20 Insight*, March, 1–5.

Srivastava, S., Singh, S., & Dhir, S. (2020). Culture and international business research: a review and research agenda. *International Business Review, 29*, 101709.

Summers, K. H., Baird, T. D., Woodhouse, E., Elisa, M., Mccabe, J. T., Terta, F., & Peter, N. (2020). Mobile phones and women' s empowerment in Maasai communities : How men shape women's social relations and access to phones. *Journal of Rural Studies, 77*, 126–137.

Suwana, F., & Lily. (2017). Empowering Indonesian women through building digital media literacy. *Kasetsart Journal of Social Sciences, 38*, 212–217.

Tay, L., Tai, H., & Tan, G. (2022). Digital financial inclusion : A gateway to sustainable development. *Heliyon, 8*, 09766.

Trivelli, C., Villanueva, C., Marincioni, M., Pels, J., Sachetti, F. C., Robino, C., & Walbey, H. (2018). Financial inclusion for women: A way forward. *Gender Economic Equity, 20*, 1–19.

Tsourela, M., & Roumeliotis, M. (2015). The moderating role of technology readiness, gender, and sex in consumer acceptance and actual use of technology-based services. *Journal of High Technology Management Research, 26*, 124–136.

Usai, A., Fiano, F., Messeni Petruzzelli, A., Paoloni, P., Farina Briamonte, M., & Orlando, B. (2021). Unveiling the impact of the adoption of digital technologies on firms' innovation performance. *Journal of Business Research, 133*, 327–336.

Verhoef, P. C., Broekhuizen, T., Bart, Y., Bhattacharya, A., Qi Dong, J., Fabian, N., & Haenlein, M. (2021). Digital transformation: A multidisciplinary reflection and research agenda. *Journal of Business Research, 122*, 9–901.

Verma, P., Kumar, V., Daim, T., Kumar, N., & Mittal, A. (2022). Identifying and prioritizing impediments of industry 4. 0 to sustainable digital manufacturing: A mixed method approach. *Journal of Cleaner Production, 356*, 131639.

Vial, G. (2019). Understanding digital transformation : A review and a research agenda. *Journal of Strategic Information Systems, 28*, 118–144.

Yang, X., Huang, Y., & Gao, M. (2022). Can digital financial inclusion promote female entrepreneurship? Evidence and mechanisms. *North American Journal of Economics and Finance, 63*, 101800.

Siti Nur Azizah is a lecturer at the Faculty of Islamic Economics and Business, Sunan Kalijaga State Islamic University, Yogyakarta, Indonesia. Her research interests include Islamic social finance, the halal industry, and women's economic empowerment. Siti Nur Azizah is the corresponding author and can be contacted at: siti.azizah@uin-suka.ac.id, **ORCID**: https://orcid.org/0000-0003-3554-9637

Suradi is a senior researcher at the National Research and Innovation Agency, Indonesia. previously worked at the Ministry of Social Affairs, with a concentration on social policy, empowerment, and social protection. The doctorate in social welfare was obtained from the Faculty of

Social and Political Sciences, Padjadjaran University, Bandung. He actively participates in activities and is a speaker at national and international seminars, especially those discussing indigenous communities and poverty. His papers have been published in various national and international journals and other media. Experience as a reviewer in national and international journals, as chairman of the editorial board of the sosio konsesia journal [welfare business journal] and sosio informa [social welfare issues journal] managed by the Indonesian Ministry of Social Affairs, as social welfare research coordinator and social welfare doctor coordinator at Padjadjaran University, Bandung.

Koeswinarno is a research professor since 2011. He has dedicated most of his research career to the transgender in Indonesia, which is still a marginalized group. He started his research on transgender when he received research grants from The Toyota Foundation (1992) and Ford Foundation (1996), and then he strengthened transgender studies in his master's program (1997). By the year of 2008, he continued his doctoral study in anthropology at Gadjah Mada University, Yogyakarta, by writing a dissertation on transgender. He began his career as a researcher at the Ministry of Communication and Informatics (1989–2001) and continued at Sunan Kalijaga State Islamic University (2001–2012). In 2012, he moved to work at the Ministry of Religious Affairs, and finally, he joined the National Research and Innovation Agency from 2022 until now.

M. Fakhruddin is a researcher at the Research Centre of Public Policy, the National Research and Innovation Agency (BRIN). Before joining BRIN in January 2021, he was a researcher at the Centre for Research and Development of Religious Public Guidance and Religious Services for the Ministry of Religious Affairs. He completed his Master of Administrative Science at Krisnadwipayana University, Jakarta.

Irma Yuliani is a lecturer at the Faculty of Islamic Economics and Business, Ponorogo State Islamic Institute, Indonesia. Her research interests include Islamic finance, gender, Islamic economic, and Sharia Banking. She is an editor and reviewer in several journals.

Online Marketing in Islamic Perspective: The Case of Halal Companies in Kosovo

Asdren Toska

Abstract This chapter aims to shed some light on the operation of halal companies in the context of online marketing in the Kosovar economic environment. Since, in the era of digitalization, it is inevitable for companies to use online marketing tools, this chapter aims to examine halal companies in the Kosovar context regarding the use of online marketing and the application of the principles of Islamic online marketing by these entities. The chapter uses practical case studies from the respective companies to reach the conclusions. The findings emphasize that halal companies in the Kosovar context carefully apply the principles of Islamic online marketing in online communications with their target customers. The chapter will also have implications in the academic community to further advance studies in this field, as in the Kosovar context, as well as in other economic environments to see the similar and distinctive elements.

1 Introduction

Today, online marketing is considered one of the most applied and successful forms that help companies achieve competitive advantages by targeting the right customers through online tools. Nowadays, it is difficult to think of a successful business operation while ignoring online marketing tools. In spite of this, it is not only necessary to apply online marketing tools, but companies must to be up to date with the changes that online marketing brings since this form of marketing is very dynamic and requires careful monitoring of these changes (Toska et al., 2023). When we talk about the need to monitor these changes carefully, the researchers argue that, specifically, the forms of direct online marketing are the most dynamic since we often have new technologies that can be introduced into the forms of direct online

A. Toska (✉)
South East European University, Tetovo, North Macedonia
e-mail: at16851@seeu.edu.mk

A. Rafiki et al. (eds.), *Strategic Islamic Business and Management*, Contributions to Management Science,
https://doi.org/10.1007/978-3-031-61778-2_6

communication. Therefore, companies must follow these trends to stay in line with the competition and dynamic environment (Kotler & Armstrong, 2011). Many studies have been carried out that emphasize the benefits of applying online marketing, seen from the perspective of the customer and companies as well. Being present to consumers at different geographical distances and providing information at any time, including the interactivity offered by direct communication, online marketing has a vital role in creating long-term and healthy relationships with customers (Sprott et al., 2009; Toska & Zeqiri, 2023a, 2023b). Likewise, online marketing enables companies to improve sales performance and increase customer engagement opportunities, which leads to creating advocates customers, as argued in Marketing 4.0 (Kotler et al., 2017).

Since the *Marketing 4.0* form is widely applied by companies today, where its necessity is evidenced by many empirical studies carried out so far, on the other hand, companies should not ignore the fact of the increase in the number of Muslim consumers who use the Internet for online shopping and who make a large market by focusing on Halal consumption in accordance with Islamic practices and values (Syafril & Hadziq, 2021). Therefore, it is suggested that companies operating in such economic environments that Muslim consumers dominate focus on Halal business practices that follow Islamic practices (Alserhan, 2011). In addition, online marketing strategies in such economic environments must be in full compliance with Islamic values, principles, and standards to ensure companies have a successful operation in these markets (Amin et al., 2013). There were doubts that online marketing is compatible with Islamic values and practices. Regarding this, some of the scientific studies extinguish these doubts. Since the Islamic religion sees business as an important source of livelihood, forms of online marketing are acceptable if they are consistent with Islamic values and principles by offering ethical business practices (Ahasanul Haque & Kabir, 2022; Tabrani & Marlizar, 2016).

The authors Ahasanul Haque and Kabir (2022) argue that it is important for the academic community to bring different empirical experiences that focus on Islamic marketing practices in different contexts, not only focusing on marketing strategies but also on its practices. The Islamic *Marketing 4.0* phase also emphasizes the need for scientific evidence from different countries and cultures to test existing theories in these countries (Alserhan et al., 2022). Islamic marketing is essential for companies since, by ethically offering products and services, they develop long-term relationships with their customers (Wilson, 2012). Referring to demographic statistics, Kosovo is dominated by a Muslim population of 85.6%; seeing this statistic, it is important for the academic community to bring scientific evidence for Islamic marketing practices applied by Kosovar companies during communications with their online customers.

This chapter aims to analyze the application of Islamic marketing principles such as the principle of truthfulness (*al-Sidq*), responsibility (*al-Mas'uliyyah*), impartiality (*al-Adl*), moderation (*al-Wasatiyyah*), and corporate social responsibility in accordance with Islamic principles. The chapter will analyze three case studies from Halal companies in the Kosovar market, bringing scientific evidence to the academic community on how Halal companies apply Islamic principles in the Kosovar economic environment, which is dominated by the Muslim population.

The chapter will have the following structure: continuing with the economic environment in Kosovo, stages of marketing development, online marketing, Islamic practices in accordance with online marketing, case studies of Kosovar companies, and finally, conclusion.

2 Economic Environment in Kosovo

Some general knowledge about Kosovo and the economic environment in Kosovo is necessary in this chapter since the focus of this chapter is on the Kosovo context.

Geographically, Kosovo is located in the Western Balkans and is considered one of the six countries of this region. Likewise, Kosovo is considered a new state since it declared its independence on February 17, 2008. The territorial surface of Kosovo is 10,887 square kilometers, which makes it the smallest state in this region, with a population of 1.9 million inhabitants. This also made this country the country with the smallest number of inhabitants in this region after Montenegro (The World Bank, 2020). Kosovo is dominated by the Albanian ethnic population at 91%, followed by the Serbian ethnic group at 3.4% and other ethnic groups at 5.6% (Mustafa & Krasniqi, 2018). Referring to the Kosovar context, it is important to bring statistics on the composition of religions in Kosovo.

Regarding the existing data, the extent of religions is as follows: Muslim 85.6%, Orthodox 10%, Catholic 5%, and other religions 0.06% (Mustafa & Krasniqi, 2018). The Muslim population in Kosovo is dominant. Therefore, companies operating in this economic environment should not ignore Islamic marketing practices since consumer behavior is closely aligned with ethnic, cultural, and religious factors.

Pristina, the capital of Kosovo, together with the Sharr Mountains, the Albanian Alps, and the area of Mitrovica are listed as some of the most attractive areas to visit and also the regions where entrepreneurship is most developed (KAS, 2020; MTI, 2014). Referring to the existing statistics in Kosovo, 182,825 businesses are active, dominated by micro-enterprises and small enterprises, which comprise a percentage of 99.7%. In contrast, medium and large enterprises comprise 0.3% (KBRA, 2022). Within the number of these businesses in Kosovo, it is reported that 85% of businesses in Kosovo are family businesses, which have an important role in the country's economy, contributing to economic growth and reducing unemployment (Gashi & Ramadani, 2013; Toska et al., 2022). Referring to the empirical data brought by Toska and Zeqiri (2023b), it is concluded that the online context is also developed in the Kosovar environment. Initially, it was argued that Kosovo had the highest internet accessibility of all Western Balkan countries, with 96% (Table 1).

It is then argued that Kosovar consumers prefer online shopping, ranging from 16.9% of purchases/orders made online during the last 3 months in 2018, this percentage increased to 23.1% in 2021.

Table 1 Purchase online order difference 2018–2021

	2018	2021
During the last 3 months	16.9%	23.1%

Source: KSA (2022)

3 Stages of Marketing Development

Above, relying on scientific arguments, we argued that marketing is dynamic and evolves very quickly. Therefore, marketing science has evolved in five since the free market economy's actualization. From what *Marketing 1.0* was in the beginning, today, researchers and companies discuss *Marketing 5.0*, which aims to use technology for human benefit. In order to see the importance of the online context, it is essential to explain these five stages briefly, what was unique about each of them, and what brought about a change in each subsequent stage.

- ***Marketing 1.0***—At this stage, companies were focused on production and product. So, the company focused on producing a larger quantity of products and services, ignoring the market demands, leading to Marketing Myopia. Companies used to assume that if you had products and staff, customers would buy (Kotler et al., 2017).
- ***Marketing 2.0***—This concept is thought of differently. As we have technological development at this stage, customers are exposed to information. Therefore, companies cannot sell what they want to customers. This stage suggests that companies should consider the value of the products and services they offer customers.
- ***Marketing 3.0***—This phase becomes even more complex regarding the company and customers. As consumers have deeper needs and wants, companies cannot simply address them. Therefore, companies should think about the sentimental issues of customers. Also, this phase suggests that companies take functional and environmental issues into account when designing marketing strategies and communicating value to target customers (Fuciu & Dumitrescu, 2018).
- ***Marketing 4.0***—At this stage, the importance of the online context is emphasized. Here, it is explained that the dynamics of life and the significant surge of technological development have caused customers to focus on online tools to fulfill their needs and wants. In contrast, companies focus on creating online platforms to communicate the value of marketing to their target customers. The phase emphasizes that the online context has created the "*advocate customer*," who in the preliminary stages was known as the "*loyal customer*."
- ***Marketing 5.0***—This stage suggests that technology should be for the benefit of humanity. So, the phase incorporates the efficient combination of technology, which can be used from the information brought by the technology for the proper design of the marketing strategy, giving superior value to their target customers. At this stage, elements of marketing that help this process are expressed, which are contextual marketing, augmented marketing, agile marketing, predictive

marketing, and data-driven marketing (Kotler et al., 2019, 2021; Reshidi & Baca, 2023).

Since Islamic marketing studies have started very late, the authors Alserhan et al. (2022) bring several phases of Islamic marketing in the context of its actualization in the academic world for both scientific studies and practical implications. Referring to this argument, the literature lists four stages of Islamic marketing. *Islamic Marketing 1.0*: It is thought that it started as early as 2007. The academic community at this stage carried out debates of scientistic value in the context of Islamic marketing, an academic institutional framework was built, and the limitations of scientistic publications were listed, thus laying the foundations of research in this field. *Islamic Marketing 2.0*: This phase recognizes scientific research in the context of Islamic marketing as unique to the academic community and encourages researchers to deal with this field as the institutional framework has already been established. This phase tests the theory from the models tested by previous studies, but now in the context of Islamic marketing. *Islamic Marketing 3.0:* In this phase, studies in the field of Islamic marketing were further advanced. What the creators of the first stages failed to achieve was that solid groups of researchers in this stage advanced the science of Islamic marketing to another level. Various prejudices at this stage were extinguished. *Islamic Marketing 4.0*: This phase argues that, so far, a set of theories has been tested in the Islamic context. It is in the interest of the academic community to bring scientific evidence from different countries and cultures in this way to have Islamic marketing results from different places, to compare these results, and to increase the scientific debate in this field.

Arguing the evolutionary phases of marketing, we can conclude that today, the *Marketing 4.0* phase and *Marketing 5.0* phase are current both for the academic community and for the practical community, focusing more on studies in the online context. Also, relying on the stage of *Islamic Marketing 4.0*, it is suggested that scientific research in different countries and cultures is more necessary to enrich the existing theory and increase the scientific debate in the context of Islamic Marketing; this chapter will contribute as little as possible literature by bringing scientific evidence for Islamic online marketing from the Kosovar context and addressing these suggestions.

4 Online Marketing

Online marketing is a permanent challenge for companies based on its dynamics, but on the other hand, it is also a real opportunity for companies to reach the target audience using online tools. Today, in a very competitive environment, companies must be careful in online communications, relying on the right strategies, specifically on proper marketing content, which attracts consumer attention and makes them react, leading to purchase intentions (Kajtazi & Zeqiri, 2020). Therefore, *online marketing* is defined as an advanced technique using the Internet that companies use to reach the market and consumers through the inclusion of online channels such as social media, websites, email, and other online tools (Garg et al., 2021). In discussions of the online context, it is crucial to distinguish between e-business, e-marketing, and e-commerce. *E-business* is an umbrella that includes e-marketing and e-commerce. Electronic business includes automating and digitizing all business processes, including raw material supply, inventory management, workforce management, distribution system management, information systems management, marketing, and other activities in digitalized companies. E-marketing also includes e-commerce, but not e-business. So, the function of e-marketing is not only to carry out transactions as in e-commerce but also to build long-term relationships with customers using online tools. Through online marketing, companies build appropriate strategies for communicating with the market and creating healthy customer relationships. Customers can now feel important to the company, as they can also be co-creators of value in the company, thanks to the co-creation that online platforms offer. Companies: *Firefox, Linux, T-shirt, Electrolux, DHL,* and *Heineken* are examples of such practices where long-term relationships with consumers were built through co-creation (Ahasanul Haque & Kabir, 2022; O'Hern & Rindfleisch, 2010; Vargo & Lusch, 2016).

Figure 1 shows the steps that companies must follow to realize online presence. So, companies should create social media networks, create a website, place online ads, and finally use email (Kotler & Armstrong, 2011). Depending on the activity, companies can also choose the type of websites to communicate with their target customers. If we are dealing with a manufacturing company, a *corporate website* would be more favorable in this case, which aims to strengthen the marketing network without offering the possibility of making a direct transaction from the end customer. *Website marketing* aims to engage customers in online platforms. The transaction is made directly by end customers, but also other marketing results from this engagement (Hwang et al., 2003).

Therefore, some forms of online communication are business-to-consumer (B2C), this form includes final customers in the company's direct sales strategy, allowing high customer-company interactivity. The main focus is on the final customer, minimizing the focus on the distribution channel. Statistics show that online purchases using the direct site globally are expected to increase from 19.4% in 2021 to about 24.5% in 2025 (eMarketer, 2021). On the other hand, business-to-business (B2B) involves selling products from business-to-business from online platforms. This strategy helps to strengthen relationships and partnerships at the business level. The differences between B2B and B2C digital marketing are highlighted in the value chain development and the different focus on strengthening relationships with end customers versus supporting the distribution network. Other forms include consumer-to-consumer (C2C), where consumers buy and sell products/services among themselves, and consumer-to-business (C2B), a form of engagement where consumers provide suggestions for product and service improvements. An example of C2B practices can be taken from the company Priceline.com (Kannan & Li, 2017; Santosa et al., 2020).

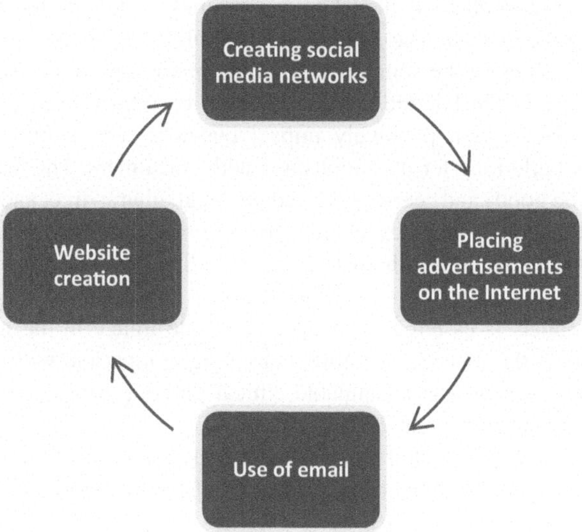

Fig. 1 Online presence

In addition to websites, online communication is also very successful through social media. Social media is considered the newest and most applied online marketing tool, where users share essential information among friends. Social media also allow interactivity between creators and users where, from a marketing perspective, they can be seen as helping consumers find information about products and services. Social media such *as Facebook*, *Twitter*, *YouTube*, *Pinterest*, *Flickr*, and *LinkedIn* are used to build relationships among friends and provide a platform for professional promotions (Hajli et al., 2017). According to Statista (2018), social media has usage of over 1/3 globally in 2021, presenting a significant opportunity for companies looking to reach customers through online tools.

Recent empirical studies emphasized that web marketing positively impacts sales performance and customer engagement (Alderete, 2019; Lorca Fernández et al., 2019). Also, some other research shows that social media brings apparent advantages for companies. Overall, through the analysis and conclusions of recent studies, we can conclude that social media positively impacts brand awareness, drives purchase intentions, increases consumer engagement, and improves sales performance in many aspects (Nisar & Whitehead, 2016; Xie et al., 2016).

5 Islamic Practices in Accordance with Online Marketing

Researchers have addressed the dilemma of whether online marketing is compatible with Islamic practices and whether it is acceptable to Islamic consumers. Based on the study of Ahasanul Haque & Kabir (2022), it is confirmed that online marketing can be considered acceptable in the Islamic community as long as it respects Islamic principles from the Qur'an. The authors make a solid contribution to the scientific literature, elaborating on the suitability of online marketing to the Islamic context. Online marketing in the Islamic context is based on ethical practices that reflect faith and morality, which positively impact business sustainability, as such an approach helps build long-term relationships with consumers. The main purpose is to ensure that the goods and services offered are halal, gharar-free, and interest-free. The cancellation policy is a key element in ensuring consumer trust. Clear and transparent communication with customers, including accurate information about products, services, and terms of sale, is another important company approach. Honesty in products is essential to create a credible image in the market. These approaches reflect the essence of online marketing compatible with Islamic practices, ensuring a reliable and sustainable ethical practice while operating in such economic environments (Amin, 2008).

In an environment dominated by an Islamic population, it is all the more necessary for companies to adopt the basic principles of Islamic marketing. These principles, which have been explained by researchers Shafiq et al. (2016), present an ethical and legal framework that helps promote marketing activities in accordance with Islamic principles. In this context, truthfulness (*al-Siqd*) in information and promotion is considered essential to protect the maqasid of property, according to Islam. So, companies should carefully create accurate and honest product descriptions during online communications with target customers. This principle is also presented in the teachings of the Qur'an, in Surah Bekare (42), where Allah commands: "*Do not mix the truth with falsehood and do not hide the truth since you know*." So with this, companies should focus on honest and trustworthy information during communications with the market and customers because this can affect healthy company–consumer relationships. The principle of impartiality (*al-Adl*) does not allow, during communications with the market, to use advertisements that underestimate others while elevating ourselves or our company in this case. Such demeaning practices toward others are contrary to the principles of Islam. Therefore, we should also be careful about others' reputations.

Moderation (*al-Wasatiyyah*) refers to caution in advertising by eliminating chastity practices. Advertising is important for businesses in their economic activity to inform consumers about products and services. Businesses should be careful in their portrayals, so campaigns should be with careful and modest clothing that is by Islamic principles. This principle fits the recommendations of the Qur'an that encourage a calm and wise attitude toward the market and its participants. Responsibility (*al-Mas'uliyyah*) rests on the principle that companies must be responsible for providing products and services in accordance with Islamic practices. So, companies offering harmful and forbidden products by Islamic principles can harm consumers' physical and moral health. Merchants are also prohibited from offering products and services, relying on lying and deceptive practices to gain a sale or transaction. Social Responsibility in Islamic marketing is presented as a moral obligation that aims to contribute to the welfare and improvement of society. The Social Responsibility of Islamic organizations is based on Responsibility toward the natural environment, community, and stakeholders (Abdullah & Ismail, 2011; Ibn'Abidin, 2012; Shafiq et al., 2016).

In the following section, through the analysis of case studies, it will be identified how many halal companies in the Kosovar context use Islamic principles to communicate with their customers in online channels. Through the analysis of activities on social media, websites, and other online platforms, we aim to find out to what extent principles such as truthfulness (*al-Siqd*), impartiality (*al-Adl*), moderation (*al-Wasatiyyah*), and Responsibility (*al-Mas'uliyyah*), and Social Responsibility are used. This analysis provides an overview of Islamic marketing practices applied by halal companies in Kosovo, focusing on how they communicate and promote products or services online to their target customers.

6 Case Studies of Kosovar Companies

Case 1: MEKA HALAL FOOD

The company "*MEKA HALAL FOOD*" is a typical example of a company in the Kosovar context that operates in the market with the principles of Islamic marketing. The company was founded in 1992 in Dragash in the Republic of Kosovo. The company offers a wide range of meat products, which are entirely halal, to Kosovar consumers.

The company has an online presence to communicate with its customers. The company's website is integrated with social media, where through these tools the company communicates value to target customers. Since the company is a manufacturer, the approach of the *corporate website* is used, where the online campaigns are intended to support the distribution channel instead of directly selling the product.

Figure 2 shows two examples of companies communicating value to target customers based on Islamic principles. Similar images have been featured in most of the company's online ads over the years. The company relies on communication based on truth (*al-Siqd*) by delivering accurate advertisements and reliable content. Likewise, it can be concluded that there is also a responsibility (*al-Mas'uliyyah*) to provide healthy and halal products according to Islamic practices. The company consistently advertises and uses images of satisfied customers to prove its sustainability and a proper *eWom*. Also, carefully used images that eliminate the practices of chastity, relying on the principle of moderation (*al-Wasatiyyah*), emphasizing the family and its value. Therefore, the advertisements feature images of modest clothing accepted in Islamic principles.

One of the most emphasized forms of social responsibility is the focus on stakeholders, but not overlooking the responsibility toward the natural environment and the community in general. Figure 3 shows a decision by the company's CEO that for each birth of children by employees in the company they will be supported with a check of 1000 euros from the company and if two children are born, 2000 euros, and so on. A very welcome and discussed decision in the Kosovar environment in a positive context, which affected the company's reputation. Also, this figure shows the

Fig. 2 Meka company—online communication with customers in social media

Fig. 3 Meka company—social responsibility

organization of Iftar by the company, both for the employees and the community where it operates, during the holy month of Ramadan. The first case study helped us to see how a halal company in the Kosovar environment uses online tools to communicate with its customers in accordance with the principles of Islamic marketing.

Case 2: Hallal Pizza

In the second case study, the company "*Halal Pizza*" which operates in the capital of Kosovo, Pristina, will be discussed. The company focuses on gastronomy, offering halal food to its target customers. Online communication is focused on social media and other forms of direct online communication, such as *Viber* and *WhatsApp*, to receive customer orders and also carry out other marketing communications. The company does not have a website, another important form of online communication that allows integration with social media. Unlike the previous case, here, the transaction from the final customer is possible since the company's concept is B2C, which does not focus on strengthening the distribution network but on the direct sale itself (Fig. 4).

The company has consistency in online communications with customers. By chronologically analyzing the advertisements in social media, an evident consistency is seen, which proves in trust toward consumers, based on appropriate and true advertisements (*al-Siqd*), the company managed to create an *eWom* aimed at this target. The company ensures that the advertisements are in accordance with Islamic principles. It can also be concluded that no element is used that undermines the principle of moderation (*al-Wasatiyyah*), where the company is careful in using appropriate portrayals that do not contain elements of chastity, images are carefully using modest clothing, in full accordance with Islamic principles. The principle of impartiality (*al-Adl*) is also evident in using advertising campaigns, which specify the business's strengths, not belittling and disparaging competitors. The figure also shows the advertisements given to customers in the Holy Month of Ramadan, enabling the delivery of halal products to customers during Syfyr. This element emphasizes the principle of responsibility (*al-Mas'uliyyah*) toward this community. The second case study helped us understand how another halal company in the Kosovar context carefully uses online marketing tools based on Islamic principles to communicate with its target customers. So, the practices create reliability and sustainability for the company in such communities.

Case 3: AGA BURGER

The last case study will focus on another company in the gastronomy sector. The *AGA BURGER* company offers fast food products and operates in Pristina, the

Fig. 4 Hallal Pizza—online communication tools

capital of the Republic of Kosovo. The company has an online presence that focuses more on social commerce, making more use of *Instagram* and *Facebook* to communicate with their customers.

Figure 5 shows some of the company's online communications. As in previous cases, the company does not use images prohibited in Islam, being in accordance with the principle of moderation (*al-Wasatiyyah*). In online communications, it can be seen that the company has credibility regarding the information it provides to consumers; since it has a high *eWom*, it is also by the principle of truthfulness (*al-Siqd*). During the *COVID-19* pandemic, the company also enabled the delivery of products to customers at home, the cause of social distancing, proving responsibility (*al-Mas'uliyyah*), a concept that continues even now.

7 Conclusion

This chapter discusses the application of Islamic online marketing principles in the context of halal companies in the Kosovar economic environment. The chapter links online marketing in general and Islamic online marketing in particular. Given that Kosovo is a country with 85.6% Muslim population, empirical evidence from such an economic environment was important and also necessary for the academic community. The five basic principles of Islamic marketing were chosen to be analyzed as the principle of truthfulness (*al-Sidq*), responsibility (*al-Mas'uliyyah*), impartiality (*al-Adl*), moderation (*al-Wasatiyyah*), and corporate social responsibility by Islamic principles. Based on case studies of concrete companies, it is concluded that halal companies in the Kosovar environment apply the basic principles of online marketing when communicating with their customers through online tools and are very careful in this aspect. This chapter offers opportunities to advance studies in the field of online marketing, as in the Kosovar environment and other

Fig. 5 Aga Burger—online communication tools

environments. Referring to *Islamic Marketing 4.0*, we find solid suggestions that marketing studies in the Islamic context are more necessary from the perspectives of different cultures and countries to see the similarities and compare the differences. Therefore, such empirical evidence will profoundly influence scientific discussions, further adding to scientific debates and leading to a more advanced level through empirical information from different contexts.

References

Abdullah, K., & Ismail, Y. (2011). Malaysian consumers' experience with deceptive marketing practices of sellers-An Islamic perspective. In O. M. Zain (Ed.), *Readings in marketing: An Islamic perspective* (pp. 247–275). IIUM Press.

Ahasanul Haque, A. K. M., & Kabir, S. M. H. (2022). E-marketing in Islamic markets. In B. A. Alserhan, V. Ramadani, J. Zeqiri, & L.-P. Dana (Eds.), *Strategic Islamic marketing: A roadmap for engaging Muslim consumers* (pp. 109–123). Springer.

Alderete, M. V. (2019). Electronic commerce contribution to the SME performance in manufacturing firms: A structural equation model. *Contaduría y administración, 64(4)*.

Alserhan, B. A. (2011). *The principles of Islamic marketing*. Gower.

Alserhan, B. A., Zeqiri, J., Ramadani, V., & Dana, L. P. (2022). An introduction to strategic Islamic. In B. A. Alserhan, V. Ramadani, J. Zeqiri, & L.-P. Dana (Eds.), *Strategic Islamic marketing: A roadmap for engaging Muslim consumers* (pp. 1–5). Springer.

Amin, H. (2008). E-business from Islamic perspectives: Prospects and challenges. *Journal of Internet Banking and Commerce, 13(3)*, 1–13.

Amin, M., Isa, Z., & Fontaine, R. (2013). Islamic banks: Contrasting the drivers of customer satisfaction on image, trust, and loyalty of Muslim and non-Muslim customers in Malaysia. *International Journal of Bank Marketing, 31*(2), 79–97.

eMarketer. (2021). Retrieved from https://www.insiderintelligence.com/content/global-ecommerce-forecast-2021

Fuciu, M., & Dumitrescu, L. (2018). From marketing 1.0 to marketing 4.0—The evolution of the marketing concept in the context of the 21st century. In *International conference knowledge-based organization* (pp. 43–48).

Garg, M., Bansal, A., & Single, K. (2021). Impact of digital marketing on consumer decision making. *International Journal for Scientific Research and Development, 8*(3), 213–218.

Gashi, G., & Ramadani, V. (2013). Family businesses in Republic of Kosovo: some general issues. In V. Ramadani & R. C. Schneider (Eds.), *Entrepreneurship in the Balkans* (pp. 91–115). Springer.

Hajli, N., Sims, J., Zadeh, A. H., & Richard, M. O. (2017). A social commerce investigation of the role of trust in a social networking site on purchase intentions. *Journal of Business Research, 71*, 133–141.

Hwang, J. S., McMillan, S. J., & Lee, G. (2003). Corporate web sites as advertising: An analysis of function, audience, and message strategy. *Journal of Interactive Advertising, 3*(2), 10–23.

Ibn'Abidin, M. A. (2012). *The book of sales: Kitab al-Buyu', Muhammad Anas al-Muhsin & Amer Bashir*. IBFIM.

Kajtazi, K., & Zeqiri, J. (2020). The effect of e-WOM and content marketing on customers' purchase intention. *International Journal of Islamic Marketing and Branding, 5*(2), 114–131.

Kannan, P. K., & Li, H. A. (2017). Digital marketing: A framework, review and research agenda. *International Journal of Research in Marketing, 34(1)*, 22–45.

KBRA. (2022). *Report on basic performance indicators for the registration of commercial companies in Kosovo for 2021*. Kosovo Business Registration Agency.

Kosovo Agency of Statistics (KAS). (2020). *Statistical Repertoire of Economic Enterprises in Kosovo Q3 2020*. Kosovo Agency of Statistics.
Kosovo Agency of Statistics (KAS). (2022). Kosovo Agency of Statistics. Retrieved from https://askdata.rks-gov.net/pxweb/sq/ASKdata/
Kotler, P., & Armstrong, G. (2011). *Principles of marketing* (14th ed.). Prentice Hall.
Kotler, P., Kartajaya, H., & Setiawan, I. (2017). *Marketing 4.0: Moving from traditional to digital*. Wiley.
Kotler, P., Kartajaya, H., & Setiawan, I. (2019). *Marketing 3.0: From products to customers to the human spirit*. Springer.
Kotler, P., Kartajaya, H., & Setiawan, I. (2021). *Marketing 5.0: Technology for humanity*. Wiley.
Lorca Fernández, P., Andrés Suárez, J., & García Díez, J. (2019). Impact of e-commerce sales on profitability and revenue. The case of the manufacturing industry. *Engineering Economics, 30*(*5*), 544–555.
Ministry of Trade and Industry (MTI). (2014). *Sector profile of tourism*. PristinaMTI.
Mustafa, M., & Krasniqi, B. A. (2018). Entrepreneurship in Kosovo. In R. Palalić, L.-P. Dana, & V. Ramadani (Eds.), *Entrepreneurship in former Yugoslavia: Diversity, institutional constraints and prospects* (pp. 51–65). Springer.
Nisar, T. M., & Whitehead, C. (2016). Brand interactions and social media: Enhancing user loyalty through social networking sites. *Computers in Human Behavior, 62*, 743–753.
O'Hern, M. S., & Rindfleisch, A. (2010). Customer co-creation: a typology and research agenda. *Review of Marketing Research, 6*(*1*), 84–106.
Reshidi, N., & Baca, G. (2023). *Strategjia e Marketingut*. Ricoh.
Santosa, S., Mudiantono, M., Murniyono, C., Hersugondo, H., & Soesanto, H. (2020). Increasing consumers to consumers (C2C) e-commerce in central Java, Indonesia. *Accounting, 6*(*5*), 753–762.
Shafiq, A., Haque, A., & Abdullah, K. (2016, September 28–29). A collection of Islamic advertising principles. In *Proceedings of the 8th international management and accounting conference* (p. 53). Universiti Kebangsan.
Sprott, D., Czellar, S., & Spangenberg, E. (2009). The importance of a general measure of brand engagement on market behavior: Development and validation of a scale. *Journal of Marketing Research, 46*(*1*), 92–104.
Statista. (2018). Retrieved from https://www.statista.com/chart/15355/socialmedia-users/
Syafril, S., & Hadziq, M. F. (2021). Islamic principles in marketing: An overview of islamic marketing mix in social-media campaign. *El-Qish: Journal of Islamic Economics, 1*(*1*), 69–82.
Tabrani, M., & Marlizar, M. (2016). The marketing concept in Islamic perspective: a literature review. *AFEBI Islamic Finance and Economic Review, 1*(*01*), 20–37.
The World Bank. (2020). *An overview of the World Bank's work in Kosovo*. The World Bank. In Kosovo.
Toska, A., Ramadani, V., Dana, L. P., Rexhepi, G., & Zeqiri, J. (2022). Family business successors' motivation and innovation capabilities: The case of Kosovo. *Journal of Family Business Management, 12*(*4*), 1152–1166. https://doi.org/10.1108/JFBM-11-2021-0136
Toska, A., & Zeqiri, J. (2023a). Analyzing online marketing forms in customer engagement: Evidence from micro-enterprises. In *10th Eurasian conference on economics and social sciences* (pp. 36–46). ECLSS.
Toska, A., & Zeqiri, J. (2023b). Does online marketing have positive impact on sales growth in family businesses? In *Twenty-eighth international conference on: "Social and natural sciences—Global challenge 2023" (ICSNS XXVIII-2023)* (pp. 37–46).
Toska, A., Zeqiri, J., Ramadani, V., & Ribeiro-Navarrete, S. (2023). Covid-19 and consumers' online purchase intention among an older-aged group of Kosovo. *International Journal of Emerging Markets, 18*(*12*), 5921–5939.
Vargo, S. L., & Lusch, R. F. (2016). Institutions and axioms: an extension and update of service-dominant logic. *Journal of the Academy of Marketing Science, 44*(*1*), 5–23.

Wilson, J. (2012). Looking at Islamic marketing, branding and Muslim consumer behaviour beyond the 7P's: The call for supportive course content and more P's please. *Journal of Islamic Marketing, 3*(3), 212–216.

Xie, K. L., Zhang, Z., Zhang, Z., Singh, A., & Lee, S. K. (2016). Effects of managerial response on consumer eWOM and hotel performance: Evidence from TripAdvisor. *International Journal of Contemporary Hospitality Management, 28*(9), 2013–2034.

Asdren Toska is the founder of Toska Electronics Company and also a part-time teaching assistant at the Faculty of Economics at Kadri Zeka University in Gjilan, Kosovo. He completed his Bachelor studies at the South East European University, Faculty of Economics, Department of Business Administration. He completed his Master's degree at the same university in the marketing field. He also completed Doctoral studies at the South East European University, Faculty of Economics, Department of Business Administration, in Tetovo, North Macedonia. Asdren has published some scientific papers in well-known international journals.

Islamic Marketing and Circular Economy to Foster Sustainability of Product Life Cycle Enhancement

Mohammad Mominul Islam, Mohamed Syazwan Ab Talib, and Nazlida Muhamad

Abstract Implementation of product life cycle management for resource optimization within the framework of the circular economy in Islamic marketing remains unstudied. This chapter aims to emphasize the sustainability of product life cycle enhancement governed by Maqasid-al Shariah. The Maqasid-al-Shariah methodology and a qualitative approach have guided this chapter to showcase sustainable consumption and production of halal products. The extensive literature on the circular economy was aligned with Islamic principles to achieve resource optimization. Data were analyzed using ATLAS.ti, resulting in thematic and content analysis. The content analysis reveals that the circular economy prioritizes resource optimization while adhering to prohibitions (Haram) and purity (Khabis). However, Islamic marketing can contribute to enhancing the product life cycle for resource optimization with Halal (permissibility) and Tayyib (purity) norms, benefiting both individuals and society. This chapter provides valuable insights for academics interested in exploring the circular economy from the perspective of Maqasid-al-Shariah and Islamic marketing. Ultimately, halal marketers can reconfigure and market their products based on circular economy principles.

1 Introduction

The product life cycle (PLC) is a concept that was first introduced by Raymond Vernon in his 1966 book titled "International Investment and International Trade in the Product Cycle," and it has since become a widely accepted model in marketing and strategic management (Vernon, 1966). The stages of PLC encompass the introduction, growth, maturity, and decline phases (Kotler & Keller, 2016). The phases

M. M. Islam (✉) · M. S. Ab Talib · N. Muhamad
School of Business and Economics, Universiti Brunei Darussalam,
Gadong, Brunei Darussalam
e-mail: 21h8961@ubd.edu.bn; syazwan.talib@ubd.edu.bn; nazlida.muhamad@ubd.edu.bn

A. Rafiki et al. (eds.), *Strategic Islamic Business and Management*,
Contributions to Management Science,
https://doi.org/10.1007/978-3-031-61778-2_7

of PLC serve as a strategic tool to guide decision-making related to product development, marketing strategies, and resource allocation (Kotler, 2016).

The PLC needs to have new avenues to face competitiveness in the market (Michelle Grantham, 1997). Also, previous studies have recognized the PLC from an End-of-Life (EOL) product return perspective to mitigate waste and foster sustainability (Ghanadpour et al., 2023). EOL underscores the importance of considering a holistic perspective, encompassing social, economic, and environmental factors that influence the behavior of both manufacturers and consumers (Mies & Gold, 2021).

Consumer behavior is shaped by factors such as perceived environmental benefits, financial incentives, and product design. Conversely, manufacturers' decisions are influenced by factors like the cost of implementing EOL product return programs, legal and regulatory requirements, and market pressures (Kumar & Putnam, 2008). Furthermore, it has become evident that understanding the interaction between consumers and manufacturers is essential in the context of EOL product return (Chen & Akmalul'Ulya, 2019). The presence of financial incentives for consumers to participate in EOL product returns may hinge on manufacturers' willingness to implement EOL (Michaud & Llerena, 2011; Nnorom et al., 2009).

EOL product return boosts product life cycle enhancement (PLCE) which is viewed from the circular economy (CE) point of view. The CE framework emerges as a prominent decision-making approach, aiming to minimize waste generation and promote resource efficiency (Agrawal et al., 2021; Hunka et al., 2021; Kirchherr et al., 2017; Lawrenz et al., 2021). Circular Economy (CE) focuses on environmentally friendly and economically beneficial business models through Product Life Cycle Enhancement (PLCE). Producers can adopt a sustainable production approach based on Islamic marketing's Halal and Tayyib principles. However, Shariah compliance in PLCE and End-of-Life (EOL) management under CE in Islamic marketing requires further exploration. This study aims to highlight the significance of PLCE governed by Islamic marketing and circular economy.

2 Literature Review

The circular economy (CE) presents a different development approach than the traditional take-make-dispose linear model (Wilson & Goffnett, 2022; Haydn et al., 2023). The CE concept has gained prominence in recent years outlining substitution instead of manufacturing resulting in cost-effectiveness (Ding et al., 2023). Researchers have expanded the principles of CE associated with environmental implications and sustainability for accelerating resource efficiencies (Geissdoerfer et al., 2017), by balancing between ecology and profit orientation in traditional business modeling (Bocken et al., 2016; Ghisellini et al., 2016). As such, CE has emerged as a vital approach to address resource scarcity, environmental degradation, and sustainability (Lynne et al., 2019).

Sustainability becomes a major issue for organizations motivated by consumer demand and environmental concerns (Wrålsen et al., 2021). Resource constraints and scarcity (Hunka et al., 2021), economic benefits (Calzolari et al., 2021), regulatory pressure (Moreno-Mondéjar et al., 2021), stakeholder and competitive forces (Jaeger & Upadhyay, 2020; Nag et al., 2021) drive organizations to embrace sustainability practices, aligning with EOL. Also, technological uncertainty (Tecchio et al., 2017), coupled with resource scarcity (Formentini et al., 2022; Kazancoglu et al., 2022), corporate reputation (Kapetanopoulou & Tagaras, 2011), limited information sharing (Kaviani et al., 2020), and shifting from linear to circular (Marques & Manzanares, 2023) are the influential factors for EOL.

Factors such as collaborating on innovations, consistently patenting new ideas, and partnering with research institutions were identified as key drivers in reducing the number of wasteful patents and boosting business success by adhering to CE principles (Marín-Vinuesa et al., 2023). Further, internal policies and practices (Shaharudin et al., 2015), effective waste management practices, and product quality (Bernon et al., 2013; Li et al., 2018; Cecchin et al., 2019) are essential drivers, contributing to environmental sustainability and EOL. Diversification of motivation and intrinsic phenomena help foster the progress of CE (Savu & Max, 2023). However, spiritual drivers from Shariah are missing in the studies cemented with IM. How the principles of IM can navigate EOL, PLCE, and sustainability remains unknown.

Implementing sustainability practices, including EOL product returns, leads to reduced environmental pollution through resource conservation, waste reduction, and enhanced resource efficiency (Agrawal et al., 2021; Hunka et al., 2021; Lawrenz et al., 2021). Sustainable manufacturing systems can optimize resource utilization and enhance environmental performance (Nag et al., 2021; Tecchio et al., 2017). As such, sustainability practices, including EOL product returns, contribute to cost savings and reduced risk (Foster, 2020). Effective sustainable practices and efficient waste management enhance organizational reputation and customer loyalty while improving efficiency (Uhrenholt et al., 2022; Zailani et al., 2017). Yet, sustainability from MS and IM and their potential outcome related to PLCE and EOL are neglected. In particular, the contribution of halal producers, consumers, and halal consumption leading to sustainability to enhance PLC and EOL has not been studied.

Digital technology can help create sustainable circular products. Involving customers through digitalization is crucial. Experts recommend a product-service system to accelerate the shift toward a circular economy (Agrawal et al., 2022) followed by big data analytics capabilities (Riggs et al., 2023) with reverse logistics (Fernando et al., 2023). Research also shows that blockchain technology can have a beneficial impact on the circular economy, as it promotes visibility, transparency, relationship management, and smart contracting (Syed et al., 2022). Notably, it is important to understand the link between Industry 4.0 (I4.0) technologies and CE practices. Nevertheless, there is still insufficient research on the possible connection between them—particularly on how different I4.0 technologies might impact various CE practices (Lei et al., 2023; Paolo et al., 2020).

Fig. 1 Content analysis of the literature on CE and PLC

Figure 1 shows the key concepts that the previous studies have focused on. Product (127 times) and its EOL (77 times) center on the CE (34 times) and economy (38 times). Consumers (45 times) and production (23 times) are associated with sustainability (43 times). To ensure sustainability, resources (43 times) require a return (63 times) recycling (9 times) by managing waste (27 times) with efficiency (9 times) through the supply chain (43 times). Hence, the key issue of CE is to enhance PLC by optimizing resources and sustainability through production and consumption. However, the implementation of PLC management in halal marketing using Maqasid-al-Shariah (MS) has not been studied yet. To improve customer value, marketers can use the PLCE approach to optimize resources. The combination of MS and IM with CE principles has not been given attention. This chapter aims to fill in the gaps and demonstrate how IM and CE can contribute to the sustainability of PLCE.

3 Methodology

3.1 Maqāṣid al-Sharī'ah

According to Imam Al-Ghazālī, the main goal of Sharī'ah is to protect people's faith, lives, intellect, offspring, and wealth. Actions that protect these aspects are beneficial while those that harm them are not (Hudaefi & Badeges, 2022). MS evaluates the compatibility of structural reforms with Islamic principles, aims, and objectives (Tubarad et al., 2022) to explore knowledge through Ijtihad. The Noble Quran entails *Ijtihad* in chapters 3:142; 29:69; 47:31 and others. Allah says:

وَٱلَّذِينَ جَاهَدُوا۟ فِينَا لَنَهْدِيَنَّهُمْ سُبُلَنَا وَإِنَّ ٱللَّهَ لَمَعَ ٱلْمُحْسِنِينَ

"As for those who struggle in Our cause, We will surely guide them along Our Way. And Allah is certainly with the good-doers" The Noble Quran, 29:69). The messenger of Allah (PBUH) said:

إِذَا حَكَمَ الْحَاكِمُ فَاجْتَهَدَ ثُمَّ أَصَابَ فَلَهُ أَجْرَانِ وَإِذَا حَكَمَ فَاجْتَهَدَ ثُمَّ أَخْطَأَ فَلَهُ أَجْرٌ

> If a judge makes a ruling, striving to apply his reasoning and he is correct, he will have two rewards. If a judge makes a ruling, striving to apply his reasoning and he is mistaken, he will have one reward (Ṣaḥiḥ al-Bukhari, 7352, Ṣaḥih Muslim, 1716).

The prime intention is to satisfy Allah, ensure human welfare, and put efforts into knowing the Shariah sources from the true ground. These are called *Maslahah (welfare intention)*, *Usuliyyun (ruling)*, and *Ijtihad (effort)*, composing the MS methodology (Fisol et al., 2019). MS methodology has guided the study to justify the related Islamic principles of sustainability in the Islamic economy, CE, IM, SC, SP, RO, and PLCE. Importantly, Shariah unveils the resources, namely the Quranic verses and Sunnah to benefit humanity. The prescriptive verses and sayings of the prophet have benefited the present work followed by the relevance and context and MS methodology to explore suitable Quranic verses.

3.2 Research Approach

A qualitative approach facilitates the researcher to understand the depth of the research problem. Notably, qualitative research focuses on the depth of the topic when the research area is new (Bryman, 2016). Since the CE from IM research phenomena is novel, this study has followed a qualitative approach inducting insights into the halal business domain.

3.3 Data Collection

This study has gathered only secondary data about CE associated with EOL to optimize resources. Then, the relevant Quranic verses have been explored to showcase the way of resolving the shortcomings of the traditional CE. As such, conventional and Shariah aspects of business have been put together to enhance inclusiveness and sustainability.

3.4 Data Analysis

This study has analyzed the content of the previous studies to focus on the key concepts and synthesized the MS methodology to resolve the current problem of CE. Besides, a thematic analysis has addressed the research objective. The conceptual framework of the study has shown the way to cement MS, IM, and CE to optimize resources and PLCE. The qualitative data analysis software ATLAS.ti has facilitated the analysis.

4 Results

4.1 Sustainability in the Islamic Economy

In a sustainable Islamic economy, the three lenses to consider are people, planet, and profit. The Quran states that people are at the center of the planet's creation (2:29) and that Allah has maintained balance in the earth's creation, so it is essential to maintain that balance (55: 7–8). By doing so, people can engage in halal business for profit without causing harm to others or being unjustified by others (2:275–279). Therefore, people must manage the planet justifiably for profit in a sustainable Islamic economy (Fig. 2).

Allah permits lawful business activities (The Noble Quran, 2:275), which may yield profit or loss. Stakeholders must engage in halal production, ensuring that profits are distributed justly (The Noble Quran, 4:29; 2:279) to prevent injustice. Equitable profit contributes to the concept of sustainability. The profit formula involves reducing costs and increasing benefits. Since usury is forbidden (The Noble Quran, 3:130), marketers must avoid interest-based practices, as they are unsustainable (The Noble Quran, 2:276). Additionally, Allah prohibits resource wastage and encourages optimal resource utilization (The Noble Quran, 17:27,

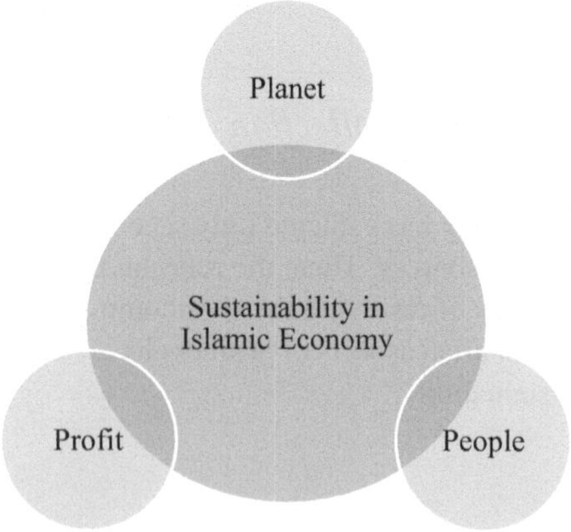

Fig. 2 Sustainability in Islamic Economy adopted from Islam (2023)

17:29), further emphasizing the importance of profit generation from halal and efficient practices.

The three pillars of the planet, people, and profit (3Ps) are integral to shaping a sustainable Islamic economy. Islamic entrepreneurship can catalyze this sustainability model, driven by the triple bottom lines of the 3Ps. The 3Ps can signify the welfare of the people guided by MS. IM governed by MS can contribute to fostering halal products, its sustainable production (SP), and sustainable consumption (SC).

4.2 Sustainable Consumption

Halal products are the primary factor in achieving sustainability in production and consumption. Allah has commanded both Muslims and non-Muslims to consume Halal and Tayyib products (The Noble Quran, 2:168, 2:172). Even the Quran instructs Allah's messengers to consume Halal and Tayyib (The Noble Quran, 23:51). Consequently, the scope of consumption of halal products is broad which can lead to sustainable consumption.

Halal consumption is encouraged in the Noble Quran, but waste is prohibited. Allah says:

"O Children of Adam! …Eat and drink, but do not misuse or abuse. Surely Allah does not like to be extravagant" (The Noble Quran, 7:31). Using extra is a sinful act and haram (prohibited). This explicit instruction must guide about two million Muslims around the globe for responsible consumption leading to sustainability. Controlling mis/abuse will accelerate resource optimization taking part in enhancing sustainability in the present time. Then, the resources will be reserved for the future generation attaining sustainability for the future time. Hence, using more than the need in the form of abuse, more use, and misuse as *Tusrifu* (تُسْرِفُوٓا) signifies the milestone of achieving resource optimization.

Sustainable consumption (SC) can also result from restricting the wastage in consumption. This is much more crucial than ab/more/misuse. Allah compares the individuals who waste with the devil. The words of Allah say: "And do not spend wastefully, Surely the wasteful are like brothers to the devils. And the Devil is ever ungrateful to his Lord" (The Noble Quran, 17:26–27). The Holy Quran outlines the word *Tubajjir* (تُبَذِّرْ), which is more stressful than *Tusrifu* (تُسْرِفُوٓا). Hence, *Tusrifu* (extra use) and *Tubajjiru* (waste) are the threats to resource optimization, sustainability, and CE. IM can motivate halal consumers to refrain from extra use and waste to achieve sustainability resource optimization goals.

Consumers can help to maximize resource optimization by being economical and rational in spending. The Noble Quran teaches how to be balanced in expenditure. It says: "Do not be so tight-fisted, for you will be blameworthy; nor so openhanded, for you will end up in poverty" (The Noble Quran, 17:29). The principle of frugality can help run the wheel of the economy in one hand, and consumer expenditure on halal products on other hands.

Muslim consumers must not consume Haram and Khabis which can stop resource use from the very beginning. Further, rational expenditure on Halal and Tayyib products can make the resource use more productive and conservative. Halal and Tayyib are positive sides and Haram, Khabis, Tubajjiru, and Tusrifu are the negative sides of consumption. Enhancing the positive and checking the negative can produce sustainable consumption from the consumers' perspective. Hence, halal consumers can contribute to fostering PLCE through reusing and recycling products being economical to derive sustainable production.

4.3 Sustainable Production

The business of halal products is ordained, as stated in The Noble Quran (2:275). It also means that halal businessmen cannot deal with Haram and Khabis. When producers do not manufacture haram products, wastage and wrong uses of resources become net zero. This ensures resource optimization in the CE. This is the first step of sustainable production. In the second stage, halal producers must check misuse in producing halal products. The Quranic rulings "Surely Allah does not like to be extravagant" (The Noble Quran, 7:31) apply to the producers in manipulating the scarce resources. Beyond excessive use, sustainable production can also be achieved by curbing wastage according to the Holy Quran (The Noble Quran, 17:26–27). The Holy Quran uses the term *Tubajjir* to describe wastefulness, which is considered more severe than *Tusrifu* (extra use). Therefore, both excessive use (*Tusrifu*) and waste (*Tubajjir*) pose threats to resource optimization, sustainability, and the concept of CE. Islamic motivation can encourage Halal producers to avoid excess use and waste, aligning with the goals of resource optimization and sustainability.

Producers' efforts for Halal and Tayyib products can lead to more efficient and conservative resource utilization. Halal and Tayyib represent positive aspects, while Haram, Khabis, Tubajjir, and Tusrifu represent negative aspects of production. Promoting the positive and curbing the negative can foster sustainable production from the producers' perspective. This can foster PLCE, resource optimization, and circular economy.

4.4 PLCE, SC, and SP

Islamic Marketing (IM) is a Shariah-based discipline that addresses the consumption of halal products, on the one hand, and halal production, on the other. IM essentially follows the Halal and Tayyib approach to produce and consume products. Products serve as a means of sustenance for consumers and a source of income. Since products require resources, optimizing resource usage is imperative.

Figure 3 depicts resource optimization (RO), MS, IM, and its role in accelerating PLCE. The green zone represents SP and SC based on Halal and Tayyib principles,

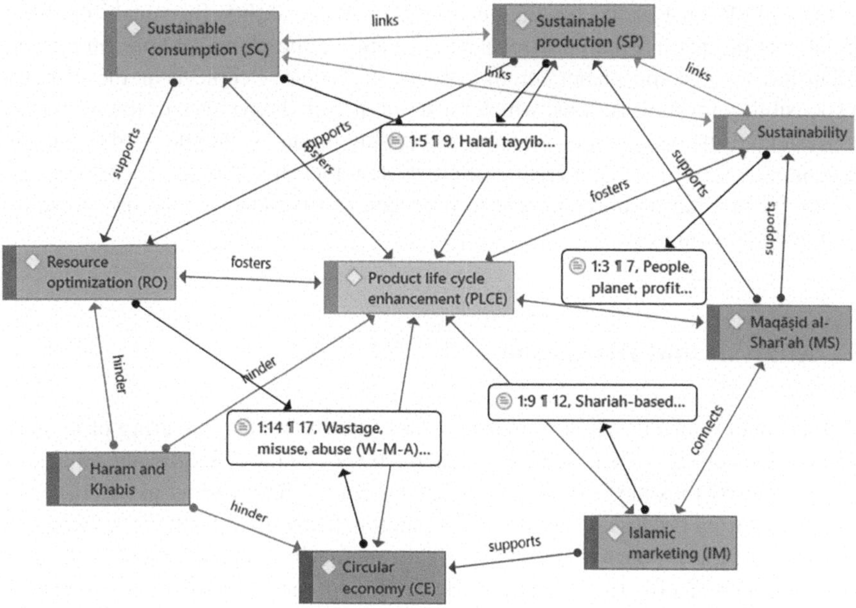

Fig. 3 CE and IM for PLCE

resulting in sustainability. IM can contribute to fostering PLCE and RO by recognizing Halal and Tayyib as the rule of thumb guided by MS. Consequently, RO can only be attained through CE when W-M-A is checked. However, Haram and Khabis with W-M-A lead to the red zone, which is dangerous for sustainability, PLCE, and CE (blue zone). Muslim producers and consumers with strong religious sensitivity can avoid the red zone to reach the blue and green zones. Hence, IM guided my MS can play a major role in supporting CE.

CE promotes RO and vice versa. However, wastage, misuse, and abuse (W-M-A) pose threats to RO. Wastage refers to the unnecessary consumption or use of resources, materials, or goods resulting in their loss or squandering. Misuse refers to the inappropriate or incorrect use of something, often contrary to its intended purpose or rules. Abuse involves intentional or unintentional actions that may lead to damage, harm, or reduced effectiveness of the item or resource. W-M-A, RO, and CE are not compatible.

To achieve RO within CE, responsible producers and consumers must take action to reduce W-M-A. Global Muslim producers and consumers believe that Allah, the Almighty and Sustainer, revealed Quranic principles to guide humanity toward responsible living. Humans are stewards of resources and bear significant responsibility for RO. The prospect of being questioned on Judgment Day motivates producers and consumers to avoid Haram and Khabis practices. According to Shariah principles, W-M-A is Haram and Khabis, fostering resource optimization in the Muslim community, leading to SC and SP.

SC and SP must be the essential elements of sustainability. Sustainability, from an Islamic perspective, emphasizes a balance among the planet, people, and profit. Allah has created the planet for the benefit of people. People bear the primary responsibility for their actions, which must not disrupt the balance of the planet, its ecosystem, or its people in consuming and selling products. People need to engage in economic activities to generate profit to sustain their livelihoods. Therefore, sustainability in Islam requires people to take care of the planet while also pursuing their economic interests.

5 Analysis and Discussion

EOL product return research primarily delves into the remanufacturing process as an environmentally conscious effort, bolstered by government support and advanced technologies (Moosmayer et al., 2020; Zhang et al., 2020; Zhu et al., 2014). The remanufacturing process rest of the PLCE approach that gives birth to the introduction stage of the product in the market. This approach is highly supported by MS and IM to ensure effective resources. The EOL approach emphasizes the management of waste generated during operation (Liu et al., 2020). The MS suggests that waste (Tubajjir) is strictly Haram and sinful. So, IM can navigate PLCE by downsizing waste and extra use of resources.

The development of EOL (de Jesus et al., 2021; Foster, 2020; Julianelli et al., 2020), modeling of EOL (Bimonte et al., 2022; Esenduran et al., 2020; Tuunanen & Cassab, 2011), theoretical development of EOL (Bigliardi et al., 2020; Debo et al., 2005; Singhal et al., 2020), interaction of buyers and sellers for EOL (Shang et al., 2017; Subramanian & Subramanyam, 2012) received attention. This study posits that SP and SC can result in better top-to-bottom and bottom-to-top approaches to link buyers and sellers to foster EOL and PLCE. Importantly, the circulation of CE, PLCE, and IM can be a model to enhance EOL.

The challenges of managing EOL include human resources, financial resources, time (Zailani et al., 2017), and product return policies (Shaharudin et al., 2015). Also, perceived risk and moral obligation to purchase refurbished products pose challenges (Alyahya et al., 2023). The present study identifies SC based on Islamic considerations that can motivate Muslim consumers to adopt PLCE in CE.

The current perception of CE is mostly descriptive and lacks practical relevance (Naoum et al., 2023), and coupled with a shortage of supply chains (Piller, 2023), logistics and innovation (Tom et al., 2023), maintenance (Kuhl et al., 2023), and practicality (Kaviani et al., 2020). Further, reverse logistics (Frei et al., 2020), sustainability (Agrawal et al., 2021; Hunka et al., 2021; Lawrenz et al., 2021), supply chain management (Jiménez-Parra et al., 2014), redistribution to reduce waste hinders EOL product return (Walsh et al., 2016). Against this backdrop, halal supply chain management (HSCM) can link the halal producers and halal consumers. IM and HSCM are coined, and MS outlines the Halal and Tayyib principles to facilitate the EOL product return.

Some CE practices are more critical and prone to consumer behavior from a conventional view, and purity and wholesomeness according to Islamic teaching. For instance, wastewater treatment plants and the process of converting human waste into energy are in practice in non-Muslim majority countries (Malgorzata et al., 2023). But this practice is questionable according to the Shariah ruling. Thus, the practice of reversing human manure into power and using human waste as bio-fertilizers in agriculture challenges the Islamic aspect of CE. MS does not allow this practice as human manure is Haram and Khabis. Similarly, IM restricts Haram and Khabis in SP and SC though human manure is worth being a component of CE and EOL product returns governed by Islamic principles.

Islamic principles underpin Islamic marketing, which is much more spiritual than materialistic (Islam, 2021a). This spirituality-based marketing targets the mass market, consisting of both Muslims and non-Muslims (Islam, 2021b), tapping into marketing opportunities beyond borders (Islam, 2022a). Hence, in addition to spirituality and materialism, Islamic marketing is also strategic (Islam, 2022b). Although Halal and Haram principles are the restrictions of Islamic marketing (Islam, 2022c; Islam et al., 2023a), these restrictions can be managed through common consensus (Islam et al., 2023b). Against the backdrop of wastage and misuse of resources tied to products, Islamic marketing becomes a torchbearer with its Halal and Haram provisions, supported by MS. Eventually, the MS, galvanized with welfare governed by the restrictions (Haram) and permissibility (Halal), can boost PLCE, RO, and CE.

6 Conclusion

The decline of a product's life cycle is not the end, but the beginning of a new phase that encourages sustainable consumption practices. Producers should follow the Halal and Tayyib norms of Islamic marketing, along with Maqasid-al-Shariah guidelines, to reduce waste and optimize resources. Combining Islamic marketing principles with a circular economy fosters sustainability and enhances the production and consumption of Halal products. Future research can expand the understanding of this intersection.

The decline stage should not be the end phase for products. Instead, the end of the PLC should mark the introduction of a new phase. Therefore, enhancing the product life cycle can promote a circular economy by encouraging consumers to engage in rational, halal, and sustainable consumption practices. In this context, producers should adopt a sustainable production approach guided by the Halal and Tayyib norms of Islamic marketing. Notably, Maqasid-al-Shariah can provide guidelines to reduce waste and excessive resource usage, involving both consumers and producers in sustainable production and consumption. Consequently, PLCE can be a viable option for marketers seeking to optimize resources and provide better customer value. Combining the principles of Maqasid-al-Shariah and Islamic marketing with the concept of a circular economy can foster sustainability within

PLCE. This study offers insights to academics interested in exploring the circular economy from the perspective of Maqasid-al-Shariah and Islamic marketing. Unlike the ontological view of the circular economy, which does not consider concepts like permissibility (Halal), wholesomeness (Tayyib), prohibition (Haram), and impurity (Khabis), a Maqasid-al-Shariah view of the circular economy can help mitigate the challenges associated with conventional approaches. Furthermore, Islamic marketing, when viewed through the lens of Maqasid-al-Shariah, can contribute to sustainability by enhancing the production and consumption of halal products.

Halal marketers can reconfigure and market their products by circular economy principles. Ensuring Shariah compliance, with a focus on strict adherence to Halal and Haram, underscores the practicality of halal marketing. Consequently, halal marketers should embrace a conservative approach to resource utilization in line with the sustainable production norms outlined in this study. It is worth noting that there is a scarcity of literature addressing the intersection of circular economy, Islamic marketing, and Maqasid-al-Shariah. Future research endeavors have the potential to expand understanding in this area, with both empirical and conceptual studies contributing to the body of knowledge.

References

Agrawal, R., Wankhede, V. A., Kumar, A., & Luthra, S. (2021). Analysing the roadblocks of circular economy adoption in the automobile sector: Reducing waste and environmental perspectives. *Business Strategy and the Environment, 30*(2), 1051–1066. https://doi.org/10.1002/bse.2669

Agrawal, R., Wankhede, V. A., Kumar, A., Upadhyay, A., & Garza-Reyes, J. A. (2022). Nexus of circular economy and sustainable business performance in the era of digitalization. *International Journal of Productivity and Performance Management, 71*(3), 748–774. https://doi.org/10.1108/IJPPM-12-2020-0676

Alyahya, M., Agag, G., Aliedan, M., Abdelmoety, Z. H., & Daher, M. M. (2023). A sustainable step forward: Understanding factors affecting customers' behaviour to purchase remanufactured products. *Journal of Retailing and Consumer Services, 70*. https://doi.org/10.1016/j.jretconser.2022.103172

Bernon, M., Upperton, J., Bastl, M., & Cullen, J. (2013). An exploration of supply chain integration in the retail product returns process. *International Journal of Physical Distribution and Logistics Management, 43*(7), 586–608. https://doi.org/10.1108/IJPDLM-03-2012-0060

Bigliardi, B., Campisi, D., Ferraro, G., Filippelli, S., Galati, F., & Petroni, A. (2020). The intention to purchase recycled products: Towards an integrative theoretical framework. *Sustainability (Switzerland), 12*(22), 1–20. https://doi.org/10.3390/su12229739

Bimonte, G., Ioppolo, G., Senatore, L., & Trincone, B. (2022). Government eco-innovation incentives in a recycling system: A Stackelberg-type model. *Business Strategy and the Environment*. https://doi.org/10.1002/BSE.3337

Bocken, N. M. P., de Pauw, I., Bakker, C., & van der Grinten, B. (2016). Product design and business model strategies for a circular economy. *Journal of Industrial and Production Engineering., 33*(5), 308–320. https://doi.org/10.1080/21681015.2016.1172124

Bryman, A. (2016). *Social research methods*. Oxford University Press.

Calzolari, T., Genovese, A., & Brint, A. (2021). The adoption of circular economy practices in supply chains—An assessment of European Multi-National Enterprises. *Journal of Cleaner Production, 312*. https://doi.org/10.1016/j.jclepro.2021.127616

Cecchin, A., Lamour, M., Davis, J. M., & M., & Jácome Polit, D. (2019). End-of-life product management as a resilience driver for developing countries: A policy experiment for used tires in Ecuador. *Journal of Industrial Ecology, 23*(5), 1292–1310. https://doi.org/10.1111/jiec.12861

Chen, C.-K., & Akmalul'Ulya, M. (2019). Analyses of the reward-penalty mechanism in green closed-loop supply chains with product remanufacturing. *International Journal of Production Economics, 210*, 211–223.

Debo, L. G., Toktay, L. B., & Van Wassenhove, L. N. (2005). Market segmentation and product technology selection for remanufacturable products. *Management Science, 51*(8), 1193–1205. https://doi.org/10.1287/mnsc.1050.0369

de Jesus, A., Lammi, M., Domenech, T., Vanhuyse, F., & Mendonça, S. (2021). Eco-innovation diversity in a circular economy: Towards circular innovation studies. *Sustainability (Switzerland), 13*(19). https://doi.org/10.3390/su131910974

Ding, P., Xia, W., Zhao, Z., & Li, X. (2023). The impact of government subsidies on build operate-transfer contract design for charging piles in circular economy. *Industrial Management and Data Systems, 123*(4), 1084–1121. https://doi.org/10.1108/IMDS-01-2022-0060

Esenduran, G., Lin, Y. T., Xiao, W., & Jin, M. (2020). Choice of electronic waste recycling standard under recovery channel competition. *Manufacturing and Service Operations Management, 22*(3), 495–512. https://doi.org/10.1287/msom.2018.0750

Fernando, Y., Shaharudin, M. S., & Abideen, A. Z. (2023). Circular economy-based reverse logistics: dynamic interplay between sustainable resource commitment and financial performance. *European Journal of Management and Business Economics, 32*(1), 91–112. https://doi.org/10.1108/EJMBE-08-2020-0254

Fisol, W. N. M., Nor, S. F., & Binti, M. (2019). The engineering of the Maqasid Shariah Theory through the scientific methodology of Al-Quran and Al-Sunnah in Islamic Finance. *Journal of Social Science and Humanities, 2*(6), 38–44. https://doi.org/10.26666/rmp.jssh.2019.6.4

Formentini, M., Secondi, L., Ruini, L., Guidi, M., & Principato, L. (2022). Enablers and barriers to circular supply chain management: a decision-support tool in soft wheat bread production. *Journal of Enterprise Information Management, 35*(3), 796–816. https://doi.org/10.1108/JEIM-02-2021-0069

Foster, G. (2020). Circular economy strategies for adaptive reuse of cultural heritage buildings to reduce environmental impacts. *Resources, Conservation and Recycling, 152*. https://doi.org/10.1016/j.resconrec.2019.104507

Frei, R., Jack, L., & Krzyzaniak, S. (2020). Sustainable reverse supply chains and circular economy in multichannel retail returns. *Business Strategy and the Environment, 29*(5), 1925–1940. https://doi.org/10.1002/bse.2479

Geissdoerfer, M., Savaget, P., Bocken, N. M. P., & Hultink, E. J. (2017). The circular economy—A new sustainability paradigm? *Journal of Cleaner Production, 143*, 757–768. https://doi.org/10.1016/j.jclepro.2016.12.048

Ghanadpour, S. H., Shokouhyar, S., & Pourabbasi, M. (2023). Effective end-of-life (EOL) products management in mobile phone industry with using Twitter data analysis perspective. *Environment, Development and Sustainability, 25*, 11337–11366. https://doi.org/10.1007/s10668-022-02529-7

Ghisellini, P., Cialani, C., & Ulgiati, S. (2016). A review on circular economy: The expected transition to a balanced interplay of environmental and economic systems. *Journal of Cleaner Production, 114*, 11–32. https://doi.org/10.1016/j.jclepro.2015.09.007

Haydn, B., Abraham, Z., & Wang, J. X. (2023). Integrating product design and supply chain management for a circular economy. *Production Planning and Control, 34*(11), 1097–1113. https://doi.org/10.1080/09537287.2021.1983063

Hudaefi, F. A., & Badeges, A. M. (2022). Maqāṣid al-Sharī'ah on Islamic banking performance in Indonesia: a knowledge discovery via text mining. *Journal of Islamic Marketing, 13*(10), 2069–2089. https://doi.org/10.1108/JIMA-03-2020-0081

Hunka, A. D., Linder, M., & Habibi, S. (2021). Determinants of consumer demand for circular economy products. A case for reuse and remanufacturing for sustainable development. *Business Strategy and the Environment, 30*(1), 535–550. https://doi.org/10.1002/bse.2636

Islam, M. M. (2021a). Segmenting, targeting and positioning in Islamic marketing. *Journal of Islamic Marketing, 12*(7), 1385–1404. https://doi.org/10.1108/JIMA-10-2018-0181

Islam, M. M. (2021b). The satisfaction of Bangladeshi pilgrims: Service gaps in spiritual tourism based on gender and expenditure. *International Journal of Religious Tourism and Pilgrimage, 9*(1), 106–117. https://doi.org/10.21427/2hw0-cd28

Islam, M. M. (2022a). Reaching halal and Islamic market segments: tapping marketing opportunities beyond the borders, Chap. 15. In G. H. Brodowsky, C. P. Schuster, & R. Perren (Eds.), *Handbook of research on ethnic and intra-cultural marketing* (pp. 200–212). Edward Elgar.

Islam, M. M. (2022b). Strategic perspectives of Islamic entrepreneurship and marketing, Chap. 13. In *Strategic Islamic marketing* (pp. 183–200). Springer.

Islam, M. M. (2022c). Evaluating negative attitudes of the students and shoppers towards halal cosmetics products. *Journal of Islamic Marketing, 13*(3), 565–585. https://doi.org/10.1108/JIMA-03-2020-0067

Islam, M. M. (2023). Institutionalization of Halal entrepreneurship for a sustainable economy. In L. Raimi, S. M. Adekunle, & M. S. Shabbir (Eds.), *Contemporary discourse of halal and Islamic entrepreneurship*. Springer. https://doi.org/10.1007/978-981-99-6427-7_5

Islam, M. M., Ab Talib, M. S., & Muhamad, N. (2023a). Challenges of halal standards and halal certification for Islamic marketing. *Muslim Business and Economic Review, 2*(1), 104–124. https://doi.org/10.56529/mber.v2i1.156

Islam, M. M., Ab Talib, M. S., & Muhamad, N. (2023b). Developing theoretical lenses for upstream halal businesses. *Journal of Islamic Marketing*. https://doi.org/10.1108/JIMA-05-2022-0156

Jaeger, B., & Upadhyay, A. (2020). Understanding barriers to circular economy: cases from the manufacturing industry. *Journal of Enterprise Information Management, 33*(4), 729–745. https://doi.org/10.1108/JEIM-02-2019-0047

Jiménez-Parra, B., Rubio, S., & Vicente-Molina, M.-A. (2014). Key drivers in the behavior of potential consumers of remanufactured products: a study on laptops in Spain. *Journal of Cleaner Production, 85*, 488–496.

Julianelli, V., Caiado, R. G. G., Scavarda, L. F., & Cruz, S. P. D. M. F. (2020). Interplay between reverse logistics and circular economy: Critical success factors-based taxonomy and framework. *Resources, Conservation and Recycling, 158*. https://doi.org/10.1016/j.resconrec.2020.104784

Kapetanopoulou, P., & Tagaras, G. (2011). Drivers and obstacles of product recovery activities in the Greek industry. *International Journal of Operations and Production Management, 31*(2), 148–166. https://doi.org/10.1108/01443571111104746

Kaviani, M. A., Tavana, M., Kumar, A., Michnik, J., Niknam, R., & Campos, E. A. R. D. (2020). An integrated framework for evaluating the barriers to successful implementation of reverse logistics in the automotive industry. *Journal of Cleaner Production, 272*. https://doi.org/10.1016/j.jclepro.2020.122714

Kazancoglu, I., Kazancoglu, Y., Kahraman, A., Yarimoglu, E., & Soni, G. (2022). Investigating barriers to circular supply chain in the textile industry from stakeholders' perspective. *International Journal of Logistics Research and Applications, 25*(4–5), 521–548. https://doi.org/10.1080/13675567.2020.1846694

Kirchherr, J., Reike, D., & Hekkert, M. (2017). Conceptualizing the circular economy: An analysis of 114 definitions. *Resources, Conservation and Recycling, 127*, 221–232. https://doi.org/10.1016/j.resconrec.2017.09.005

Kotler, P. (2016). *Marketing management: Analysis, planning, implementation, and control*. Pearson.

Kotler, P., & Keller, K. L. (2016). *Marketing management*. Pearson.

Kuhl, C., Skipworth, H. D., Bourlakis, M., & Aktas, E. (2023). The circularity of product-service systems: the role of macro-, meso and micro-level contextual factors. *International Journal of Operations and Production Management, 43*(4), 619–650. https://doi.org/10.1108/IJOPM-01-2022-0055

Kumar, S., & Putnam, V. (2008). Cradle to cradle: Reverse logistics strategies and opportunities across three industry sectors. *International Journal of Production Economics, 115*(2), 305–315.

Lawrenz, S., Leiding, B., Mathiszig, M. E. A., Rausch, A., Schindler, M., & Sharma, P. (2021). Implementing the circular economy by tracing the sustainable impact. *International Journal of Environmental Research and Public Health, 18*(21). https://doi.org/10.3390/ijerph182111316

Lei, Z., Cai, S., Cui, L., Wou, L., & Liu, Y. (2023). How do different Industry 4.0 technologies support certain Circular Economy practices? *Industrial Management and Data Systems, 123*(4), 1220–1251. https://doi.org/10.1108/IMDS-05-2022-0270

Li, G., Reimann, M., & Zhang, W. (2018). When remanufacturing meets product quality improvement: The impact of production cost. *European Journal of Operational Research, 271*(3), 913–925. https://doi.org/10.1016/j.ejor.2018.05.060

Liu, H., Long, H., & Li, X. (2020). Identification of critical factors in construction and demolition waste recycling by the grey-DEMATEL approach: a Chinese perspective. *Environmental Science and Pollution Research, 27*(8), 8507–8525. https://doi.org/10.1007/s11356-019-07498-5

Lynne, E. M., Devin, S., & William, M. M. (2019). *Resource recovery from wastes: Towards a circular economy. Royal Society of Chemistry*. https://doi.org/10.1039/9781788016353

Malgorzata, J., Kacprzak, I., & Iwona, K. (2023). The specificities of the circular economy (CE) in the municipal wastewater and sewage sludge sector—local circumstances in Poland. *Clean Technologies and Environmental Policy, 25*, 519–535. https://doi.org/10.1007/s10098-021-02178-w

Marín-Vinuesa, L. M., Portillo-Tarragona, P., & Scarpellini, S. (2023). Firms' capabilities management for waste patents in a circular economy. *International Journal of Productivity and Performance Management, 72*(5), 1368–1391. https://doi.org/10.1108/IJPPM-08-2021-0451

Marques, L., & Manzanares, M. D. (2023). Towards social network metrics for supply network circularity. *International Journal of Operations and Production Management, 43*(4), 595–618. https://doi.org/10.1108/IJOPM-02-2022-0139

Michaud, C., & Llerena, D. (2011). Green consumer behaviour: an experimental analysis of willingness to pay for remanufactured products. *Business Strategy and the Environment, 20*(6), 408–420.

Michelle Grantham, L. (1997). The validity of the product life cycle in the high-tech industry. *Marketing Intelligence and Planning, 15*(1), 4–10. https://doi.org/10.1108/02634509710155606

Mies, A., & Gold, S. (2021). Mapping the social dimension of the circular economy. *Journal of Cleaner Production, 321*(128960). https://doi.org/10.1016/j.jclepro.2021.128960

Moosmayer, D. C., Abdulrahman, M. D. A., Subramanian, N., & Bergkvist, L. (2020). Strategic and operational remanufacturing mental models: A study on Chinese automotive consumers buying choice. *International Journal of Operations and Production Management, 40*(2), 173–195. https://doi.org/10.1108/IJOPM-12-2018-0684

Moreno-Mondéjar, L., Triguero, Á., & Cuerva, M. C. (2021). Exploring the association between circular economy strategies and green jobs in European companies. *Journal of Environmental Management, 297*. https://doi.org/10.1016/j.jenvman.2021.113437

Nag, U., Sharma, S. K., & Govindan, K. (2021). Investigating drivers of circular supply chain with product-service system in automotive firms of an emerging economy. *Journal of Cleaner Production, 319*. https://doi.org/10.1016/j.jclepro.2021.128629

Naoum, T., Tomás, S. H., & Jagjit, S. S. (2023). Digital supply network design: a Circular Economy 4.0 decision-making system for real-world challenges. *Production Planning and Control, 34*(10), 941–966. https://doi.org/10.1080/09537287.2021.1980907

Nnorom, I. C., Ohakwe, J., & Osibanjo, O. (2009). Survey of willingness of residents to participate in electronic waste recycling in Nigeria—A case study of mobile phone recycling. *Journal of Cleaner Production, 17*(18), 1629–1637. https://doi.org/10.1016/j.jclepro.2009.08.009

Paolo, R., Claudio, S., Andrea, U., Davide, C., & Sergio, T. (2020). Assessing relations between Circular Economy and Industry 4.0: a systematic literature review. *International Journal of Production Research, 58*(6), 1662–1687. https://doi.org/10.1080/00207543.2019.1680896

Piller, L. W. (2023). Designing for circularity: sustainable pathways for Australian fashion small to medium enterprises. *Journal of Fashion Marketing and Management: An International Journal, 27*(2), 287–310. https://doi.org/10.1108/JFMM-09-2021-0220

Riggs, R., Roldán, J. L., Real, J. C., & Felipe, C. M. (2023). Opening the black box of big data sustainable value creation: the mediating role of supply chain management capabilities and circular economy practices. *International Journal of Physical Distribution and Logistics Management, 53*(7/8), 762–788. https://doi.org/10.1108/IJPDLM-03-2022-0098

Savu, R., & Max, F. (2023). What motivates entrepreneurs into circular economy action? Evidence from Japan and Finland. *Journal of Business Ethics, 184*, 71–91. https://doi.org/10.1007/s10551-022-05122-0

Shaharudin, M. R., Zailani, S., & Tan, K. C. (2015). Barriers to product returns and recovery management in a developing country: Investigation using multiple methods. *Journal of Cleaner Production, 96*, 220–232. https://doi.org/10.1016/j.jclepro.2013.12.071

Shang, G., Pekgün, P., Ferguson, M., & Galbreth, M. (2017). How much do online consumers really value free product returns? Evidence from eBay. *Journal of Operations Management, 53–56*, 45–62. https://doi.org/10.1016/j.jom.2017.07.001

Singhal, D., Tripathy, S., & Jena, S. K. (2020). Remanufacturing for the circular economy: Study and evaluation of critical factors. *Resources, Conservation and Recycling, 156*. https://doi.org/10.1016/j.resconrec.2020.104681

Subramanian, R., & Subramanyam, R. (2012). Key factors in the market for remanufactured products. *Manufacturing and Service Operations Management, 14*(2), 315–326. https://doi.org/10.1287/msom.1110.0368

Syed, A. R. K., Zhang, Y., Salman, S. D. I. G., Sumeela, A., & Sobia, S. (2022). The role of block chain technology in circular economy practices to improve organisational performance. *International Journal of Logistics Research and Applications, 25*(4-5), 605–622. https://doi.org/10.1080/13675567.2021.1872512

Tecchio, P., McAlister, C., Mathieux, F., & Ardente, F. (2017). In search of standards to support circularity in product policies: A systematic approach. *Journal of Cleaner Production, 168*, 1533–1546. https://doi.org/10.1016/j.jclepro.2017.05.198

Tom, B. J., Coenen, K. V., & Leentje, V. (2023). A systemic perspective on transition barriers to a circular infrastructure sector. *Construction Management and Economics, 41*(1), 22–43. https://doi.org/10.1080/01446193.2022.2151024

Tubarad, C. P. T., Mohammed, N. F., & Mohamad, M. (2022). Maqasid Shariah Performance Index conceptualisation: Examples from Malaysia and Indonesia. *International Journal of Academic Research in Business and Social Sciences, 12*(12), 1146–1152. https://doi.org/10.6007/IJARBSS/v12-i12/15852

Tuunanen, T., & Cassab, H. (2011). Service process modularization: Reuse versus variation in service extensions. *Journal of Service Research, 14*(3), 340–354. https://doi.org/10.1177/1094670511413912

Uhrenholt, J. N., Kristensen, J. H., Rincón, M. C., Jensen, S. F., & Waehrens, B. V. (2022). Circular economy: Factors affecting the financial performance of product take-back systems. *Journal of Cleaner Production, 335*. https://doi.org/10.1016/j.jclepro.2021.130319

Vernon, R. (1966). International investment and international trade in the product cycle. *The Quarterly Journal of Economics, 80*(2), 190–207. https://doi.org/10.2307/1880689

Walsh, G., Albrecht, A. K., Kunz, W., & Hofacker, C. F. (2016). Relationship between online retailers' reputation and product returns. *British Journal of Management, 27*(1), 3–20. https://doi.org/10.1111/1467-8551.12120

Wilson, M., & Goffnett, S. (2022). Reverse logistics: Understanding end-of-life product management. *Business Horizons, 65*(5), 643–655. https://doi.org/10.1016/j.bushor.2021.10.005

Wrålsen, B., Prieto-Sandoval, V., Mejia-Villa, A., O'Born, R., Hellström, M., & Faessler, B. (2021). Circular business models for lithium-ion batteries—Stakeholders, barriers, and drivers. *Journal of Cleaner Production, 317*. https://doi.org/10.1016/j.jclepro.2021.128393

Zailani, S., Govindan, K., Shaharudin, M. R., & Kuan, E. E. L. (2017). Barriers to product return management in automotive manufacturing firms in Malaysia. *Journal of Cleaner Production, 141*, 22–40. https://doi.org/10.1016/j.jclepro.2016.08.160

Zhang, Q., Wang, L., & Zhou, D. (2020). Remanufacturing under energy performance contracting—an alternative insight from sustainable production. *Environmental Science and Pollution Research, 27*(32), 40811–40825. https://doi.org/10.1007/s11356-020-10116-4

Zhu, Q., Sarkis, J., Lai, K., & hung. (2014). Supply chain-based barriers for truck-engine remanufacturing in China. *Transportation Research Part E: Logistics and Transportation Review, 68*, 103–117. https://doi.org/10.1016/j.tre.2014.05.001

Mohammad Mominul Islam is a PhD Candidate at the School of Business and Economics, Universiti Brunei Darussalam (UBD), under the prestigious UBD Graduate Scholarship. Previously, he taught at EXIM Bank Agricultural University, Bangladesh. He received an MBA and BBA in Marketing from Rajshahi University, Bangladesh and an MSc in Business and Economics from Sweden. His research interests focus on Islamic marketing, Islamic entrepreneurship, Halal certification, Sustainable marketing, Halal cosmetics, Spiritual tourism, Halal supply chain management, and Marketing pedagogies. His scholarly works were published in the *Journal of Islamic Marketing, International Journal of Islamic Accounting and Business Research, International Journal of Islamic Marketing and Branding, Social Business, International Journal of Religious Tourism and Pilgrimage*, Springer, and Edward Elgar Publishers, USA. Mohammad Mominul Islam received "The Bangladesh Education Leadership Awards 2017."

Mohamed Syazwan Ab Talib is a Senior Assistant Professor of Logistics Management at the School of Business and Economics, UBD. Prior to UBD, he was a Senior Lecturer in Marketing at Azman Hashim International Business School, Universiti Teknologi Malaysia, and a Lecturer in Logistics Management at the Faculty of Business and Information Science, UCSI University. His research and teaching focus primarily on logistics and supply chain management with the inclusion of halal principles and Islamic distribution. He has published numerous academic papers and presented at international conferences on various subjects, particularly on halal, halal food, halal management standards, halal logistics, and halal food supply chains.

Nazlida Muhamad is an Associate Professor of Marketing at the School of Business and Economics, UBD. She received PhD in Marketing from The University of Western Australia in 1999. She has taught marketing modules in Malaysian, Saudi, and Australian public universities, prior to working at UBD. Her research interests and publications are in the area of consumer behavior—with a special interest in issues pertaining to Islam, and its role in Muslim consumer behaviors in the marketplace. Among others, her works have been published in *Psychology and Marketing, the Journal of Islamic Marketing, the Journal of Business and Policy Research*, and *the Journal of Computer-Mediated Communication.*

Organizational Culture Practices in Sustaining Shariah-Based Quality Management System MS1900 at Zakat Collection Center

Musaiyadah Ahmadun, Siti Arni Basir, Masnih Mustapa, and Nur Ilma Ambo

Abstract Organizational culture is one of the factors that influence the effectiveness of Quality Management System practices in any organization. A weak organizational culture deteriorates the implemented quality management system. Organizational culture describes how an organization operates, makes decisions, manages customers, and manages performance. Therefore, this article aims to analyze organizational culture practices in sustaining Shariah-based Quality Management System MS 1900 at a Zakat collection center in Malaysia. This study employed a qualitative method involving the Pahang Zakat Collection Center (PZCC) located on East Coast of Malaysia as a case study organization. Semi-structured interviews were used as a data collection method and were supported by observation and document review. A total of fifteen informants consisting of top management and officers of the Zakat center were interviewed to obtain data and information related to the topic studied. The thematic analysis method was used with the help of Atlas.ti version 7 to analyze the data collected. The research findings revealed that there are five organizational cultures practices in sustaining the Shariah-based Quality Management System MS 1900 at Pahang Zakat Collection Center (PZCC), namely management commitment, customer focus, performance measurement, partnership with suppliers, and emphasis on excellence. The positive culture found helps to the effective and sustainable implementation of MS 1900.

M. Ahmadun (✉) · M. Mustapa
Academy of Islamic Contemporary (ACIS), Branch Sabah Campus, Kota Kinabalu, Sabah, Malaysia
e-mail: musaiyadah@uitm.edu.my; masnihmustapa@uitm.edu.my

S. A. Basir
Academy of Islamic Studies, Universiti Malaya, Kuala Lumpur, Malaysia
e-mail: sitiarni@um.edu.my

N. I. Ambo
Academy of Islamic Contemporary (ACIS), Campus Shah Alam, Shah Alam, Selangor, Malaysia
e-mail: nurilma@ums.edu.my

A. Rafiki et al. (eds.), *Strategic Islamic Business and Management*, Contributions to Management Science,
https://doi.org/10.1007/978-3-031-61778-2_8

The results of this study prove that work culture facilitates sustainable quality management system implementation and further accelerates organizational competitiveness in a challenging management and economic environment.

1 Introduction

Organizational culture is a behavior that is constantly present in an organization. Organizational culture is a set of beliefs shared by organizational members that influence how they interact with and respond to environmental and external circumstances (Limaj & Bernroider, 2019). Each organization has a unique pattern and culture based on its objectives, vision, and mission (Paais & Pattiruhu, 2020). Organizational culture also serves as a social controller, reinforcing dominant behavior and beliefs in order to influence organizational relationships (Ahmadun, 2019). Organizations typically encounter obstacles in preserving a specific culture, such as integrating employees into a team and efficiently adapting to the external environment (Akpa et al., 2021).

As a result, there is an important relationship between an organization's culture and its quality management. Organizational culture is also an essential component of the quality management system. This is because, with a strong organizational culture, a quality management system-based organization can improve employees work efficiency and strategy effectiveness (Koh & Low, 2008; Watson & Howarth, 2011). In this regard, understanding organizational culture becomes vital to improving an organization's quality (Nukic & Huemann, 2016). As a result, three major aspects must be emphasized and paid attention to: the holistic aspect, the factor perspective, and the cognitive perspective, in order for organizational culture to contribute to the sustainability of an organization quality management system (Asbari et al., 2020; Bhaduri, 2019; Sackman, 1991).

Previous researchers conducted research on organizational culture and quality management, including elements of organizational culture and quality management in the construction industry (Patyal et al., 2020); Perspectives on organizational culture and leadership (Khan et al., 2020; Maamari & Saheb, 2018); organizational culture perspective and customer focus (Famiyeh et al., 2018; Kao et al., 2020). However, the studies conducted are primarily conceptual and quantitative, with fewer researchers applying to conduct research on organizational cultural practices and their relationship with the quality management system, particularly in Zakat Collection Centers. As a result, the purpose of this article is to examine organizational cultural practices related to the maintenance of the MS1900 Sharia-based management system in Malaysian Zakat Collection Centers.

2 Literature Review

Organizational culture has a significant impact on the way an organization operates, makes decisions, and manages staff behavior and performance. As a result, some researchers, such as Basir et al. (2011), Mannion and Devis (2018), Qi and Chau (2018), Soomro and Shah (2019), and Wetzel and Tint (2019), have dispersed the cultural elements that organizations require in order to improve quality, namely top management commitment, customer focus, performance measurement, supplier partnership, and focus on excellence.

Top Management Commitment

Organizational culture is an essential element of an organization's management system. In line with that, top management must understand how to manage and set a strong emphasis on cultural norms that are compassionate and based on Islamic Sharia. Previous research has indicated that transformational leadership is one type of leadership that is required to guarantee the successful operation of an embracing quality culture in an organization.

Transformational leadership is leadership that encourages its followers to do more than what is considered ordinary tasks by changing the vision, providing an example, and encouraging and stimulating employees to change for the better (Iqbal, 2021; Sartika et al., 2020). In this regard, to successfully implement this concept, top management must collaborate actively in attempts to increase organizational quality by empowering employees through training requirements (Source: Rodriguez & Walters, 2017). This is consistent with a study conducted by Niati et al. (2021), which found that the more frequently the leader provides training to the employee, the more motivated the worker will be. Next, management must ensure that the results of the team are always aligned with the organization's quality policies and objectives (Javed, 2015; Tovmasyan, 2017). This is because the quality objective can determine the progress of improvement activities and enable the organization to measure the effectiveness of the quality management system (Bugdol & Jedynak, 2022). Next, top management not only plays the role of managing the staff within an organization, but the responsibility of top management is also to cover the planning, organization, and supervision processes in order to determine the goals of the organization through the use of existing resources (Havidz & Suprapto, 2021). According to Kabeyi (2019), planning is a tool for promoting the direction, objectives of organization through three stages: strategy formulation, implementation, and evaluation. Instead of organizational processes, top management acts to gather and organize all the necessary resources, especially human resources, finance, and infrastructure, so that the work that is being carried out meets the organizational goals (Marolah et al., 2018). Top management must also conduct the monitoring process on a regular basis by establishing a method of measuring performance, determining whether the organization's performance meets the established standards, and taking corrective action if management is insufficient (Manyuchi & Sukdeo, 2021).

Focus on the Customer

The customer is one of the organizational assets that has a significant impact on the effectiveness of the quality management system. Customer focus is one of the critical concepts in adopting a quality management system, allowing the quality of a product or service served by an organization to be enhanced and evaluated based on customer feedback (Nor et al., 2014). A previous study has demonstrated that service quality can influence customer satisfaction and loyalty to an organization (Anwar et al., 2019; Hsieh et al., 2018). According to Bahri and Herawan (2019), in order to provide quality services to customers, the organization must focus on three main areas: service strategy, human resources, and service systems.

A strategy is a plan that is implemented to face challenges and capitalize on opportunities by integrating the ideas of various units within an organization (Desfitrina et al., 2019). As a result, the organization must use a range of service techniques to ensure that the customer stays the primary emphasis. First, the organization must have a core strategy that specifies the organization's goals and describes the function and direction of the service. The second is the customer strategy, in which the organization strives to provide a full-service delivery system for the customer by providing simple procedures, a complete infrastructure, and a user-friendly system. Third, a supervision plan in which the organization constantly monitors and records customer complaints in an efficient and systematic method. Fourth, the organization must have a cultural strategy in which management promotes a collective and positive work culture among employees (Asya et al., 2024).

Next, human resources have a significant role in customer satisfaction and loyalty to an organization. This is because human resources are the front line of an organization's ability to carry out day-to-day operational tasks (Widyatmoko & Wibowo, 2020). At the same time, management must develop and equip employees with fundamental values including responsiveness, punctuality, credibility, respect, friendliness, fairness, and empathy (Mampu, 2008). This is supported by a study conducted by Bahadur et al. (2018), who discovered that empathy among staff while interacting with customers can assist organizations to enhance service strategies and increase customer loyalty (Ling et al., 2023).

Performance Measurement

It is essential for an organization to develop a systematic performance measuring system since it can aid in ensuring the successful implementation of the quality management system (Daud & Samsudin, 2016). As a result, several models have been developed to assess and enhance an organization's work and service performance. These include key performance indicators (KPIs), balanced scorecards (BSCs), and business excellence models (BEM). These models provide a foundation for achieving performance and allow critical evaluation of the quality management system's implementation success (Neyestani & Juanzon, 2016).

The performance measurements used vary and are determined by the industry, company, department, or individual being assessed. In addition, organizations can evaluate such performance in financial or non-financial terms, qualitatively or quantitatively, and across a wide variety of disciplines such as sales, customer service, operations, marketing, and human resources (Aparanji, 2023). When performance reduction is undertaken within an organization, it can determine the success of a project, assist the organization in understanding the working process, assess whether the demands of stakeholders and customers are satisfied, and suggest changes (Gunasekaran & Kobu, 2007; Parker, 2000).

As a result, organizations must focus on several key factors throughout the performance measuring process. First, do a timely assessment. This is because an assessment that is not completed on time would be ineffective for the employee (source: Podolinsky, 2013). Second, the appraisal should consider the staff's scope. If the assessment is relevant to the staff's job, it makes it easier for employees to complete and discuss their behavior based on that assessment (Yazid et al., 2017). Third, the evaluator should provide more precise reviews to the staff, as non-specific reviews are a waste of time for both the assessor and the employees. Fourth, the judgment must be objective, measurable, and not subjective, i.e. based solely on the judge's sentiments (Midor et al., 2020). Fifth, reward employees when the assessment is completed to keep them motivated to execute their jobs (Aparanji, 2023). Sixth, performance metrics must be careful and complete, with assessments addressing team dynamics, personal development, knowledge base development, and asset acquisition.

Sharing with Suppliers

Sharing with suppliers has also been a crucial culture in quality management. The provider is one of the entities that is continually in relationship with organizations and service organizations. The relationship between suppliers and organizations has a significant impact on an organization's operational journey and progress. According to previous researchers such as Jack and Powers (2015), Kim et al. (2015), and Shakeel et al. (2014), better supplier relationships might have a significant impact on the success of supply chain strategic implementation, product and service quality, financial performance, and product innovation.

In this context, efficient communication must be conducted in order to sustain the relationship between the organization (the buyer) and the supplier. This is because effective communication between organizations (buyers) and suppliers can reduce misconceptions and clarify the aims that both sides seek to achieve. In this regard, Islam has created communication ethics that must be followed throughout the transaction, namely honesty in conversation and deeds, as well as the ability to carry out the trusts granted (Citra Andanari et al., 2023). Second, accuracy in information delivery is essential in collaboration with the supplier, whose data and information must be precise in order to prevent damage to both stakeholders (Andriani, 2021). Third, it is the nature of a business in which organizations (customers) and suppliers are always active in price fixing, bidding, and sale purchase agreements (Shahriman et al., 2022).

Suppliers must also sustain momentum for the ongoing improvement of the products they supply to the organization. Logistics, procurement, storage, transportation, and handling of goods and services are all factors that must be continuously examined and improved. This is consistent with Barata et al.'s (2022) vision that, in order to deal with a fast-changing business environment, suppliers must be competitive in increasing product and service quality through more effective supply chain management.

Focus on Excellence

A culture of excellence is an initiative that focuses on good values for the effective conduct and execution of work at all levels of service, whether formal or informal. A brilliant organization requires efficient and effective management. This is consistent with Selvanathan et al. (2012), who claim that in order to assess an organization's level of excellence, various characteristics must be considered, including effectiveness, efficiency, and productivity. To declare that an organization is efficient, management must reinforce the elements that drive efficiency, such as human resource management, continuous improvement, and innovation in the organization.

In a competitive environment, an organization's excellence is determined by the quality of its employees (Aman & Shukor, 2015). As a result, managing and developing human capital is an important component that must be prioritized for an organization to move forward (Mansor & Tengku Kasim, 2015). In general, organizations must distinguish between five essential components of human capital management: knowledge (Cisneros & Perlines, 2018), skills (Engelman et al., 2017), productivity (Ismail & Ahmad, 2020), education (Ornek & Ayas, 2015), and health (Nor & Selamat, 2020).

Next, an organization that emphasizes excellence must always look at continuous improvement as a management approach to providing quality services (Wijayanti et al., 2020). In this aspect, in the process of improvement, the organization requires a process-oriented strategy or culture, such as Kaizen. Kaizen strives to enhance three key aspects of organizational management: quality, cost, and delivery. Kaizen also aims to improve the quality of goods or services while lowering costs at every level (Prayuda, 2020). The elements of the Kaizen continuous improvement strategy are plan-do-check-act (PDCA) and standardize-do-check-act (SDCA). PDCA is a four-stage, recurrent approach for enhancing portions, goods, and services while also solving problems. This procedure necessitates a methodical evaluation of potential solutions, an analysis of the outcomes, and the implementation of successful solutions. Organizations may ensure that their workers are always improving toward a better work culture while achieving excellent results by implementing this across the service process (Ahmad et al., 2022).

3 Methodology

The study employs a qualitative study design with case studies as a study strategy (Ormston et al., 2014), which is ideal for examining something rare and novel because it allows for the collection of detailed data and information about emotions, feelings, motivations, certain behaviors, social circumstances, or specific events

(Chua, 2011). Based on the study's objectives, the researchers selected an intrinsic category case study to gain a more in-depth understanding of the organizational culture that can support the implementation of the MS 1900 Quality Management System at PKZP.

Meanwhile, the researcher employed a semi-structured interview method to get current information by allowing respondents to express their opinions on the questions posed. For semi-structured interviews, the researcher must provide an interview guide containing a list of questions to be asked, which are based on discovered themes. The question list becomes a reference for conducting research during the interview process (Awang, 2009). Furthermore, researchers used observational methods to gain a thorough understanding of the organizational culture at PKZP. In addition, these observation methods can assist researchers in gathering more specific data on staff behavior, environment, and working environment at PKZP.

The study subjects were then chosen via purposive sampling. Only PKZP employees with certain qualities are selected as informants. This study's respondents who are experienced and actively involved in the implementation of MS1900 demonstrate two primary characteristics. Similarly, fifteen informants were interviewed for this study, including top management, Sharia officers, audit officials, ISO officers, and administrative officials who are actively involved in the MS1900 implementation. The researchers used Atlas.ti software to do analyze of themes on the data. Thematic analysis is a type of qualitative analysis that examines at data-related topics. The thematic analysis allows academics to characterize data in detail through the interpretation process (Boyatzis, 1998).

The researchers utilized a thematic analysis methodology, revising the interview findings based on the methodologies and systems related. Next, the researchers create a rough copy of the collected data, conduct comparisons, and divide the data into numerous segments based on the objectives. Finally, researchers developed themes and specific codes to help them produce intriguing, high-quality papers.

4 Result and Discussion

The study's findings revealed that five organizational cultures influence the adoption of the MS1900 Quality Management System at PKZP:

Top Management Commitment

According to the findings of the researchers' interviews and observations, the top management has long prioritized this commitment, and it is a crucial factor in guaranteeing the successful implementation of MS 1900 in PKZP. This is evident when PKZP's top management ensures that policies and quality objectives are consistently followed and connected with the staff's day-to-day work activities. According to Bugdol and Jedynak (2022), quality objectives can guide improvement operations and allow organizations to assess the performance of their quality management system. Furthermore, the informants examined argued that management is continually reviewing every aspect of each employee's activity while considering the five main areas that have been planned: income, savings, business, advertising and promotion, and ICT and social media applications.

The PKZP administration is continually in discussions with all authorities and support workers about how to set new quoting guidelines to improve the quality of PKZP's zakat services. Likewise, several informants interviewed said that PKZP administration is continually planning, monitoring, controlling, and organizing information and statistics regarding the production of financial allocations throughout the year. This is consistent with prior research, which found that all of these were created to support the organization's direction, objectives, and goals (Havidz & Suprapto, 2021; Manyuchi & Sukdeo, 2021; Marolah et al., 2018). The management of PKZP is also very focused on the needs of the employees, and they are always fully prepared with the necessary resources and infrastructure for staff and support staff training in order to successfully execute the MS 1900 Quality Management System. This is consistent with Rodriguez and Walters' (2017) opinion that in order to increase the quality of the organization, management must reinforce the staff by requiring training because the more frequently the staff is trained, the more motivated they will be (Niati et al., 2021).

Next, practically all informants in PKZP management always ensure that the organization's value is generated and applied in every staff member's work movement. The organization's ideals include perseverance, trust, and compassion. They continually remind all employees to follow and promote the pure ideals defined by management. This is consistent with a study conducted by Iqbal (2021) and Sartika et al. (2020), which discovered that transformational leadership encourages its employees to change for the better. This is because it seeks to train employees to always work for God's mercy and redemption.

Focus on the Customer

The customer's demands or requirements should be given primary focus in the implementation of MS 1900 (Hasan et al., 2018). The study's findings indicate that management is concerned with providing the best, most responsible, trustworthy, efficient, and customer-friendly service possible. This is close to study conducted by Anwar et al. (2019) and Hsieh et al. (2018), who discovered that service quality may affect customer happiness and loyalty to an organization. According to Bahri and Herawan (2019), to give quality services to customers, three critical factors must be addressed: service strategy, human resources, and service systems. This approach is consistent with the study's findings, which show that the PKZP is always offering services to customers through programs and activities that bring the PKZP closer to them. Among these include the establishment of supermarket zakat counters, the provision of previews in government, private, and school departments, and the organizing of zakat carnivals. Furthermore, based on investigators' observations, the PKZP has always taken the initiative immediately and decisively in cases regarding customer needs. For example, they supply appropriate counter-zakat and extend service hours during peak seasons such as the end of the year and Ramadan.

In addition, to determine whether the customer is satisfied with the services provided by PKZP, according to the informants approached, the organization provided a survey form each year for obtaining feedback on PKZP's external customer satisfaction with the services provided. The study's findings revealed that five major areas are prioritized in PKZP customer satisfaction studies: office environment, counter service, skilled information, service, and professionalism. Two of the most common customer complaints received by PKZP concern inadequate public facilities and the unpleasant attitude of a few PKZP employees during their daily work. Nonetheless, according to sources contacted, the PKZP is always working to resolve issues quickly and efficiently. This is consistent with a study conducted by Asya et al. (2024), which indicated that organizations should always efficiently and methodically monitor and respond to customer complaints.

Performance Measurement

Next, performance measurement is an element of organizational culture that determines the application of the MS1900 Quality Management Standards. The informant feels that the purpose of performance assessment at PKZP is not just to improve the organization's quality and performance, but additionally for the sake of Allah s.w.t. This is related to a study conducted by Daud and Samsudin (2016), who discovered that role performance measures ensure effectiveness and can help enhance the implementation of the quality management system.

In this regard, the interview findings revealed that PKZP management conducted performance measurement assessments for all officers and provided confident and transparent assistance to employees. In this regard, the informant claimed that PKZP's performance measurements consider workflow and operations, human resource requirements, environment and infrastructure, finances, expectations, and customer feedback. The study's findings also revealed that PKZP has executed four major steps: defining KPIs and setting performance targets, measuring and assessing performance, monitoring performance levels, and enhancing the performance of employees and support staff. This revenue aligns with the Aparanji field (2023) in that performance evaluations take many forms, including customer service, operations, marketing, and human resources.

The PKZP has used quality objectives to identify and define indicators of performance that are in line with the PKZP's standards. This is consistent with Midor et al.'s (2020) study, which found that assessments must be objective and quantitative rather than subjective. According to the informant, all department heads are expected to fill out a performance measurement assessment form each year, which includes two primary parts: staff job specifications and key values in the PKZP. Performance measurements are evaluated twice a year: in the middle and at the end of the year. Each department head will conduct a performance measurement assessment by analyzing the activities, operations, and services provided by PKZP's staff and support employees. The gained information is then collected, and the department director will analyze the work movement's achievements and staff service. Staff with poor performance will be monitored regularly, with meetings held to discuss methods to find an essential solution to the difficulties they encounter.

This implies that the PKZP conducted a detailed and comprehensive performance measurement, with an assessment of dynamic variety, as indicated by Aparanji (2023).

Sharing with Suppliers

The study's findings indicate that the provider is one of the stakeholders or partners who assist the PKZP in improving the efficiency and effectiveness of the service delivery system. According to one of the informants interviewed, the most important aspect of sharing with suppliers is developing trust. This is consistent with prior research Jack & Powers, (2015); Kim et al., (2015); Shakeel et al., (2014), findings that better supplier relationships are critical to the success of strategic supply chain deployment. Furthermore, PKZP management has a heavy emphasis on supplier selection, with the administrative department managing the entire process and the selection criteria fully guided by the MS1900 Quality Management System to ensure that the products and services provided by suppliers meet Shariah requirements. Furthermore, other criteria such as Islamic suppliers and zakat payers were prioritized in order to encourage more entrepreneurs in Pahang to practice zakat.

Next, management always ensures that any products or services given to PKZP are original and comply with the specified requirements. According to the informants, the methods defined by MS 1900 often involve three major purchases: secrets, printed documents, and gold weighers, for which the PKZP has selected suppliers. Furthermore, to avoid supplier monopolies, the PKZP established an electronic data system via which suppliers can make purchases and selections. This is consistent with a study conducted by Andriani (2021), which discovered that accuracy in giving information is critical in the connection between suppliers and companies to prevent losses on both sides.

The PKZP is likewise very careful about paying invoices to vendors; management has set a payment SOP of no more than fourteen working days. This has been highlighted in sales communication, where suppliers and buyers must be honest in their discussions and activities (Citra Andanari et al., 2023). Following that, the PKZP will appraise the products or supplies provided by the supplier. The evaluation criteria consider product pricing, quality, and delivery time. If the product demand does not meet the specifications, the PKZP will undertake a review and reconsideration with the supplier. According to Barata et al. (2022), suppliers must compete by enhancing the quality of their products and services through more effective supply chain management.

Focus on Excellence

To enhance and preserve the culture of excellence at PKZP, management has taken appropriate action by implementing the MS1900 Quality Management System. This is in line with findings of study suggestion by Hasan et al. (2016) that is MS 1900 implementation facilitates to achieve excellence in quality management system certified organization. The MS1900 Quality Management System can assist the company in developing new ideas that will improve the efficiency of PKZP and meet client needs. Simultaneously, PKZP has made various improvements, one of which is to reinforce and increase the efficiency of ICT use, with management and support staff working together to provide some of the latest technologies and services for PKZP customers. The resulting inventions include Zakat4u, Zakatklik, online Zakat, E-Zakat pahang, and Bizzakat.my. According to PKZP staff, innovation and originality are the primary drivers of organizations that provide systematic services, leading to creative, efficient, and competitive employees. This is consistent with Selvanathan et al. (2012), who claim that management should increase the variables that drive efficiency, including innovation and human resources.

In terms of human resource development, the PKZP intends to streamline its present workforce. Each employee is trained to be able to perform multiple duties at once. It is because, in a competitive environment, an organization's success is determined by the quality of its people (Aman & Shukor, 2015). As a result, the PKZP has stressed three key exercises for enhancing staff skills and competence: the direction course, the staff skills course, and Shariah basic training. These have boosted workers' competencies and helped to start the day-long work movement during the deployment of the MS1900 Quality Management System. is consistent with earlier research, which has identified several essential components that must be considered throughout the development of human capital, including knowledge, skills, productivity, education, and health (Cisneros & Perlines, 2018; Engelman et al., 2017; Ismail & Ahmad, 2020; Nor & Selamat, 2020; Ornek & Ayas, 2015).

5 Conclusion

Overall, this study has identified and explored organizational cultures that facilitate in the further implementation of MS1900 in PKZP. At the same time, the researchers are attempting to make some recommendations that would help in the establishment of a sustainable organizational culture in the PKZP, as well as the more effective execution of MS1900 standards. In this regard, management and staff should have a deep belief in ownership and responsibility for the MS1900 implementation at PKZP. MS1900 cannot be implemented solely by one party. With this culture, organizations may make better decisions while maintaining the value of their reputation.

At the same time, a culture of learning and quality improvement should be a key pillar throughout the implementation of the MS1900 Quality Management System. A culture of quality improvement is a continuous process, not a seasonal program. This culture needs to be integrated into the implementation of MS 1900 so that quality aspects are considered at every stage and can produce quality services. Some of the key aspects that need to be addressed in the improvement process are infrastructure development, identifying specific areas that require improvement, implementing improvement projects, and establishing a project group responsible for implementing and completing any improvement required by the organization.

In addition, all PKZP employees should cultivate a culture of open communication. In the context of management and citizenship, an open attitude can demonstrate the ability to receive constructive feedback, the willingness to hear and receive opinions, the ease with which any issue can be discussed, and the willingness to share information about work movements carried out within the organization. After all, a positive or strong organizational culture can ensure that the deployment of MS1900 continues to thrive. However, a poor organizational culture will be unable to assist the organization in achieving its stated goals. Therefore, the management should put stern efforts to nurture a positive work culture so that it increases effectiveness of MS 1900 implementation in organization involved.

References

Ahmad, M. F., Islamil, M. H., Rahman, N. A. A., & Ahmad, A. N. A. (2022). Pelaksanaan TQM dalam Syarikat Pengilang Logam: Kajian Kes. *Research in Management of Technology and Business, 3*(2), 240–250.

Ahmadun, M. (2019). *Pelaksanaan MS1900 Sistem Pengurusan Kualiti Berasaskan Syariah Di Pusat Kutipan Zakat Pahang*. Tesis Ijazah Doktor Flasafah, Akademi Pnegajian Islam, Universiti Malaya.

Akpa, V. O., Asikhia, O. U., & Nneji, N. E. (2021). Organizational culture and organizational performance: A review of literature. *International Journal of Advances in Engineering and Management, 3*(1), 361–372.

Aman, Z., & Shukor, K. A. (2015). Faktor-faktor yang Mempengaruhi prestasi Kerja dalam Kalangan Staf Sokongan Kolej Universiti Antarabangsa Selangor. In *Proceeding of the 2nd international conference on management and Muamalah* (pp. 276–286).

Andriani, N. (2021). Etika Komunikasi Bisnis Dalam Perspektif Islam. *Jurnal Al-Hikmah, 19*(1), 13–26.

Anwar, S., Min, L., & Dastagir, G. (2019). Effect of service quality, brand image, perceived value on customer satisfaction and loyalty in the Chinese banking industry. *International Journal of Business, Economics and Management Works, 6*(3), 24–30.

Aparanji, P. P. (2023). Role key performance indicators in performance appraisal. *International Journal of Research Publication and Reviews, 4*(4), 1705–1710.

Asbari, M., Purwanto, A., Maesaroh, S., Mustofa, M., Hutagalung, D., Mustikasiwi, A., Ong, F., & Andriyani, Y. (2020). Impact of hard skills, soft skills and organizational culture: Lecturer innovation competences as mediating. *Journal of Education, Psychology and Counseling, 2*(1), 101–121.

Asya, N. F., Dwiputrianti, S., Fitriani, L., & Harahap, A. S. (2024). Service triangle model to improve the quality of service for prospective workers in Surabaya Job Training Center during the new normal era. In *ICAS* (pp. 289–299).

Awang, I. (2009). *Penyelidikan Ilmiah: Amalan Dalam Pengajian Islam*. Kamil dan Shakir Sdn.

Bahadur, W., Aziz, S., & Zulfiqar, S. (2018). Effect of employee empathy on customer satisfaction and loyalty during employee-customer interaction: The mediating role of customer affective commitment and perceived service quality. *Cogent Business & Management, 5*, 1–21.

Bahri, R. R., & Herawan, E. (2019). The improving of customer satisfaction: Service triangle approach. In *Advances in social science, education and humanities research* (pp. 400–403).

Barata, F. A., Febrianto, G. N., & Yasin, M. (2022). *Supply chain management strategy in building a competitive advantage through the implementation of logistic 4.0, INSYMA* (pp. 371–377). Atlantis Press.

Basir, S. A., Davies, J., & Rude, A. (2011). The element of organizational culture which influence the maintenance of ISO 9001: A theoretical framework. *African Journal of Business Management, 5*(15), 6029–6035.

Bhaduri, R. M. (2019). Leveraging culture and leadership in crisis management. *European Journal of Training and Development, 43*(5/6), 534–549.

Boyatzis, R. E. (1998). *Transforming qualitative information: Thematic analysis and code development*. Sage.

Bugdol, M., & Jedynak, P. (2022). Quality objectives in management systems-their attributes, establishment and motivational function. *International Journal of Quality & Reliability Management, 39*(1), 115–136.

Chua, Y. P. (2011). *Keadah Penyelidikan: Kaedah dan Statistik Penyelidikan*. McGraw-Hill.

Cisneros, M. A. I., & Perlines, F. H. (2018). Intellectual capital and organization performance in the manufacturing sector of Mexico. *Management Decision, 56*(8), 1818–1834.

Citra Andanari, M., Abubakar, A., & Basri, H. (2023). Prinsip Kejujuran Dalam Jual Beli Menurut Perspektif al-Quran. *Jurnal Cakrawala Ilmiah, 2*(6), 2535–2545.

Daud, H. M., & Samsudin, M. E. (2016). Kajian Empirikal Pengukuran Prestasi Firma: Data Skunder Firma Tersenarai di Bursa Malaysia. *E-Academia Journal UiTM, 5*(2), 159–173.

Desfitrina, D., Zulfadhli, Z., & Widarti, W. (2019). Good service strategies affect competitive advantage. *International Review of Management and Marketing, 9*(6), 135–144.

Engelman, R. M., Fracasso, E. M., Schmidt, S., & Zen, A. C. (2017). Intellectual capital, absorptive capacity and product innovation. *Management Decision, 55*(3), 474–490.

Famiyeh, S., Asante-Darko, D., & Kwarteng, A. (2018). Service quality, customer satisfaction and loyalty in the banking sector the moderating role of organizational culture. *International Journal of Quality & Reliability Management, 35*(8), 1546–1567.

Gunasekaran, A., & Kobu, B. (2007). Performance measures and metrics in logistics and supply chain management: A review of recent literature (1995–2004) for Research and Applications. *International Journal of Production Research, 45*, 2819–2840.

Hasan, A. M., Ab Mumin, A. G., Siti Arni, B., Sharifah Hayaati, S. I., & Ilhaamie, A. G. A. (2016). Pelaksanaan Sistem Pengurusan Kualiti Berteraskan Syariah MS 1900, Penerbit Universiti Pertahanan Nasional Malaysia (UPNM).

Hasan A. M., Siti Arni, B., John, D., Sharifah Hayaati, S. I., Ilhaamie, A. G. A., Syahrul Nizam, K., & Musaiyadah, A. (2018). Manual Pelaksanaan Sistem Pengurusan Kualiti ISO 9000 di Sektor Perkhidmatan, Penerbit Universiti Pertahanan Nasional Malaysia (UPNM).

Havidz, H. B. H., & Suprapto, E. (2021). The role and function of management in global organizations. *Dinasti International Journal of Digital Business Management, 2*(4), 774–753.

Hsieh, S.-W., Lu, C.-C., & Lu, Y.-H. (2018). A study on the relationship among brand image, service quality, customer satisfaction, and customer loyalty—Taking 'the Bao Wei Zhen Catering Team' as an empirical study. *KnE Social Sciences, 3*(10), 1768–1781.

Iqbal, M. (2021). Kepimpinan Transformasional dalam Upaya Pengembangan Sekolah/Madrasah. *Journal Pendidikan, 10*(3), 119–129.

Ismail, S. H., & Ahmad, N. (2020). Hubungan di antara Persekitaran Kerja dan Produktiviti Tenaga Kerja dalam Sektor Pembuatan. In *Kajian Kes di Malaysia* (pp. 11–23).

Jack, E. P., & Powers, T. L. (2015). Managing strategic supplier relationships: antecedents and outcomes. *Journal of Business & Industrial Marketing, 30*(2), 129–138.

Javed, S. (2015). Impact of top management commitment on quality management. *International Journal of Scientific and Research Publications, 5*(8), 1–5.

Kabeyi, M. J. B. (2019). Organizational strategic planning, implementation and evaluation with analysis of challenges and benefits. *International Journal of Applied Research, 5*(6), 27–32.

Kao, C.-Y., Tsaur, S.-H., & Huang, C.-C. (2020). The scale development of organizational culture on customer delight. *International of Contemporary Hospitality Management, 32*(10), 3067–3090.

Khan, M. A., Ismail, F., Hussain, A., & Alghazali, B. (2020). The interplay of leadership styles, innovative work behavior, organizational culture and organizational citizenship behavior. *SAGE Journals, 10*(1), 1–16.

Kim, J., Kim, K. H., Garrett, T. C., & Jung, H. (2015). The contributions of firm innovativeness to customer value in purchasing behavior. *Journal of Product Innovation Management, 32*(2), 201–213.

Koh, T. Y., & Low, S. P. (2008). Organizational culture and TQM implementation in construction firm in Singapore. *Construction Management and Economics, 26*(3), 237–248.
Limaj, E., & Bernroider, E. W. N. (2019). The role of absorptive capacity and culture balance for exploratory and exploitative innovation in SMEs. *Journal of Business Research, 94*, 137–153.
Ling, T. H., Logendran, T., & Mariappan, L. (2023). The role of communication skill in persuading customers to purchase product. *International Journal of Research Publication and Reviews, 4*(10), 277–288.
Maamari, B. E., & Saheb, A. (2018). How organizational culture and leadership style affect employees performance of genders. *International Journal of Organizational Analysis, 26*(4), 630–651.
MAMPU. (2008). *Panduan Pengurusan Perhubungan Pelanggan*. MAMPU.
Mannion, R., & Devis, H. (2018). Understanding organizational culture for healthcare quality improvement. *BMJ*. https://doi.org/10.1136/bmj.k4907
Mansor, F., & Tengku Kasim, T. S. A. (2015). Pembangunan Modal Insan Dari Perspektif Pengurusan Islam. *Journal of Al-Tamadun, 3*(1), 276–286.
Manyuchi, M. M., & Sukdeo, N. (2021). Importance of the control function in effective operational planning and rational decision making in mining organisations. In *Proceedings of the international conference on industrial engineering and operations management* (pp. 407–411).
Marolah, A., Ujianto, & Suhermin. (2018). Management functions on the implementation of community health care. *International Journal of Advanced Research (IJAR), 6*(6), 336–342.
Midor, K., Sujova, E., Cierna, H., Zarebinska, D., & Kaniak, W. (2020). Key performance indicators as a tool to improve product quality. *New Trends in Production Engineering, 3*(1), 347–254.
Neyestani, B., & Juanzon, J. B. P. (2016). Developing an appropriate performance measurement framework for total quality management (TQM) in construction and other industries. *International Journal of Technology & Engineering, 5*(2), 32–44.
Niati, D. R., Siregar, Z. M. E., & Prayoga, Y. (2021). The effect of training on work performance and career development: The role of motivation as intervening variable. *Budapest International Research and Critics Institute-Journal, 4*(2), 2385–2393.
Nor, F. M., Sabri, S. A., & Khalid, K. (2014). Tahap Kepuasan Pelanggan Terhadap Kualiti Perkhidmatan Pengangkuta Awam di KUIS. In *Proceeding of the international conference and management and Muamalah* (pp. 337–347).
Nor, M. S. M., & Selamat, M. N. (2020). Hubungan antara Sistem Kerja Ergonomik dan Prestasi Keselamatan dan Kesihatan Pekerjaan. *Jurnal Wacana Sarjana, 4*(1), 1–13.
Nukic, I. S., & Huemann, M. (2016). Organizational culture of the Croatian construction industry. *Engineering, Construction and Architectural Management, 23*(2), 237–260.
Ormston, R., Spencer, L., Barnard, M., & Snape, D. (2014). The foundations of qualitative research. In J. Ritchie, J. Lewis, C. M. Nicholls, & R. Ormston (Eds.), *Qualitative practice: A guide for social science student and researchers* (2nd ed.). Sage.
Ornek, A. S., & Ayas, S. (2015). The relationship between intellectual capital, innovative work behavior and business performance reflection. *Procedia—Social and Behavioral Sciences, 195*, 1387–1395.
Paais, M., & Pattiruhu, J. R. (2020). Effect of motivation, leadership and organizational culture on satisfaction and employee performance. *Journal of Asian Finance, Economics and Business, 7*(8), 577–588.
Parker, C. (2000). Performance measurement. *Work Study, 49*(2), 63–66.
Patyal, V. S., Ambekar, S., & Prakash, A. (2020). Organizational culture and total quality management practices in Indian construction industry. *International Journal of Productivity and Performance Management, 69*(5), 866–913.
Podolinsky, M. A. (2013). *Productivity: managing, motivating, maximizing teams in Asia*. McGraw Hill.
Prayuda, R. Z. (2020). Continuous improvement through Kaizen in an automotive industry. *Journal of Industrial Engineering & Management Research, 1*(1), 37–42.

Qi, C., & Chau, P. Y. K. (2018). Will enterprise social networking systems promote knowledge management and organizational learning? An empirical study. *Journal of Organizational Computing and Electronic Commerce, 28*(1), 31–57.
Rodriguez, J., & Walters, K. (2017). The importance of training and development in employee performance and evaluation. *World Wide Journal of Multidisciplinary Research and Development, 3*(10), 206–212.
Sackman, S. (1991). *Culture knowledge in organization*. Sage.
Sartika, N. S., Subroto, D. E., Mauladaniyati, R., Rosdianwinata, E., Rifai, R., Sujana, A., Abidin, Z., Priadi, M. D., & Purwanto, A. (2020). Effect of pedagogic, professional competency and work motivation toward Indonesian primary school teachers performance. *Systematic Reviews in Pharmacy, 11*(9), 617–626.
Selvanathan, M., Ali, A. J., Balasubramaniam, S., & Thanaraju, P. (2012). Kesan Sikap Pekerja Ke Arah Kualiti Kerja Cemerlang. *Jurnal Kemanusiaan, 20*, 130–146.
Shahriman, W. H., Bakar, R. A., & Ismail, C. Z. (2022). Perbincangan dan Penghayatan Konsep Al-Ihsan Dalam Al-Quran Serta Muamalah. *Journal of Islamic, Social, Economics and Development, 7*(43), 208–220.
Shakeel, S. J. M., Ali Brah, S., Zahoor Hassan, S., & Kannan, R. V. (2014). An examination of product innovation and buyer-supplier relationship in Pakistani firms. *International Journal of Productivity and Performance Management, 63*(8), 1031–1045.
Soomro, B. A., & Shah, N. (2019). Determining the impact of entrepreneurial orientation and organizational culture in job satisfaction, organizational commitment and employee's performance. *South Asian Journal of Business, 8*(3), 266–282.
Tovmasyan, G. (2017). The role management in organizations: Psychological aspect. *Business Ethics and Leadership, 1*(3), 1–26.
Watson, P., & Howarth, T. (2011). *Construction quality management*. Spoon Press.
Wetzel, R., & Tint, B. (2019). Using applied improvisation for organizational learning in the Red Cross Red Crescent Climate Centre. In E. Antonacopolou & S. Taylor (Eds.), *Sensuous learning for practical judgment in professional practice* (Palgrave studies in business, arts and humanities). Palgrave Macmillan.
Widyatmoko, W., & Wibowo, D. A. (2020). Peranan Kualitas Kerja dan Pelayanan Karyawan Terhadap Kepuasan Pelanggan CV Pare Centre Kediri. *Jurnal Inovasi Penelitian, 1*(7), 1505–1512.
Wijayanti, T. C., Setini, M., Darma, D. C., Purwadi, P., & Alka, P. R. (2020). Connectivity continuous improvement program and employee performance. *International Review of Management and Marketing, 10*(1), 1–9.
Yazid, Z., Abdullah, N. A., & Baharom, R. (2017). Konflik dalam Proses Penilaian Prestasi dalam Organisasi. *Jurnal Pengurusan, 49*, 41–52.

Musaiyadah Ahmadun is a senior lecturer at the University of Technology MARA (UiTM), Sabah Campus. She has expertise in the field of Shariah, quality management, and halal standards. She is active in research and writing related to quality management, halal, and education. In addition, she has won several research awards at national and international level. She can be contacted at musaiyadah@uitm.edu.my

Siti Arni Basir is an Associate Professor in the Department of Siasah Syar'iyyah, Academy of Islamic Studies, University of Malaya. She actively conducts research and publishes articles related to the implementation of MS 1900 Shariah-Based Quality Management System and organizational management. She also produced publications related to issues of quality management and Islamic administration in public and corporate sectors in Malaysia. The focus of her latest research is relationship between the implementation of quality programs and excellence as well as innovation. In addition, she also studied halal management and the development of halal detection instruments for companies producing halal products in Malaysia. Associate Professor Dr. Siti Arni has won various research awards at national and international levels. Based on her outstanding achievements, she was accepted as a member of British Publishing House in 2020. She was also listed in BritishPedia: Successful Person in Malaysia 2nd edition (2020), published by British Publishing House. She can be contacted at sitiarni@um.edu.my

Masnih Mustapa is a lecturer at Universiti Teknologi MARA (UiTM), Sabah campus. She has expertise in the field of Islamic Studies, leadership, and Management. She is active in research and writing in Islamic Studies, leadership, and Management. She can be contacted at masnihmustapa@uitm.edu.my

Nur Ilma Ambo is a lecturer at Universiti Malaysia Sabah, and she has expertise in the fields of Shariah and halal management. She is active in research and writing related to quality halal and education. She can be contacted at nurilma@ums.edu.my

Sri *Sukuk* to Achieve SDGs with a Special Reference to the Legal and Regulatory Framework in Malaysia

Omar Kachkar and Marwa Al Fares

Abstract Malaysia has the leadership in developing Islamic socially responsible investing (SRI) and green finance. This study aims to highlight the importance of issuing SRI sukuk to achieve sustainable development and point out the incentives which attract investors to participate in the SRI sukuk within its legal and regulatory framework. This study has adopted a qualitative approach by analysing previous literature, as well as examining Malaysian SRI regulatory framework provisions. This study found that Malaysia is the key actor in provision of an environment requisite for issuing Sukuk aligned with SDGs such as SRI sukuk and SDG Sukuk. Promoting financing for SRI projects through issuing SRI sukuk will meet the United Nations' Sustainable Development Goals (SDGs). In order to attract more investors to subscribe to SRI sukuk, SC Malaysia is determined that SRI sukuk should be highly mandated to invest in green, social, sustainable, and waqf dimensions. Moreover, the Government of Malaysia (GoM) provides Tax incentives for SRI sukuk issuers, investors as well as brokers. For instance, the SRI sukuk issuer enjoys tax concessions on SRI sukuk issuance. Finally, SDG Sukuk Framework explains the Malaysian Government's intention to utilize the SDG Sukuk to fund projects that will generate environmental and social benefits and also conform to sustainable development goals.

O. Kachkar (✉)
Department of Management, Ibn Haldun University, Istanbul, Türkiye
e-mail: omar.kachkar@ihu.edu.tr

M. Al Fares
Department of Islamic Economics and Finance, Istanbul Sabahattin Zaim University, Istanbul, Türkiye

A. Rafiki et al. (eds.), *Strategic Islamic Business and Management*, Contributions to Management Science,
https://doi.org/10.1007/978-3-031-61778-2_9

1 Introduction

1.1 *Overview of* Sukuk *Significance in the Economy*

Islamic law is a comprehensive system that always offers alternatives conforming to Shariah for the things that the latter prohibits. Consequently, the prohibition of interest is associated with the permissibility of the sale contract. Also, bonds, as interest-based loans, are prohibited from Shariah perspective, on the other hand, bonds are an essential means of investment in the capitalist system, therefore, it is necessary to find a *shariah*-compliant alternative to allow Muslims to invest their money in the permissible way in order to benefit as much as possible from the Muslims money who refrain from dealing in forbidden bonds and as a means to attract their savings. Hence the importance of *sukuk* is that it is an alternative to conventional bonds that are prohibited by Shariah (IFSB, 2007).

Sukuk began to spread widely in the Muslims and international markets, as one of the innovative modern financial tools for Islamic financial engineering. The process of issuing and trading *sukuk* passes through technical procedures and is subject to *shariah* principles, starting from subscription, and then trading procedures ending with the *sukuk* maturity.

Studies like Rafay et al. (2017) find that Sukuk is an opportunity for individual investors to afford low-risk bond. Additionally, it is also a cheap alternative for companies in need of debt. The international sukuk market is led by several regions such as the GCC and South East Asian (SEA) countries. In the GCC region in particular the Sukuk market is led by countries including KSA, UAE, Qatar, and Kuwait. On the other hand, Malaysia is generally known as the earliest contender in this market. The SEA region has other leaders like Indonesia and Singapore as well. Lastly, the global market also has other countries with considerable share in the sukuk market including countries like Pakistan, Japan, USA, UK, Germany, Turkey, Egypt, and Gambia, to name a few (Zolfaghari, 2017).

The International Monetary Fund reported that a large number of investors switched to invest in *sukuk* instead of traditional debt instruments, from 2001 to 2007 the *sukuk* market witnessed a significant boom in which the financial market achieved more than 50 billion Malaysian ringgits due to the efficiency of the decisions of the SAC of SC which contributed to the product development and structuring of innovative financial instruments. As the volume of *sukuk* issuance doubled between 2004 and 2006 from 7.2 to 27 billion dollars, in 2008, there was a notable deviation in the *sukuk* growth industry due to the global financial mortgage crisis (IIFM, 2021). In 2016, the issuance of *sukuk* increased marginally compared to 2015, due to the difficult conditions in the market, especially in emerging markets, and the drop in oil prices in 2014. Malaysia alone contributed $45 billion to the total *sukuk* issued globally $116.4 billion in 2014 as the Central Bank of Malaysia (BNM) is one of the largest issuers of *sukuk* in the world (eliktisad.com, 2015). In 2015 BNM cessation of *sukuk* issuance led to a 42.5% decrease in *sukuk* issuance compared to the same period in 2014. In 2019, the Global *Sukuk* market recorded gorgeous issuances and gains of around 18.32% (US$ 145.70 billion) which is considered a positive growth of the *Sukuk* market internationally. The growth of sukuk market accelerated in year 2020 when it jumped to US$ 174.6 billion recording an increase of almost US$ 30 billion in a single year. Thereon, the growth decelerated and finally fell towards the end of year 2022. The market's total volume expanded to US$ 188.1 billion in 2021 but finally fell to US$ 182.7 billion in year 2022.

Moreover, 2019 *wakalah Sukuk* was more dominant compared to the *Ijarah Sukuk* and other *sukuk* as the share of *wakalah sukuk* by issuers in the international market was around 56% (US$ 75.21 billion), i.e. an increase of about 6% (US$ 5.3 billion). *Ijara sukuk* share has declined to 11% (US$ 4.21 billion), followed by 9% *Mudaraba sukuks* (US$ 3.5 billion), while the hybrid *Murabaha/Mudaraba sukuks* and *Ijara*/hybrid *Murabaha sukuks* rose to 8% (US$ 3 billion) and 12% (US$ 4.6 billion), respectively. The share of *Wakala* and *Mudaraba* hybrid *sukuks*, as well as *Murabaha sukuks*, amounted to 1% (US$ 0.30 billion) and 3% (US$ 1.10 billion) (Fig. 1).

In 2020, the total global issuance valued to US$ 174.641 billion which is considered the highest value of yearly *Sukuk* issuances to date. The 1-year period from 2019 to 2020 witnessed a significant increase in the Sukuk issuance that increases from 145.7 billion $ to 174.6 billion $, i.e. registering an increase of 19.84% over 1 year. The increasing pattern of sukuk issuance continued thereon to reach an overall maximum of almost US$ 188.1 billion in year 2021; however, it finally fell to US$ 182.7 billion in 2022.

The global market for Sukuk is led by Malaysia; however, other notable contenders that follow her include countries like Saudi Arabia, Turkey, Indonesia, and UAE. 2020 was a promising year for the local markets when the sovereign issuances increased their share to 60%. Next came the institutions and quasi-sovereign types with 14% and 8% shares, respectively. The share of Islamic financial institutions however was significant and reached the 18% mark in terms of share. In general, corporate sector in terms of the share in the volume of issuance performed lower

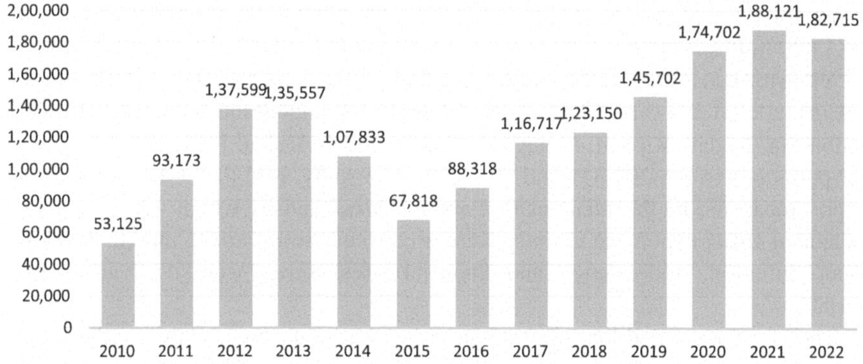

Fig. 1 Total global *Sukuk* issuances (Jan 2001–Dec 2022)—all tenors, all currencies, in USD millions. Source: Adapted by authors from *Sukuk* IIFM database

than expected. This however was an exception to the case of Malaysian government. The quasi-sovereign sukuk have shown an improved performance in 2020, reaching US$ 19.27 billion, or 45% of the market. The sovereign issuances however fell from 30% in 2019 to 23% in 2020. Financial institutions there were at 6.27 billion US$, in 2019 it increased to US$ 9.55 billion, i.e. by 23%. The corporate issuances, on the other hand, witnessed a negative turn and reached US$ 3.85 billion from US$ 4.93 billion, hence decreasing in share from 13 to 9% in 1 year.

The combined domestic and international sukuk issuance of Malaysia reached $800 billion landmark making it the global leader in the sukuk industry. Saudi Arabia is another contender with significant market share and US$ 174.60 billion issuances. Indonesia, UAE, and Turkey follow with 124.90, 98.39, and US$ 66.90 billion sukuk issuances. Malaysia has a strong capital market that helps to maintain its leading position in the market over other players from Asia and GCC regions. This has not been the case for some other countries including Bahrain, Saudi Arabia, Indonesia, the United Arab Emirates, Turkey, etc. whose increasing share in the global sukuk markets has squeezed Malaysia's total share.

In the year 2020, Ijara sukuk was outperformed by the agency structure in terms of the issuers in the international market and had an overwhelming share of 51% amounting to US$ 21 billion. Ijarah sukuk is second in the list and enjoys a significant 18% share with value around US$ 7.5 billion. Mudaraba sukuk are next in the list with a relatively lesser share of around 6% or US$ 2.7 billion. Murabaha sukuk may have marginally decreased over the time; however, it still enjoys a very significant share of 29%, a decrease of 1% over 1 year while ijarah sukuk over-performs it with 41% market share (IIFM, 2021).

Table 1 shows selected leaders in Global *Sukuk* issuances during 2022. It is clearly evident from the table that though Malaysia may lead the global sukuk market a lot of countries and individual issuers are growing their share in this market fast. In 2022 around 5500 million $ sukuk was issued by Government of Turkey followed by US$ 3250 million by Government of Indonesia and 2500 million $ by Saudi Arabia. This was an exception to the issuances in their local currencies. Islamic Development Bank alone issued sukuk for almost US$ 2600 million. Other major issuers included Al Rajhi Bank with SAR 2667 million and Infracorp with US$ 900 million sukuk.

Yıldırım et al. (2020) have conducted a study to measure the relationship between *sukuk* market development and growth by analysing data of nine countries "Brunei, Indonesia, Jordan, Kuwait, Malaysia, Nigeria, Saudi Arabia, Pakistan, and Turkey for periods between 2014Q1 and 2017Q4". The study used the "GDP, total *sukuk* export measured by the *sukuk* market and the *sukuk* density which was considered as annual *sukuk* export per country were used to determine *sukuk* market development. Inflation, trade deficit and financial stress series were used as control variables".

The aforementioned long-term relationship was confirmed by this study. In the long run, growth was positively affected by both volume and density of sukuk. A single unit increase in the former accelerated growth by 0.5%; however, a similar increase in the latter leads to an almost 1.7% increase in growth. No considerable

Table 1 Selected value leaders—global *Sukuk* issuances and trends (US$ 500 million or >, tenor > 1 year)

Issue year	Issuer	Issuance currency	USD million or equivalent	Tenor (years)
2022	Al Rajhi Bank	SAR	2667	Perpetual
2022	Arabian Drilling Company (ADC)	SAR	533	5
2022	Boubyan Bank	USD	500	5
2022	Dubai Islamic Bank	USD	750	5
2022	Dubai Islamic Bank (Sustainability Sukuk)	USD	750	5
2022	First Abu Dhabi Bank	USD	500	5
2022	Government of Sharjah	USD	750	8
2022	Government of Indonesia	IDR	2570	2
2022	Government of Indonesia (Sustainability Sukuk)	USD	1750	5
2022	Government of Indonesia (Sustainability Sukuk)	USD	1500	10
2022	Government of Malaysia (Sustainability Sukuk)	MYR	1020	15
2022	Government of Pakistan	USD	1000	7
2022	Government of Pakistan	PKR	880	5
2022	Government of Qatar	QAR	1350	3
2022	Infracorp—Real Estate Agency (Sustainability Sukuk)	USD	900	Perpetual
2022	Islamic Development Bank	USD	1600	5
2022	Islamic Development Bank	USD	1000	5
2022	Ministry of Finance Saudi Arabia	SAR	3520	10
2022	Ministry of Finance Saudi Arabia	SAR	2681	10
2022	Ministry of Finance Saudi Arabia	USD	2500	6
2022	Ministry of Treasury and Finance of Türkiye	USD	3000	5
2022	Ministry of Treasury and Finance of Türkiye	USD	2500	3
2022	Ministry of Treasury and Finance of Türkiye	TRY	1008	5
2022	PLUS MALAYSIA BHD	MYR	1235	1.4
2022	Riyad Bank (Sustainability Sukuk)	USD	750	Perpetual
2022	Riyad Bank (Private Placement)	SAR	1000	Perpetual

Source: IIFM *Sukuk* database, 10th, 2021

relationship was found between these variables in the short term. On the other hand, growth was also affected positively by the sukuk export in the long run.

Financial inclusion should be implemented in Islamic finance in order to attain SDGs by supporting activities that aim to property protection of (Hifdh al-Mal) wealth maximizing in a halal manner, adherence of the tenets of justice, supporting real economic activity, fair distribution of wealth to prevent concentration in few hands. Islamic finance has rich instruments such as *Sukuk*, Takaful, *Mudharaba*, *Murabaha*, and *Musharakah* which can cater to Shariah's objectives as well as SDGs if these instruments are used with their substance. Particularly, *sukuk* can serve social impact investments such as clean and green environment financing for sustainable development (Khan, 2020). Consequently, green and sustainable *Sukuk* can significantly fill the investment gap by directing the investments to more climate-friendly investment initiatives (Alawode et al., 2018).

The annual SDG investment gap in developing countries is quickly reaching the estimated US$ 4 trillion half of which is expected to be in the clean energy sector. As Fig. 2 shows, this estimated amount is obtained as the sum of the investment gaps derived for each SDG sector individually. The energy sector leads in its share, almost 50% followed by Water and sanitation, and Infrastructure respectively. Education and Health sectors, on the other hand, have the least share.

Malaysia also takes a leading role in the green finance and socially responsible investment sectors. Several green finance projects have commenced from Malaysia in recent years including the environmental, social, and governance (ESG) index FTSE4Good and Bursa Malaysia's Sustainability Reporting Framework, to name a few. If we exclude Japan from the analysis, Malaysia turns out to be the biggest Asian market for sustainable investment (Alawode et al., 2018). In 2011, the Securities Commission of Malaysia launched Master Plan 2 for the stock market which aims to promote investment within the scope of social responsibility. The former Malaysian Prime Minister in 2014's Budget Speech announced his intention of intensified efforts for promotion of the country as a market for SRI *sukuk* (Wahab & Naim, 2019). Later in 2014, in order to integrate fresh requirements for SRI

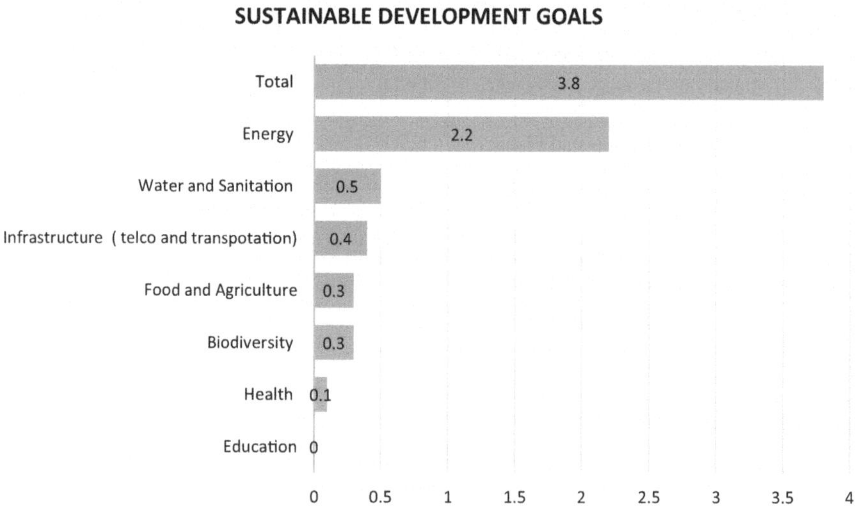

Fig. 2 Estimated annual investment gap in key SDG sectors, capital expenditure, developing countries (trillions of dollars). Source: Adapted by authors from SDG Investment Trends Monitor, UNCTAD (Sept 2023)

sukuk revised guidelines were also issued by SC Malaysia. The *sukuk* guidelines had shed the light on the framework of SRI eligible projects. Moreover, it mentioned that the proceeds will be used to finance projects aimed to preserve and protect the environment as well as to similar other causes. The guidelines also specify the disclosure requirements that the issuer is obligated to disclose under paragraph 7.05, through disclosure documents (SC, 2002) or the subscription agreement:

1. The intention of the issuer is to achieve the overall objectives of SRI project.
2. All possible details of the SRI project.
3. An undertaking of compliance with environmental, social, and governance standards by the issuer in the context of the SRI project.

Recently, achieving the SDGs has attracted the attention of the business community with an increase in demand for more climate-friendly investments, wealth maximizing in a halal manner, adherence to the tenets of justice, supporting real economic activity, fair distribution of wealth to prevent concentration in few hands. This approach pushed the financial institutions to put more effort into developing new financial instruments, which could pave the way towards SDGs achieving. SRI *sukuk* was the new and pioneer method to serve this noble goal. The most notable SRI *sukuk* issuance in Malaysia was the Ihsan. Ihsan Sukuk by Khazana was the first *sukuk* to be approved under the Securities Commission of Malaysia's SRI *Sukuk* framework. Its value was RM1 billion at face value and has a term of 25 years to maturity. Ihsan *Sukuk* by Khazana is a new and pioneer method to finance Yayasan Trust schools by financing from the capital market. In 2017, Malaysia became the first country to launch the ASEAN Green Bond Standards (AGBS). The Securities Commission Malaysia Chairman said: "ASEAN must grow in a responsible and sustainable manner. Growth cannot come at the expense of our future generations" (Global Islamic Economy Gateway, 2017). Later, in 2017, the Securities Commission announced the issuance of Malaysia's first green sukuk based on the framework of Sustainable and Responsible Investment (SRI).

When the proceeds of an Islamic bond are used for financing environmentally sustainable infrastructure projects the underlying bonds are called SRI sukuk. An example is building of a renewable energy generation facility (SC, 2017). The first Sukuk as socially responsible investment was issued by CAGAMAS Berhad in 2020 that amounted to 450 million RM. This included 350 million in Islamic medium-term notes while the remaining 100 million comprised a 3-year SRI sukuk. The framework of this initiative focused on "affordable housing, employment generation via small to medium size enterprises and five green solutions (renewable energy, energy efficiency, green buildings, low-carbon transportation, sustainable water and wastewater management)" (IIFM, 2021). In 2018, Indonesia became the global leader in issuance of green *Sukuk* (Akin Gump, 2021).

Following the Malaysian and Indonesian experience, currently, the IsDB issued the Largest Sustainability *Sukuk* ever public issuance to date, 2021 and raised US$ 2.5 billion. The IsDB issuance gained rating "Aaa/AAA/AAA by Moody's, S&P, and Fitch (all stable outlook) and priced the 5-year Trust Certificates under its US$ 25 billion Trust Certificate Issuance Programme. It was priced at par with a profit rate of 1.262%, payable on a semi-annual basis". Many parts participated in managing and organizing this issuance, for example, Société Générale, Citi, HSBC, Goldman Sachs International, NATIXIS, etc. The proceeds of this issuance will be divided (10%) into finance/refinance green while (90%) will be channelled to finance social development projects for eligible projects published by the International Capital Market Association (ICMA) (IsDB, 2021).

In line with the growing desire and appetite for sustainable financing, Turkish participation banks are planning to promote SDGs through striving to issue green, for instance, in September 2021, KFH Capital announced the issuance of Sustainability *Sukuk* for KFH-Turkey with the value of US$ 350 million and an Order book of US$ 4 billion (Akin Gump, 2021). In addition, in 2020, Türkiye Sınai Kalkınma Bankası (TSKB) had announced the first sustainable *Sukuk* issuance that aligns with UN Sustainable Development Goals (SDGs) with 45 TL million-worth issuances. This *Sukuk* was issued based on the framework of the TL 400 million-worth SDG *Sukuk* issuance programme created on behalf of Zorlu Energy Company, that issuance was the first-ever in the world SDG related *Sukuk* on behalf Zorlu Energy company. Consequently, Zorlu Energy will contribute to various SDGs, for instance, "SDG 7: Affordable and Clean Energy, SDG 12: Responsible Consumption and Production and SDG 13: Climate Action through the renewable energy investments to be made in the field of wind and solar energy" (TSKB, 2021). GIFIIP also known as the Global Islamic Finance and Impact Investing Platform was founded by UNDP's IICPSD. It was done in collaboration with the Islamic Development Bank (IsDB). GIFIIP's main work focused on supporting Islamic finance through encouragement of innovative financing tools, e.g. Sustainable and Green *Sukuk*, for SDG implementation. In Turkey, several meetings and workshops were held under this platform in order to familiarize audience with innovative finance. These forums also helped in mobilization of stakeholders to start projects that have positive social and environmental impact (GIFIIP, 2020).

Many studies have highlighted the importance of SRI *sukuk* utilization through various sectors; for instance, a study conducted by Zain and Sori's (2020) suggested that SRI *sukuk* should be employed as a tool to develop *waqf* assets, and to develop waqf assets and to include them as one of the important categories of SRI projects within a comprehensive plan for *waqf* revival. So using SRI *sukuk* to develop *waqf* assets will offer the investors a new way to devote their wealth towards sustainability which brings lasting benefit to society. Their study attempted to propose using SRI *Musharakah Sukuk* for the development of *waqf* properties/assets in Malaysia, as Malaysia has a rich history in issuing SRI *sukuk*, the most remarkable SRI *sukuk* issuance in Malaysia was the Ihsan *Sukuk*.

1.2 Research Objectives

Highlight the importance of issuing green *sukuk* to achieve sustainable development. The issuance of SRI and the legal and regulatory framework for this.

The incentives which attract investors to participate in the green *sukuk* within the legal and regulatory framework of issuing green sukuk.

1.3 Research Question

1. What is the importance of issuing SRI *sukuk* to achieve sustainable development?
2. What is the legal and regulatory framework for issuing SRI *sukuk* in Malaysia?
3. Are the legal and regulatory frameworks of issuing green *sukuk* encouraging investors to participate in it?

2 An Overview of SRI Sukuk Framework

2.1 Definition of SRI Sukuk

Under investing, it represents a category of financial instruments, such as green bonds and social impact bonds. In the past few years, SRI sukuk has experienced substantial growth in terms of both assets and the number of investors. Various sources have provided differing interpretations of the acronym, including "Responsible, Sustainable, and Impactful investments" according to The Centre for Islamic Wealth Management (2015), "Socially Responsible Investment" and "Sustainable and Responsible Investment" according to the Securities Commission Malaysia (2014). It serves as a comprehensive term encompassing investment that aligns investors' financial objectives with environmental and social problems (Moghul & Safar-Aly, 2014).

2.2 Malaysia SRI Sukuk *Background*

The IIFM (2021) report considers that when investment mandates take into account the environmental, social, and governance dimensions as well, it will lead to an increase in sukuk offering both SRI and green. Many countries have shown an increased interest in SRI projects for instance, USA, Canada, Japan, New Zealand, and Australia to name a few. In the USA until mid-2014, the SRI represented US$ 6.57 trillion, almost one-sixth of funds in the assets including private equity and hedge funds and several mutual funds (Joan & Thomas, 2015).

In the Global Sustainable Investment Review (GSIA) of 2020, it was reported that the year saw a substantial growth in global sustainable investment, reaching a staggering US$ 35.3 trillion across five major markets. This represented a significant 15% increase over the preceding 2 years until 2020. Notably, sustainable investment assets are undergoing noteworthy developments, with Canada witnessing the most substantial absolute increase in the past 2 years, boasting an impressive 48% growth. The USA has also experienced remarkable growth with a 42% increase, followed by Japan at 38% and Australasia at 25%, all measured from 2018 to 2020. In contrast, Europe faced a contrasting scenario, where a 13% decrease in the growth of sustainable investment assets was observed over 2 years' time.

Figure 3 gives an overall picture of worldwide issuance of green bonds from year 2014. The issuance has been gradually increasing from US$ 36.7 billion since 2014. A first jump was observed in 2019 when the green bonds issuance recorded a growth of almost US$ 100 billion in a single year. After a slight increase in 2020 it again observed a record increase and hit all time high of US$ 582.4 billion in 2021; however, it then declined to US$ 487.1 in the following year.

In 2017, Malaysia made history by issuing the first green sukuk in the world, earning recognition from the World Bank as a pioneer in supporting the United Nations' Sustainable Development Goals (SDGs) by utilizing the capital markets. Building on this momentum, Malaysia allocated RM5.3 billion in 2020 to finance eligible Socially Responsible Investment (SRI) projects. This was a move that targeted an increase in the development of SRI sukuk and ASEAN-labelled green, social, and sustainability bonds (cumulative April 2020). Notably, US$ 680 million was earmarked to support the CIMB SDG Bond issued under the ASEAN Sustainability Bond Standards in 2019. Since the inception of the sukuk framework in 2014 up to 2020, Malaysia has witnessed approximately 13 SRI sukuk issuances, with seven of them earning dual recognition under the ASEAN standards. Specifically focusing on green SRI sukuk, Malaysia has successfully issued RM4.3 billion, constituting over 90% of SRI sukuk issuances. These funds have played a pivotal role in financing of around 500 MGWs of renewable energy projects in the country (bixmalaysia.com, 2020).

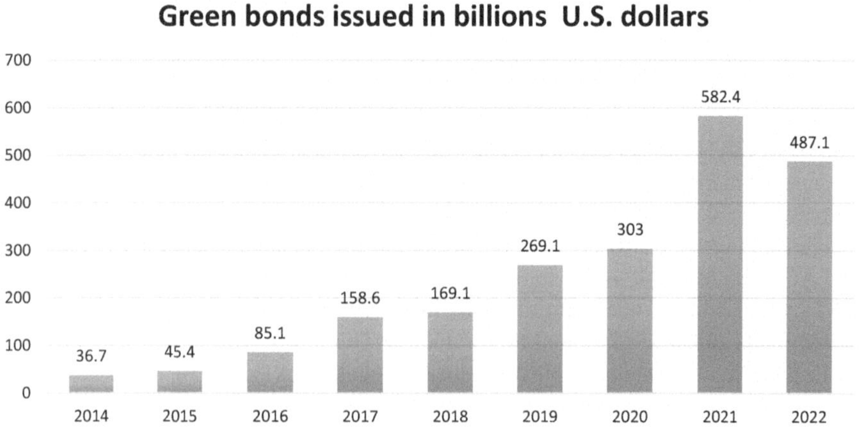

Fig. 3 Value of green bonds issued worldwide (2014–2022) in billion US$. Source: Adapted by authors from Statista 2023

In 2017, Malaysia achieved a historic milestone by pioneering the issuance of the world's inaugural green sukuk, gaining acknowledgment from the World Bank for its ground-breaking use of capital markets to support the United Nations' Sustainable Development Goals (SDGs). Building upon this achievement, Malaysia allocated RM5.3 billion in 2020 to fund several Socially Responsible Investment (SRI) projects, aiming to spur the growth of SRI sukuk and ASEAN-labelled green, social, and sustainability bonds (cumulative April 2020). From the inception of the sukuk framework in 2014 until 2020, Malaysia witnessed around 13 SRI sukuk issuances, with seven receiving dual recognition under the ASEAN standards. Focusing on green SRI sukuk, Malaysia successfully raised RM4.3 billion, constituting over 90% of SRI sukuk issuances. These funds have played a pivotal role in financing 482 megawatts of renewable energy projects in Malaysia (bixmalaysia.com, 2020) (Figs. 4 and 5).

The United Nations Conference on Trade and Development (UNCTAD) estimated that in order to achieve SDGs the world would need between US$ 5.0 trillion and US$ 7.0 trillion investment per annum until 2030 (bixmalaysia.com, 2020).

2.3 *Malaysian Experience in Issuing SRI to Enhance SDGs*

2.3.1 The World's First Ringgit-Denominated SRI Sukuk: Khazanah Ihsan Sukuk (Malaysia)

Ihsan, sponsored by Khazana, marked the initial SRI sukuk approval under the Securities Commission of Malaysia's SRI Sukuk framework in 2015. This sukuk had a face value of RM1 billion and a 25-year maturity period. The Ihsan Sukuk by

Fig. 4 Breakdown of SRI *Sukuk* issuances in Malaysia as in April 2020. Source: Adapted by authors from Sukuk IIFM database

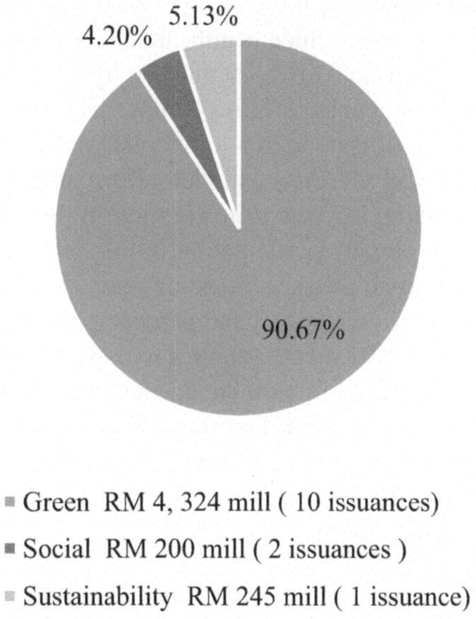

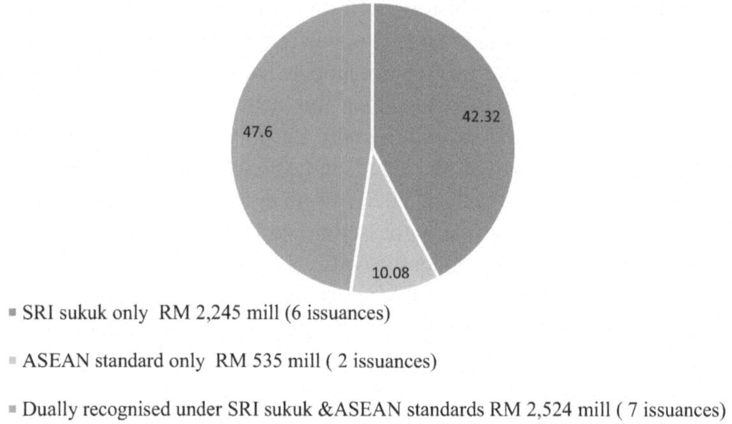

Fig. 5 Total Issuances under the SRI *Sukuk* Framework and ASEAN Green, Social, and Sustainability standards in Malaysia as in April 2020. Source: Adapted by authors from Sukuk IIFM database

Khazana pioneered an innovative approach to fund Yayasan Trust schools through the capital market. Khazanah Nasional Berhad established a Special Purpose Vehicle (SPV) called the Ihsan Sukuk Company in early 2015 to issue RM1 billion (Ghani, 2015), boasting an AAA(s) rating from RAM Rating Services Berhad. The first issuance occurred in June 18, 2015, followed by a second issuance in August 8, 2017, successfully raising $100 million each. The term was prescribed as 7 years, recurring, while distribution rate 4.3%. The funds generated are earmarked for the Yayasan AMIR Trust School Programme, with both versions aiming to enhance access to quality education in Malaysia. According to Khazana's Dato' Mohamad Azani Ghani, the executive director of investment, potential segments for these funds may include healthcare and affordable housing (Star Online, 2015).

Yayasan AMIR, established by Khazanah, is a non-profit organization (NPO). It is dedicated to elevating Malaysian public schools in terms of their standard of education. It is done as a joint collaboration with Ministry of Education. Similarly, the program serves as a model, prioritizing a comprehensive transformation of schools spanning a 5-year time duration. Before the introduction of SRI Sukuk, 30 schools were already participating in this project, which extended its impact to both rural and urban areas, influencing over 20,000 schools. The stakeholders of this project range from students and parents to teachers, principals, and the wider community.

The SRI *sukuk* complies with the Islamic principle of "Wakala Bil Istithmar". Its basis was an investment agency. Ihsan was appointed as the investment representative or agent in the first stage. Khazanah was then appointed as its sub-representative. The profits generated from the investments will be channelled to finance the schools under the Trust Program and are distributed to *sukuk* holders till the maturity date (CIMB, 2015).

2.3.2 Green *Sukuk*

Green sukuk is inherently linked to Socially Responsible Investment (SRI) sukuk, focusing on environmental conservation. Green sukuk share resemblances with conventional bonds concerning tradability and structure. They permits the issuer to borrow funds, committing to repayment, often with interest, within a specified timeframe. What sets it apart from regular bonds is its unique feature, allowing investors to participate in funding projects dedicated to addressing climate or environmental concerns.

In 2017, Malaysia became the first country to launch the ASEAN Green Bond Standards (AGBS). The Securities Commission Malaysia Chairman Tan Sri Ranjit Ajit Singh said: "ASEAN must grow in a responsible and sustainable manner. Growth cannot come at the expense of our future generations" (Global Islamic Economy Gateway, 2017). Later, in 2017, the Securities Commission announced the issuance of Malaysia's first green sukuk as an innovative tool to address global financing gaps in green finance, this issuance was based on the framework of Sustainable and Responsible Investment (SRI) for sukuk (SC, 2017).

2.3.3 Noteworthy Green/Transition Issues

To harness this potential, Malaysia's CIMB is actively contributing to the advancement of Renewable Energy Financing for SMEs. They provide comprehensive support by offering 100% financing to cover the expenses associated with the installation of solar photovoltaic systems. Additionally, CIMB Islamic has introduced the Green Biz Ready program, aiming to expedite the growth of sustainable businesses in Malaysia (CIMB, 2019).

In 2020, HSBC Amanah took a significant step by launching the first Green Sukuk (SRI) that was denominated in the local currency and amounted to 260 million Malaysian Ringgit. This Sukuk Wakalah has been employed to fund two solar power projects in the country (HSBC, 2020) (Table 2).

Notably, Malaysia continues its tradition of leading initiatives in the sukuk markets by introducing the Prihatin sukuk, marking the country's inaugural Sukuk, the digital one. The funds raised from this sukuk will be directed to a post COVID-19 project, namely Kumpulan Wang. This initiative was part of the measures intended for economic recovery. The Prihatin sukuk offers a fully online subscription process for several banks (IIFM, 2021).

2.4 *The Importance of SRI* Sukuk *to Achieve SDGs*

There are 17 SDG goals and 169 achievement targets which are expected to be achieved in 2030. Through analysing the SDGs, many eligible SRI projects can be touched in the SDGs, especially the goals of alleviating poverty and hunger and

Table 2 Anchor green Sukuk issuances

Date	Issuer	Country	Sector	Amount	Structure
March 2021	The Islamic Development Bank (IsDB)	Saudi Arabia	Finance and refinance green and social development projects	US$ 2.5 billion	n/a
September 2020	Saudi Electricity Company (SEC)	Saudi Arabia	Energy efficiency and renewable energy	US$ 1.3 Billion	n/a
2020	Etihad Airways	UAE	Climate reduction projects to reduce carbon emissions	US$ 600 million	n/a
November 2019	The Islamic Development Bank (IsDB)	Saudi Arabia	Renewable energy, clean transportation, energy efficiency, pollution prevention and control, environmentally sustainable management of natural living resources and land use	EUR 1 billion	Sukuk Al Wakalah
October 2017	Quantum Solar Park (Semenanjung) Sdn Bhd	Malaysia	Finance the construction of the largest solar power project in Southeast Asia	US$ 236.5 million	Murabaha (via a Tawarruq arrangement)

Source: Adapted by authors from Sukuk IIFM database

reducing inequality by redistributing wealth. The aforementioned goals overlap with several aspects of the five objectives of *Sharich* which are known as Maqasid al Shariah, i.e. preserving of the faith (*din*), human life- self- (*nafs*), intellect (*'aql*), posterity (*nasl*), and wealth (*mal*). Thus, *Maqasid* al Shariah are all about "promoting benefits and repelling harms". Therefore, *Maqasid* al-*Shari'ah* and SDGs have common targets such as calling for establishment of justice, poverty alleviation, and ensuring fairness.

Consequently, promoting financing SRI projects through issuing SRI *sukuk* helps in meeting the Sustainable Development Goals (SDGs). Moreover, using SRI *sukuk* will drive towards investor diversification. Thus, in order to attract more investors towards SRI *sukuk*, the SC Malaysia in 7.08 subparagraph determined that *sukuk* should be highly mandated to invest in:

- **Green**: "Renewable energy, Energy efficiency, Pollution prevention and control, environmentally sustainable management of living natural resources and land use, which are aligned with 3, 7, 8, 9, 11, 12, 13 of SGs".
- **Social**, for example: "Affordable basic infrastructure, Access to essential services, Affordable housing", which are aligned with 1, 2, 3, 4, 5, 6, 7, 8, 9, 10, 11, 13 of SGs.
- **Sustainable** and ***waqf*** properties/assets, which are aligned with 1, 3, 4, 7, 8, 9, 10, 11 of SGs (SC, 2019).

2.5 *Eligible SRI Projects*

Guidance to subparagraph 7.07 the SC Malaysia specified the criteria and conditions that must be met by the project to be eligible for a project that falls under SRI. Such project is required to achieve at least one or multiple objectives from the ones mentioned below (SC, 2019):

- Preserving and protecting the environment and natural resources
- Conserving the use of energy
- Promoting the use of renewable energy
- Reducing greenhouse gas emission
- Addressing or mitigating a specific social issue or seeking to achieve positive social outcomes especially but not exclusively for a target population or
- Improving the quality of life of the society

3 Brief Overview on the Legal and Regulatory Framework of SRI *Sukuk* Issuance in Malaysia

As a crucial component of the Malaysian government's strategic plan to market as a primary centre for sustainable finance within ASEAN, the SC introduced the SRI Sukuk framework in 2014. This framework aligns with both Malaysian and global endeavours to establish an environment conducive to sustainable and responsible investing, catering to the needs of SRI investors and issuers (SC, 2020).

The SRI Sukuk Framework emerged subsequent to the ICMA Green, Social, and Sustainable Bond Principles/Guidelines. In January 2021, the Securities Commission expanded the SRI Sukuk and Bond Grant Scheme, covering all Sukuk and bonds issued under the SRI Framework and ASEAN Green, Social, and Sustainability Bond standards. Notably, these pivotal initiatives have prompted 99 of the top 100 companies to release sustainability reports, indicating a growing recognition among local businesses regarding the significance of integrating Environmental, Social, and Governance (ESG) considerations into corporate planning. Additionally, in 2010, the Ministry of Finance introduced the Green Technology Financing Scheme (GTFS), serving as a significant catalyst for accelerating green investments. The GTFS provides a 60% government guarantee on the financing amount and a 2% interest/profit rate rebate by financial institutions. In its latest iteration, GTFS 2.0, a total of RM2.0 billion was allocated by the end of 2020, with a similar allocation for GTFS 3.0 projected to generate RM4.0 billion in green investments and facilitate the creation of 2500 job opportunities.

Bank Negara Malaysia, in alignment with EU categorization, has divided eligible green investments into several categories. This classification assists institutions in identifying economic activities that contribute to climate change objectives. The Government of Malaysia (GoM) SDG Sukuk Framework articulates the Malaysian government's intent to utilize SDG Sukuk for funding projects that generate environmental and social benefits in harmony with Sustainable Development Goals

(SDGs). SDG Sukuk issued under this framework adheres to the Social Bond Principles (SBP), Green Bond Principles (GBP), and Sustainability Bond Guidelines (SBG) of the International Capital Markets Association (ICMA). Additionally, they align with the Social Bond Standards (SBS), Green Bond Standards (GBS), and Sustainability Bond Standards (SuBS) of the ASEAN Capital Markets Forum (ACMF).

3.1 The Main Elements in SRI Sukuk Framework

1. **Utilization of proceeds**
   Any Proceeds raised from SRI *sukuk* issuance must be used only to fund the activities or transactions relating to the eligible SRI projects, as described in paragraph 7.08
2. **Project evaluation and selection**
   Set out an internal process for evaluation and selection of the eligible SRI projects
3. **Management of proceeds**
   Proceeds assigned for the eligible SRI projects are placed and accredited into a designated account or otherwise tracked in an appropriate manner
4. **Reporting**
   Annual reporting for:
   (a) "The original amount allocated for the eligible SRI projects
   (b) Amount utilized for the eligible SRI projects
   (c) Unutilized amount and where such unutilized amount is placed or invested pending utilization and
   (d) List of the eligible SRI projects in which the proceeds have been allocated to as well as brief description of the said eligible SRI projects and their impact or expected impact"
5. **External review**
   "An external reviewer may be employed to assess and provide report on the eligible SRI projects or the issuer's compliance with the relevant requirements as set out in the SC's guidelines"

3.2 Incentives for the Investors to Issue SRI Sukuk

Investors who are willing to issue SRI *sukuk* will enjoy tax deduction on SRI issuance as following:

3.3 Legal and Regulatory Framework for the Issuance of Islamic **Sukuk** *in Malaysia*

Overseeing the Malaysian debt securities and sukuk market, the Securities Commission and Bank Negara Malaysia play a dual role in regulating intermediaries and market activities. Their responsibilities extend to supporting and developing a robust market infrastructure, while also promoting and facilitating an environment conducive to the market's effective functioning.

Established on January 26, 1959, Bank Negara Malaysia is the central bank of Malaysia with the following objectives:

1. Issuing currency and maintaining reserves, safeguarding the value of the currency.
2. Acting as a banker and financial advisor to the government.
3. Promote monetary stability and sound financial structure.
4. Impact the credit situation in favour of the country.

3.3.1 Securities Commission Malaysia

Established on March 1, 1993, under the Securities Commission Act 1993, the Securities Commission (SC) was formed with the mandate to foster fair, efficient, secure, and transparent stock and futures markets. Additionally, its role extends to facilitating the systematic growth of an innovative and competitive capital market. Operating as a self-funding legal entity, the SC possesses investigative and enforcement powers to uphold its regulatory responsibilities.

The many organizational functions undertaken by the Authority include (Bank Negara Malaysia and Securities Commission, 2009):

- "To advise the Minister on all matters relating to the securities and futures industries"
- "Regulation of matters relating to securities and futures contracts"

  As an approving authority;

- Acting as a registration authority for corporate publications other than unlisted leisure clubs
- Supervising the operations of the exchange, clearing and central depository services
- Regulating acquisitions and mergers of companies
- Ensuring the proper functioning of market institutions and authorized persons

3.3.2 Government Securities Market Regulation

Capital Markets and Services Act 2007 sets out the rules for the regulation of the securities market. Typically, dealers handling government securities are banking institutions that operate under the licensing and regulation of Bank Negara Malaysia. Moreover, commercial banks, Islamic banks, and investment banks participate in the interbank market for government securities. In a parallel context, Bank Negara Malaysia issues The Malaysian Code of Conduct that assumes a crucial role in establishing the optimal market practices, principles, and standards. The central bank is also responsible for issuing rules that govern the issuance, allotment, payment of interest, redemption, and settlement of securities under a fully automated Issuance/Tendering System (FAST[1]) and Real-time Electronic Transfer of Funds and Securities[2] (RENTAS).

3.3.3 Corporate Debt Securities and Corporate *Sukuk* Market Regulation

Since July 1, 2000, the Malaysian corporate debt securities and corporate sukuk market have solely been regulated by SC. It also administers and oversees the Capital Markets and Services in accordance with the 2007 Act (CMSA). In this capacity, the SC exercises control over operations in the domestic corporate debt securities and corporate sukuk market. The SC has actively pursued the establishment of a comprehensive disclosure-based regulatory system, evident in the issuance of guidelines governing the offering of private debt securities. Guidelines related to the issuance of corporate debt securities and corporate sukuk have been issued, along with the supervision of trading activities in the secondary market. The SC also engages in collaborative assessment and inspection initiatives for investment banks, all of which are conducted in cooperation with Bank Negara Malaysia. Thus, the general purpose of setting up the guidelines for Islamic *sukuk,* i.e. its issuance, is to organize the issuance of *sukuk* and to clarify the general requirements in the framework of preparation and issuance, and the responsible parties (Bank Negara Malaysia and Securities Commission, 2009).

[1] A fully automated Issuance/Tender (FAST) system was first launched by Bank Negara Malaysia (BNM) in September 1996 to automate the bidding procedure for Government Securities/Bank Negara Papers, which are issued through the PD Network. FAST is expanded to include commercial papers and medium-term notes that are issued through tender / private investment (coordination with private investors without the need to issue sukuk to the public)

[2] Real-time Electronic Transfer of Funds and Securities System

3.4 Malaysian Guidelines on Unlisted Capital Market Products Under the Lodge and Launch Framework SC-GL/4-2015 (R7-2020)

To establish a robust capital market, the SC issued guidelines for unlisted capital market products that became effective as of June 15, 2015, to govern the preparation and issuance of such products. These guidelines specify the requirements that must be followed to make capital market products that are unlisted available to (SC, 2020):

(a) Investors in Malaysia or
(b) Individual investors from abroad

These guidelines consist of the following four sections:

Section A: determines the "general requirements on the lodge and launch framework that apply to all unlisted capital market products and responsible parties"
Section B: determines the "specific requirements for each type of unlisted capital market products' structure, the roles and responsibilities of the responsible party and their continuous obligations, and revision to information lodged"
Section C: determines the "additional requirements that apply to Shariah-compliant unlisted capital market products, as applicable" and
Section D: determines the "transitional provisions that apply to unlisted capital market products which have been approved or authorized by the SC prior to the Lodge and Launch framework"

3.4.1 Parties in a Debt Securities or *Sukuk* Issue Instruments

Lead Arranger/Principal Advisor, Advisor, Underwriter, Facility Agent, Paying Agent, Legal, Counsel, Shariah Advisor, Trustee, Credit Rating Agency, Financial Guarantee Institution.

(a) **The Malaysian Debt Securities and *Sukuk* Market IN**
"Corporate or Private Debt Securities, Debt Securities Listed and Traded on Bursa Malaysia, Government and BNM Securities, Listed Debt Securities and *Sukuk* under Exempt Regime, Futures Contracts".
(b) **Issuing Debt Securities and *Sukuk* in Malaysia**
"Requirements for a Ringgit-denominated Issue, Additional Requirements for a *Sukuk* Issue, Requirements for a Foreign Currency-denominated Issue, Operational Requirements, Listing of Debt Securities and *Sukuk* under Exempt Regime".

The issuance of Islamic securities or *sukuk* falls within the ambit of the Guidelines on the Offering of Islamic Securities (IS Guidelines). For these issues, an issuer must appoint either of the following:

(a) "An independent Shariah advisor approved by the SC, and that meets a specific criterion"
(b) "The Shariah Committee of an Islamic bank or a licensed institution approved by Bank Negara Malaysia to carry out Islamic Banking Scheme, to advise on all aspects of Islamic securities, including documentation, structuring, investment, and other administrative as well as operational matters in relation to Islamic securities"

3.5 The Facilities Provided by the Malaysian Government to Develop and Increase Islamic Sukuk *Issuance*

In addition to the strong regulations and conductive environment which have been issued by the Malaysian government, the Malaysian government has provided facilitative issuance framework, incentives, and tax exemptions, which are considered one of the most important motives that helped to increase the *sukuk* issuances in domestic corporations and attracted a large number of issuances from foreign corporations. The following are some of these incentives (Ibid, 2009):

3.5.1 Facilitating Regulatory Issues

Malaysia has sought to provide convenient regulatory solutions by establishing a sturdy issuance and trading foundation for *sukuk* in order to stimulate the issuance of *sukuk* (Ibid, 2009, Appendix 4):

1. Malaysia worked to provide a set of solutions in line with the types of *sukuk* by providing a basis for issuance and trading of *sukuk* characterized by:
   Sukuk are Islamic tools that can be trusted—A strong legal and regulatory structure—Its subject to the provisions and principles of Shariah—Availability of specialized and trained human resources.
2. Malaysia has worked to provide the tools of the Islamic financial infrastructure industry (such as Islamic banks, Takaful insurance, Islamic money market, Islamic financial market), which supports and promotes the *sukuk* market, also the presence of multiple parties in the Islamic financial system helps to manage the issuance and distribution of *sukuk*. The presence of a wide range of investors in addition to the innovation element helps to increase demand for *sukuk*.
3. Provided hedging and risk management tools such as Islamic futures contracts.
4. Malaysia has established an information centre about the Malaysian financial market, similar to the Malaysia International Financial Centre.
5. Facilitated the dealing and trading procedures in the stock market, through the following information systems:
   (a) RENTAS: Real-time Electronic Transfer of Funds & Securities

(b) FAST: Full Automated System for Issuance/Tender in (Bond)
(c) Information and Dissemination System (BIDS)

Sukuk Market Establishment
The SC has established a special Islamic *sukuk* market, in which *sukuk* are traded whether it's issued locally or from external parties, such as governments and foreign companies. Additionally, this market has been established to provide information related to new issuances and *sukuk* prices to be traded among the investors as well as to be independent of the other stock markets whether Islamic or conventional.

3.6 Tax Incentives

Reducing the tax burdens for the investors, issuers, brokers, and market workers has contributed significantly to the growth of the issuance of *sukuk* in the Malaysian *sukuk* market, and that was as follows:

3.6.1 Tax Incentives for Issuers (Angdon, 2002)

- "Issuance for PDS is exempted from stamp duty".
- "All instrument related to issuance and transfer of PDS approved by the Securities Commission are exempted from stamp duty with effect from 1 July 2000".
- "All instruments related to *sukuk* issuance approved by the SC are exempted from stamp duty".
- Taxes deduction on *sukuk* issuance expenses incurred up to the assessment year MASTER PLAN (2010).
- "Tax-neutral for the originator and SPV in a securitization", the tax is not applied to the issuers and the SPV in case of securitization.
- Extension of tax incentives for *sukuk*, postponing the payment of reduced taxes on the *sukuk* issued based on *Musharaka*, *Mudaraba*, Leasing, and *Istisna*, until the evaluation year in 2010.
- Tax deductions on the special purpose vehicle, which is established in purpose to issue *sukuk*, by exempting SPV from the income tax, effective 2007. In this case, SPVs established for Islamic financing objective are not required to comply with the administrative requirements under the Income Tax Act 1967.

3.6.2 For Investors

- "Resident individual's unit-trust companies and listed closed-end fund companies are exempted from income tax on interest income and profits for *sukuk* issued in the local currency (ringgit), sovereign bonds, as well as PDS and *sukuk* approved by the SC".

- "Non-resident investors are exempted from income tax on interest income and profit earned of sovereign bonds from ringgit-denominated debt securities issued by the Malaysian Government as well as PDS or *sukuk* approved by the SC".
- "Exemption on interest income from foreign currency-denominated *sukuk*".
- Resident investors are exempted from paying taxes on income earned from profits obtained from *sukuk* issued in foreign currency within Malaysia.
- Non-resident investors and *sukuk* holders issued in foreign currency within Malaysia, their profits and incomes are exempt from paying tax.
- There are no stamp rights when transferring the *sukuk*, in the *sukuk* approved by the Malaysian Securities Commission, Malaysia is also distinguished by the fact that it does not impose taxes on capital gains.

3.6.3 For the Brokers

Tax exemptions are not limited to investors and issuers only, but also brokers included. There are tax exemptions for financial brokers working in the stock exchange. In this regard, Malaysian government agreed, in its 2003 budget, to give tax incentives with regard to structuring processes and preparing documents necessary for the issuance of sukuk. The Malaysian Minister of Finance also stated, regarding the 2004 budget, that they worked on putting several tax measures to provide necessary facilities to encourage local issuers to offer issuances compliant with Shariah. Thus, ensure the tax treatment of conventional bonds is similar to sukuk. Therefore, Malaysian government worked to remove the tax obstacle that prevents the development and growth of sukuk. The most important incentives for the brokers (SC, 2002):

- Exemption from fees payable on trading sukuks, guaranteeing risks, and distributing sukuks issued in foreign currencies (other than the ringgit) outside Malaysia.
- Exemption from tax on profits obtained from bonds issued in foreign currency.

Malaysia have developed a strong and solid regulations for sukuk issuance, for instance, comprehensive regulatory framework that regulates the issuance process, as well as many facilities and tax exemptions which encourage the issuance of sukuk.

4 Conclusion

Malaysia has always been keen to create an easy and solid environment for the issuance of *sukuk*. This environment is represented by Islamic banks, Takaful (insurance), the Islamic money market, and the Islamic financial market. Malaysia also has a strong base for issuing and trading *sukuk*, which is characterized by: they are Islamic instruments that can be trusted, have a solid legal and regulatory framework, are subject to the provisions and principles of Shariah, provide specialized and trained human competencies, and provide hedging and risk management tools such as Islamic futures contracts. This strong legal and regulatory framework has enabled Malaysia to issue any type of *sukuk*, including SRI *sukuk*. The strength and durability of the legal and regulatory framework, as we mentioned earlier, is evident through:

1. Bank Negara Malaysia (BNM)
2. Securities Commission Malaysia
3. Islamic money market and Islamic capital market
4. Regulating corporate debt securities and the corporate *sukuk* market

Bank Negara Malaysia is the central bank of Malaysia with an important role in acting as a banker and financial advisor to the government and promoting monetary stability in the country. Bank Negara Malaysia issues rules and guidelines governing the issuance, allocation and payment of interest, redemption and settlement of non-performing securities under a fully automated system of issue/tendering (FAST), Fully Automated System for Issuance/Tendering (Bond), (RENTAS), Real-time Electronic Transfer of Funds & Securities, and Information and Dissemination System (BIDS).

Consequently, Malaysia has developed strong and solid regulations for sukuk issuance, the comprehensive regulatory framework that regulates the issuance process, many facilities, and tax exemptions to encourage the issuance of sukuk. This comprehensive framework has paved the way towards sustainable responsible investment. Malaysia currently has the largest market for sustainable responsible investment sukuk. Malaysia has set a strong framework for issuing SRI *sukuk* characterized with utilization of proceeds, project evaluation and selection, management of proceeds, reporting and external review. The (SRI) *Sukuk* framework came after the ICMA Green, Social, and Sustainable Bond Principles/Guidelines. Currently, in January 2021, the Securities Commission developed and expanded the SRI *Sukuk* and Bond Grant Scheme to cover all *Sukuk* and bonds issued under the SRI Framework and ASEAN Green, Social, and Sustainability Bond standards.

Malaysia is the key factor in providing enabling environment for issuing Sukuk aligned with SDGs such as SRI sukuk and SDG Sukuk. For example, Bursa Malaysia's Sustainability Reporting Framework and the environmental, social, and governance (ESG) index FTSE4Good promoting financing SRI projects through issuing SRI sukuk will meet the United Nations' Sustainable Development Goals (SDGs), in order to attract more investors to subscribe to SRI sukuk, SC Malaysia is determined that SRI sukuk should be highly mandated to invest in green, social, sustainable, and waqf. Moreover, the government of Malaysia (GoM) provides tax incentives for SRI sukuk issuer, investors as well as brokers. Since reducing the tax burdens for the investors, issuers, brokers, and market workers will contribute significantly to the growth of the issuance of *sukuk* in the Malaysian *sukuk* market. For instance:

- Taxes deduction on sukuk issuance expenses incurred up to the assessment year MASTER PLAN (2010).

- "Tax-neutral for the originator and SPV in a securitization", the tax is not applied to the issuers and the SPV in case of securitization.
- Extension of tax incentives for sukuk, postponing the payment of reduced taxes on the sukuk issued based on Musharaka, Mudaraba, Leasing, and Istisna, until the evaluation year in 2010.
- Tax deductions on the special purpose vehicle, which is established in purpose to issue sukuk, by exempting SPV from the income tax, effective 2007. In this case, SPVs established for Islamic financing objective are not required to comply with the administrative requirements under the Income Tax Act 1967.
- The SRI sukuk issuer will enjoy tax deductions on SRI sukuk issuance such as:
  - Grant: Green SRI Sukuk Grant Scheme which extended to issuers to fund an external review cost relating to the green SRI sukuk issuance under the framework.
  - Claim: Issuer can claim for the Grant based on an issue or programme.
  - Should there be more than one external review, the issuer is able to claim on that review cost incurred.
  - Claim amount 90% of the actual external review cost subject to a maximum of RM300,000.

Finally, SDG Sukuk Framework explains the Government of Malaysia ("GoM") intention to utilize the SDG Sukuk to fund projects that will generate environmental and social benefits which are in close alignment with SDGs. SRI sukuk aligned not only with SDGs but also with *Maqasid* al Shariah, *Maqasid* al Shariah aims to preservation and safeguarding of faith (*din*), human life- self- (*nafs*), intellect (*'aql*), posterity (*nasl*), and wealth (*mal*). Thus, *Maqasid* al Shariah is all about promoting benefits and repelling harms, so *Maqasid* al-*Shari'ah*, SDGs as well as eligible SRI projects all have common goals that calls for establishment of justice, elimination of unfairness and alleviation of destitution.

References

Akin Gump. (2021, October 12). *ESG Sukuk: A next step in the evolution of the global sukuk market, environmental, social and governance alert*. Akin Gump Strauss Hauer & Feld LLP 2.

Alawode, A., Hafiz, A., Azizana, A., & Aviles, M. (2018). Sustainable development goals and the role of Islamic finance. *World Bank Blogs*. Retrieved from https://blogs.worldbank.org/eastasiapacific/sustainable-development-goals-and-role-islamic-finance.

Angdon, N. (2002). *Sukuk development in Malaysia*. Research presented to the symposium on Islamic Sukuk—The Malaysian experience—King Fahd Convention Center, Saudi Arabia (p. 22).

Bank Negara Malaysia and Securities Commission. (2009). Malaysian Debt Securities and Sukuk Market: a guide for issuers & investors. Op Cit, Appendix 4.

bixmalaysia.com. (2020). SRI Sukuk: The journey towards sustainable and responsible investment. Retrieved from https://www.bixmalaysia.com/Learning-Center/Articles-Tutorials/SRI-Sukuk-The-Journey-Towards-Sustainable-Respo

Centre for Islamic Wealth Management. (2015). *Special edition 2013-2015 bulletin*. Centre for Islamic Wealth Management (Ed.). INCEIF and BNP Paribas.

CIMB. (2015). *Ihsan Sukuk Berhad: Information memorandum*. CIMB.

CIMB. (2019). CIMB allocates RM100 million for renewable energy financing for SMES. Retrieved from https://www.cimb.com/en/newsroom/2019/cimb-allocates-rm100-million-for-renewable-energy-financing-for-smes.html.

Eliktisad.com. (2015). اصدارات-الصكوك-الاسلامية-ستشهد-انخفاضا-حادا Retrieved from www.eliktisad.com/news/show/183981/2015-اصدارات-الصكوك-الاسلامية-ستشهد-انخفاضا-حادا

Ghani, M. I. (2015). *The world's first ringgit-denominated SRI Sukuk*. Thomson Reuters-RFI Responsible Finance Report.

GIFIIP. (2020). Turkey Issues its First Sustainable Sukuk. Retrieved from http://www.gifiip.org/turkey-issues-its-first-sustainable-sukuk/.

Global Islamic Economy Gateway (2017). Malaysia first in region to launch ASEAN Green Bond Standards. Retrieved from https://www.salaamgateway.com/en/story/malaysia_first_in_region_to_launch_asean_green_bond_standards-SALAAM08112017070919/

Global Sustainable Investment Alliance (GSIA). (2020). Global sustainable investment review 2020.

HSBC. (2020). HSBC Amanah arranges first ASEAN green sustainable and responsible investment (SRI) Sukuk for Leader Energy The debut sukuk of MYR260 million for Leader Energy will fund two solar photovoltaic power projects in Kedah. Retrieved from file: https://www.about.hsbc.com.my/-/media/malaysia/en/news-and-media/200812-hsbc-amanah-arranges-first-aseangreen-sri.pdf.

IIFM. (2021). Sukuk report, July 2021 | 10th Edition.

Islamic Development Bank Issues Largest Sustainability Sukuk (IsDB). (2021). https://www.isdb.org/news/islamic-development-bank-issues-largest-sustainability-sukuk-ever

Islamic Financial Services Board (IFSB). (2007). Capital adequacy requirements for Sukuk securitizations and Real Estate Investments draft, Dec.4/2007 (p. 3).

Joan, J., & Thomas, D. B. (2015). Socially responsible investing: A review of the critical issues. *Journal Managerial Finance, 41*(11), 1176–1201.

Khan, F. (2020). Sustainable development for real economy: Some lessons from the 12th ICIEF, Daily Sabah. Retrieved from https://www.dailysabah.com/opinion/op-ed/sustainable-development-for-real-economy-some-lessons-from-the-12th-icief.

Moghul, U. F., & Safar-Aly, S. H. K. (2014). Green Sukuk: The introduction of Islam's environmental ethics to contemporary Islamic finance. *The Georgetown International Environmental Law Review, 27*(1), 20–23.

Rafay, A., Saidq, R., & Ajmal, M. (2017). Uniform framework for sukuk al-Ijarah—A proposed model for all madhahib. *Journal of Islamic Accounting and Business Research, 8*(4), 420–454.

SDG Investment Trends Monitor. (2023). The annual SDG investment gap in developing countries. UNCTAD. Retrieved from https://unctad.org/system/files/official-document/diaemisc2023d6_en.pdf

Securities Commission Malaysia (SC). (2002). Securities Commission annual report 2002, Op Cit (p. 2.28).

Securities Commission Malaysia (SC). (2014). *SC introduces sustainable and responsible investment Sukuk framework*. Securities Commission Malaysia.

Securities Commission Malaysia (SC). (2017, July 27). Malaysia's first green Sukuk under SC's sustainable responsible investment Sukuk framework, Kuala Lumpur. Retrieved from https://www.sc.com.my/post_archive/malaysias-first-green-sukuk-under-scs-sustainable-responsible-investment-sukuk-framework/.

Securities Commission Malaysia (SC). (2019). Sustainable and responsible investment Sukuk framework an overview.

Securities Commission Malaysia (SC). (2020). Guidelines on unlisted capital market products under the lodge and launch framework SC-GI/4-2015 (R7-2020) (pp. 2–3, 69–74).

The Government of Malaysia SDG Sukuk Framework. (2021). Retrieved from https://www.mof.gov.my/pdf/ekonomi/sustainability/sukuk/

The Star Online. (2015). Khazanah's RM100m sustainable Sukuk priced at 4.3%. Retrieved June 4, 2015, from https://www.thestar.com.my/Business/Business-News/2015/06/04/Khazanah-RM100m-sustainable-Sukuk/?style=biz

Türkiye Sınai Kalkınma Bankası (TSKB). (2021). TSKB arranged the first Sukuk issuance according to UN sustainable development goals framework. Retrieved from https://www.tskb.com.tr/web/307-4715-1-1/tskb-site-en/en-hakkimizda/tskbden-haberler-en/tskb-arranged-the-first-sukuk-issuance-according-to-un-sustainable-development-goals-framework .

Wahab, M. Z., & Naim, A. M. (2019). Malaysian initiatives to support sustainable and responsible investment (SRI) especially through Sukuk approach. *Journal of Emerging Economies & Islamic Research, 7*(3), 44–54.

Yıldırım, S., Yıldırım, D. C., & Diboglu, P. (2020). Does Sukuk market development promote economic growth? *PSU Research Review, 4*(3), 209–218. Retrieved from https://doi.org/10.1108/PRR-03-2020-0011

Zain, N. S., & Sori, Z. M. (2020). An exploratory study on *Musharakah* SRI Sukuk for the development of Waqf properties/assets in Malaysia. *Qualitative Research in Financial Markets, 12*(3), 301–314.

Zolfaghari, P. (2017). *An introduction to Islamic securities (Sukuk)* (Working Paper 2017:2). Uppsala Faculty of Law.

Dr. Omar Kachkar, is a Trainer and Islamic finance expert with over 15 years of work experience as a consultant and academician in English and Arabic. Registered Shariah Advisor at Securities Commission Malaysia. Associate of Association of Shariah Advisors in Islamic Finance – ASAS. Certified Zakah Accountant, and Certified Shari'ah Advisor and Auditor (CSAA) of AAOIFI. Assistant professor at Ibn Haldun University, deputy director of Ibn Haldun University Center for Islamic Economics and Finance.

Marwa Al Fares is currently a PhD candidate at Istanbul Sabahattin Zaim University in Islamic Economics and Finance. She holds a Master's degree from the University of Malaya in Kuala Lumpur. She has an excellent background in Islamic law with excellent command of the Arabic and English languages. Marwa is equipped with excellent research skills. Her research interests include Waqf, Zakat, social finance, Green Sukuk, alternative finance, and Credit Guarantee.

Proposing Strategic Maqashid Management Framework (SMMF) to Sustainable Islamic Business: Integrating Maqashid Principles with Strategic Management Theories

Muhammad Majdy Amiruddin

Abstract Strategic Maqashid Management Framework (SMMF) is a comprehensive model designed to integrate Maqashid principles with established strategic management theories. This study is library research. This framework encompassed key components, derived from the exploration of key Maqashid principles by Ibn Asyur, including Al Rawaj, Al Wuduh, Al Hifz, Al Tsabat, and Al Adl. This framework also integrated SWOT Analysis and VRIO as established strategic management. This paper proposed the applicability of the mentioned theories in the context of Maqashid-based strategic management, such as Circulation Excellence Alignment (CEA), Circulation Resource Optimization (CRO), Transparency Enhancement Strategy (TES), Transparent Resource Advantage (TRA), Preservation Value Maximization (PVM), Sustainable Preservation Edge (SPE), Legal and Ethical Strategic Alignment (LESA), Legitimacy Resource Integration (LRI), Justice Enhancement Initiative (JEI), and Equitable Advantage Framework (EAF). SMMF is anticipated to play a substantial role in assisting organizations in attaining a harmonious sustainability by integrating economic objectives with the preservation of ethical and Sharia principles. This research significantly advances the comprehension and enhances the implementation of Maqashid in the realm of strategic management.

1 Introduction

Strategic management is crucial for the smooth operation of a business (Von Pock, 2007). However, traditional techniques have not entirely adapted to incorporate the features of Sharia, particularly in Islamic finance institutions. The importance of a

M. M. Amiruddin (✉)
Institut Agama Islam Negeri Parepare, Parepare, South Sulawesi, Indonesia
e-mail: muhammadmajdyamiruddin@iainpare.ac.id

A. Rafiki et al. (eds.), *Strategic Islamic Business and Management*, Contributions to Management Science,
https://doi.org/10.1007/978-3-031-61778-2_10

business continuity based on the principles of Maqashid al-Shariah has increased in the economic and social growth of Islamic civilization (Erba & Nofrianto, 2022). Although there has been extensive research on strategic management at Islamic institutions, a thorough understanding of how to integrate Sharia principles into strategic management practices has not yet been achieved. Socioeconomic disparity necessitates the creation of a novel framework that integrates these two parts, merge the second element with the existing one.

Upon reviewing the literature, it is evident that numerous studies have examined various aspects of strategic management in the framework of Sharia. However, studies still tend to focus on specific elements, such as the implementation of Sharia policies in financial institutions or the measurement of company performance based on Maqashid. A study conducted by Priyatno et al. (2022) on the performance of Islamic banks in Indonesia demonstrated the utilization of an approach for measuring performance. It demonstrated that utilizing the Maqashid Sharia Index approach for performance evaluation offers a comprehensive assessment of the extent to which Sharia banks may attain Sharia goals and principles. Pratiwi et al. (2023) conducted a study to evaluate the dedication of Sharia banking using the Shariah Maqashid Index. Their research revealed that Islamic banking has not completely met the Sharia objective of synthesis. However, their study only provided a partial description of the connection between Maqashid and management practices. Erba and Nofrianto (2022) had undertaken research that elucidated the effectiveness of applying Maqashid Syariah in the banking, cooperative, and Sharia insurance sectors. However, a comprehensive synthesis of this information was still absent. However, Muna and Hapsari's research (2015) demonstrated the effectiveness of BMT Amanah Ummah in implementing sharia strategic management. In contrast, Toha and Rozikin (2020) emphasized the shortcomings of conventional strategic management concepts and suggested an alternative approach by implementing Maqashid al-Shariah. However, a thorough synthesis that integrated the ideas of Maqashid al-Shariah with a validated strategic management model had not been developed yet.

The objective of this research is to develop a Strategic Maqashid Management Framework (SMMF), which is a comprehensive framework that merges essential components from Maqashid with established strategic management theories. This research addresses a gap in the existing literature by developing a model that integrates Maqashid with strategic management. Additionally, it offers a fresh perspective on how these two concepts might be integrated. SMMF is anticipated to play a substantial role in assisting organizations in attaining a harmonious sustainability by integrating economic objectives with the preservation of ethical and Sharia principles.

This research makes a significant contribution by not only offering a practical framework but also positively influencing the literature and practice of strategic management. SMMF provides practical solutions to address current deficiencies, establishing a solid basis for using Maqashid concepts in strategic management. Furthermore, this study presents a compelling case for the significance of incorporating Maqashid al-Shariah into strategic management as an essential measure toward achieving sustainable and ethically equitable Islamic business practices. Therefore, this research significantly advances the comprehension and enhances the implementation of Maqashid in the realm of strategic management.

2 Literature Review

2.1 Islamic Strategic Management

The term "strategy" originates from the Greek word "stratēgos," which is associated with the concept of planned activity to achieve one or more objectives within an organization. This is also regarded as a strategic approach to achieving "harmony" between the internal and external environments of a business (Assauri & Assauri, 2011). According to Muafi (2021a), the strategy takes into account the planning and emphasizes the significant relevance between the internal and external environments of the firm. The interpretation of the term "strategy" in the field of management varies depending on the perspective. Different strategies are employed in field management, and there are diverse interpretations of what the term "strategy" means. Strategy is perceived as a multifaceted, ever-changing, and adaptive process, wherein a corporation may encounter errors and then learn from them. Strategy is commonly understood as the allocation of resources and capabilities by a corporation to specified areas and sources. The power was compelled to act in accordance with the essential conditions. In order to get a higher level of performance and quality that is more competitive. Effective for organizational purposes, the manager will identify and outline potential strategies implemented through the use of strategic management approaches. Strategic success management is a field that is based on core principles that allow for the consideration of different perspectives while still maintaining consistent qualities (Muafi, 2021a).

The objective of the strategy, as outlined in the study conducted by Hasim et al. (2019), is to anticipate and consider future possibilities in a futuristic framework where a comprehensive understanding of the strategy is necessary for evaluating both external and internal factors in the industry. According to research, the examination of the external environment involves two components: the macro environment and the micro environment. The macro environment refers to numerous factors that have an impact on the entire industry, while the micro environment refers to the specific industry in which the organization competes. The second environment needs to be examined in order to identify the elements that will impact the success of the organization, including developments and projections. In order to achieve goals and objectives in a strategic organization, it is necessary to examine the second environment before establishing a functional plan (As-Shodiq, 2020). The formation of corporate strategy is contingent upon the efficacy of the environmental scanning process and the understanding of macro and micro elements affecting organizations (Assauri & Assauri, 2011). In order to accomplish outstanding strategic system management, every firm, including the industry of tourism and hospitality, should build and assess both the external and internal environments with proficiency. This information is crucial for establishing a foundation for strategic objectives, goals, and company strategy. The conclusions indicate that strategic management is a continuous process that involves analyzing and overseeing the markets in which a company operates. It entails evaluating competitors, setting objectives, and devising strategies to outperform both current and potential rivals. Additionally, periodic assessments are conducted to determine the effectiveness of the strategy and whether it needs to be replaced. Strategic management involves the formulation and communication of the strategy that managers should implement in order to attain superior performance and competitive excellence for their firm. An organization claims to have a competitive advantage if its profitability is higher than the average profitability in the industry. Innovation, rapid technological advancements, and customer expectations drive companies to strategically consider and make smart decisions for continued success (Poister & Streib, 1999; Sujova & Rajnoha, 2012).

Islamic strategic management, as defined by Muafi (2021b), refers to a set of processes in Islamic management that involve formulating, implementing, and evaluating strategic decisions of an organization. These decisions aim to achieve both worldly and spiritual organizational goals. Islamic strategic management possesses four important qualities that set it apart from traditional (non-Islamic) strategic management. The four characters are evaluated based on many aspects: The four components are: (1) Principles, (2) Orientation, (3) Motivation, and (4) The method itself.

When organizations/companies use Islamic strategic management, they must establish their vision, mission, and goals based on the premise of monotheism/divinity. Their orientation should encompass both worldly and spiritual aspects, and their motivation should be driven by the desire for the blessings and pleasure of Allah SWT. The article authored by Muafi (2021a) provides a definition of Islamic strategic management as a set of processes involving the formulation, implementation, and evaluation of strategies to achieve organizational goals. In this context, Islamic values serve as the fundamental basis for all organizational activities, characterized by the principles of monotheism, worldly ukhrawi orientation, and mardatillah motivation. Sharia-compliant management apps will differ significantly from traditional strategic management that does not adhere to Islamic principles. These distinctions are evident in the process of defining vision, mission, and objectives, assessing the internal and external environments, conducting sharia analysis to guide strategy selection, and executing strategies.

Fitri et al. (2021) did a study that specifically examined the elements that contribute to the growth of income in rural areas. The strengths that have been identified include effective government policies and the strong dedication of the village head. On the other hand, the shortcomings that have been identified are insufficient sanctions and the leakage of user fees. Potential for growth exists in the progress of public services facilitated by information technology, but challenges arise from the issues of illicit garbage dumping and inadequate waste collection stations. Muna and Hapsari (2015) evaluated the effective implementation of Sharia strategic management at BMT Amanah Ummah. The study emphasized the efficient procedures involved in developing, executing, and assessing strategies. The process adheres to Fred R. David's philosophy, integrating Sharia principles into strategic management, with a foundation in Maqashid Sharia.

2.2 *Maqashid Al-Shariah*

The study of Maqashid Sharia is a relatively recent scientific discipline in the history of Islamic knowledge. It emerged as a result of the development of the scientific discipline of Usul al-Fiqh. However, the application of Maqashid Sharia has been practiced by the companions of the Prophet, the Tabi'in, and following generations up to the present day (Ningsih, 2021). The evolution of Maqashid Sharia may be categorized into three distinct periods. The first period involves the application of Maqashid Sharia by scholars, who utilize it to address various issues. During this time, scholars leverage current events or enact laws that promote previously undefined benefits. The usage of this practice is derived from the teachings of Rasulullah, S.A.W, as well as later generations, including scholars of madzhab. However, its foundation lies in the field of Ushul Fiqh, specifically in the context of maslhahat (benefit) and madharat (harm).

During the second phase, known as the codification period of Maqashid Sharia, its application expanded significantly for issuing fatwas, although it still remained a component of the field of Ushul Fiqh. In the third period, Maqashid Sharia emerged as an independent discipline. The term "The Maqashid of Sharia" was initially introduced by Abu Abdillah bin Ali al-Tirmidhi in his book, which elucidated the rationale for prayer (Firdaus, 2018):

The term "Maqashid Syariah" is derived from the word "maqsad," which refers to something aimed at or a goal to be achieved (Lifia & Rofiq, 2023). It can also be traced back to the word "qasada," which means toward, aim, desire, and deliberate (Lifia & Rofiq, 2023). In English, it is referred to as "ends," in Greek as "telos," in French as "finalite," and in German as "zweck." On the other hand, "kat al-Syariah" denotes the place where water flows. Terminologically, it refers to the objectives, principles, and advantages that are to be attained by the disclosure of sharia law on a global scale and in a comprehensive manner. Originally, the concept of Maqashid Syaruiah was proposed, but Jasser Auda rejected it, arguing that there is a distinction between Maqashid and the science of Ushul Fiqh. Ushul Fiqh primarily emphasizes the literal interpretation of texts, whereas Maqashid places greater emphasis on the underlying meaning of the texts. This viewpoint aligns with the opinion of Al-Tahir Ibn Ashur (Asyur, 1999).

Imam Syatibi believed that Allah revealed his Shari'a with the intention of promoting benefits and preventing evil. On the other hand, Ibn Asyur's perspective is that Maqashid al-Tasyri' al-ammah refers to the purpose and wisdom behind the Shari'a, which can be understood by observing the various conditions of the Shari'a, rather than being limited to a specific type of sharia law (Ningsih, 2021). The ultimate objective of Maqashid Syariah is to achieve the application of Shari'a in order to generate benefits and promote goodness for all beings on earth, hence leading to positive outcomes in the hereafter.

The information contained in the Al-Qur'an and Hadith is finite, whereas societal problems are always evolving and arising. As a result, there are instances where the textual guidance provided in the Al-Qur'an and Hadith may not explicitly address the specific problems that occur. To effectively address these issues, it is crucial to comprehend the underlying principles and essence of the text. The essence of Islam is embodied in the concept of Maqashid Sharia, which aims to achieve maslhahat as a legal objective (Maharani et al., 2022) Therefore, the presence of Maqashid Sharia is crucial when the laws in Islam cannot be derived through a linguistic analysis (Muawaffaq, 2021). Consequently, the discourse on Maqashid cannot be separated from the discussion on maslhahat, as the essence of Maqashid, according to Ibn Asyur, is elucidated.

The general objectives of Islamic law, known as Maqashid al-tasyri', refer to the underlying meaning and wisdom that the sharia makers possess. These objectives can be understood by carefully observing the various conditions of the Sharia, rather than being limited to a specific form of Sharia law (Fauzan, 2023). The ultimate objective of Maqashid Syariah is the attainment of benefit and goodness for all beings on earth by the application of Shari'a, resulting in subsequent benefit and goodness in the hereafter.

The information included in the Al-Qur'an and Hadith is limited, but societal problems are constantly evolving and rising. As a result, there are instances where the textual sources of the Al-Qur'an and Hadith do not explicitly address the problems that occur. To address these issues, it is crucial to comprehend the essence and underlying principles of the text. The essence of Islam is embodied in the concept of Maqashid Sharia, which aims to achieve maslhahat as a legal objective (Karim & Sahroni, 2015). The presence of Maqashid Sharia is crucial in cases where Islamic law cannot be derived solely from a linguistic analysis (Utama et al., 2019). Therefore, discussions on Maqashid cannot be detached from discussions on maslhahat, as welfare is an integral part of Sharia (Maharani et al., 2022).

The ulama classifies Sharia aims into three basic categories: urgent goals known as dharuriyat, goals that fulfill human daily necessities and requirements (hajiyat), and goals that are supplementary and complimentary (tahsiniyat). The essential needs, or dharuriyat, encompass the soul, religion, reason, property, and lineage. If these requirements are not fulfilled, difficulties may arise, such as challenges in terms of rukhshah law. Additionally, the complementary needs, or tahsiniyat, must align with moral values and principles (Asyur, 1999).

Al-Syatibi was credited with being the first person to systematically and methodically write about Maqashid Sharia in his work al-Muwafaqat Fi Usul al-Shariah. He was considered the pioneer and the foremost teacher in the field of Maqashid Sharia, often referred to as the father of Maqashid Shari'ah (al mu'allim al awwal) (Hafidz, t.t., p. 2). The work of Al-Muwafaqat by Al-Syatibi did not receive support from society due to the prevailing intellectual crisis during that time, compounded by the collapse of the city of Granada. However, this significant work started to be studied by the subsequent generation in 1884 AD. Later in the twentieth century, a prominent figure named Muhammad Thair ibn 'Asyur (1(Maharani et al., 2022) 879–1973 AD) emerged as the pioneer of contemporary Maqashid and is widely recognized as its founding father. Prior to Al-Tahir Ibn Asyur, several scholars, such as Imam Al-Juwaini, Imam Al-Ghazali, Imam Asy-Syatibi, Izzudin bin Abdussalam, Ibn Taimiyyah, and Ibn Al-Qayyim Al Jauziah, have contributed to the field of Maqashid Sharia. However, this study will focus solely on the ideas put forth by Imam Al-Juwaini and Imam Asy-Syatibi, without delving into the thoughts of the other scholars (Lifia & Rofiq, 2023).

Toha and Rozikin contended that traditional strategic management is insufficient, prompting the investigation of Maqashid Al-Shariah as a viable alternative. The research highlighted the inadequacy of traditional management in controlling behavior for the purpose of enhancing organizational performance, proposing that the adoption of Maqashid Al-Shariah is a more effective alternative. The study conducted by Erba and Nofrianto (2022) in *AL-FALAH Journal* demonstrated the effective application of Maqashid Syariah in Sharia-compliant commercial transactions, encompassing banking, cooperatives, and insurance. The study demonstrated the positive impact of Islamic banks on social welfare by adhering to Maqashid principles and avoiding any detrimental effects. Bakhri et al. (2022) argued in the *Tasharruf Journal* that Islamic financial institutions are motivated by spirituality and ethical principles based on Maqashid al-Shari'ah. The article establishes the correlation between these institutions and Maqashid Shariah, highlighting their contribution to the advancement of the halal business. The study conducted by Amalia and Mawasti (2022) in *Realita Journal* focused on the empowerment of a social enterprise called PT Kampung Marketerindo Empowered, within the context of Maqashid Shariah. The study unveiled novel strategies that are in line with Maqashid ideas, such as providing education and training to young individuals, establishing social companies, and facilitating the connection between skilled workers and urban firms. Hadi's (Hadi, 2019) research in the *I-ECONOMICS Journal* emphasized that Sharia cooperatives encompass six Maqashid, which contribute to the accumulation of capital, macroeconomic expansion, employment prospects, camaraderie among capital owners, organizational progress, and adaptable transactions. Salycasanda et al. (2022) investigated the execution of efficient zakat allocation by PZU KLP Pameungpeuk in the Bandung area. The research exhibited conformity with Maqashid Shariah principles, specifically in the preservation of religion, life, and wealth. The study conducted by Priyatno et al. (2022) in the *Journal of Scientific Islamic Economics* employed the Maqashid Sharia Index to evaluate the effectiveness of Islamic banks in Indonesia. The investigation revealed diverse levels of performance, underscoring the significance of attaining Shariah objectives and principles. The research conducted by Pratiwi et al. (2023) in the *International Journal of Current Science Research and Review* assessed the dedication of Islamic banks using the Shariah Maqashid Index. The findings indicated a deficiency in the dedication to carrying out Sharia tasks, indicating a requirement for more stringent adherence in corporate operations.

3 Methodology

This study used a qualitative research methodology to investigate the comprehension of concepts and the integration between Maqashid principles and strategic management theory. The study methodology encompasses several phases in the creation of the Strategic Maqashid Management Framework (SMMF). The research primarily examines established strategic management theories and Maqashid concepts through literature analysis. Next, the subsequent stage entails conducting a SWOT analysis and utilizing the Vrio Framework to assess the organization's strengths, weaknesses, opportunities, and threats, as well as its precious, rare, inimitable, and valuable resources.

The development process of SMMF incorporates the incorporation of Imam Ibn Asyur's Maqashid Syariah doctrines, namely Al Rawaj, Al Wuduh, Al Hifz, Al Tsabat, and Al Adl, within the strategic management framework. The Maqashid approach incorporates essential components from established strategic management theories. During this stage, crucial players, including managers of Islamic boarding schools, teacher councils, and other relevant parties, are also engaged.

4 Results

This study examines the integration of Maqashid Ibn Ashur with the SWOT and VRIO analyses frameworks to construct the Strategic Maqashid Management Framework (SMMF). The main focus of the findings was on integrating the draft Maqashid in SWOT and VRIO analyses, resulting in the creation of SMMF as a holistic framework for strategic management.

During the SWOT analysis, each principle of Maqashid Ibn Ashur is assessed individually. Al Rawaj (Circulation) is enhanced by a culture of collaborative work and competent teamwork. However, it is necessary to identify and rectify any obstacles that hinder the circulation process in order to address its temporary weaknesses. Al Wuduh achieved robustness by enhancing transparency in reporting and communications. However, it requires the identification of weaknesses and the implementation of reform methods pertaining to transparency. Al Hifz, a preservation group, is seeking a financial opportunity that aligns with its core principles. However, it is currently facing a temporary threat to its continuity and must find a way to overcome it. Al Tsabat focuses on the legality of corporate operations, ensuring compliance with both legal and ethical standards. Additionally, it involves recognizing and developing solutions to address potential legal or ethical challenges. Al Adl, which refers to justice, is enhanced by the implementation of corporate ethics, fairness, and equitable distribution. Wealth is spread evenly, but it is important to identify and remedy any required weaknesses.

Within the VRIO Framework study, each principle analyzed by Maqashid Ibn Ashur is evaluated based on its worth and rarity. Al Rawaj (Circulation) establishes a strong reputation by effectively utilizing internal resources to gain a competitive advantage. Transparency, known as Al Wuduh, enhances the arrangement of marks by promoting clarity and openness. This quality is challenging to replicate, making it a key factor in achieving long-term supremacy and sustainability. Al Hifz (Preservation) enhances efficiency by conserving energy resources and promoting sustainable business practices, hence creating long-lasting excellence. Al Tsabat (Legality) enhances one's score by ensuring compliance with Islamic law and ethics, hence creating a competitive advantage. Al Adl, which means justice, enhances one's mark through practicing corporate fairness, hence creating a competitive advantage.

Aspect	Dimensions	Aspect	Description
Al Rawaj (circulation)	SWOT	Strength	Enhancement circulation is positive through culture work collaboration and team competent.
		Weakness	Identification and repair of obstacle circulation negative, like imbalance distribution not quite enough answers or weak internal communication.
Al Wuduh (transparency)	SWOT	Strength	Increase strength through transparency in reporting and communication, building trust holder interest.
		Weakness	Identification of weakness in transparency and strategies for improving it, for example, in reporting finance or communication external.
Al Hifz (preservation)	SWOT	Opportunity	Look for opportunity business in accordance with Al Hifz principles, such as investment in project preservation environment or continuity.
		Threat	Prevent threats to continuity organization by defending and maintaining mark ethics and morals are appropriate Maqashid.
Al Tsabat (legality)	SWOT	Opportunity	Explore opportunities legal and ethical businesses in accordance with Al Tsabat principle.
		Threat	Identification threat law or ethics and designing strategies for minimizing risk.
Al Adl (justice)	SWOT	Strength	Increase the strength organization through practice of business fairness and distribution of equal wealth.
		Weakness	Identification of weakness in justice organization and its improvement strategies, such as policy more compensation fair.
Al Rawaj (circulation)	VRIO	Mark	Build rotation mark positive with utilize source internal power increases efficiency and effectiveness organization.
		Scarcity	Evaluate whether circulation is positive difficulty imitated by competitors, creating superiority competitive.

Aspect	Dimensions	Aspect	Description
Al Wuduh (transparency)	VRIO	Mark	Increase mark organization through transparency in utilization source power and policy, help build reputation positive.
		Ability imitated	Ensure a level of high transparency difficult imitated by competitors, creating superiority sustainability.
Al Hifz (preservation)	VRIO	Mark	Increase mark organization with conserve source power and engage practice sustainable and friendly business environment.
		Organization	Organize and manage source power in a way ethical in accordance with the principles of Al Hifz, creating superiority sustainable.
Al Tsabat (legality)	VRIO	Mark	Increase mark organization with ensure that source business power and strategy in accordance with Islamic law and ethics.
		Ability imitated	Create strategies and policies difficult law imitated by competitors, creating superiority competitive.
Al Adl (justice)	VRIO	Mark	Increase mark organization through practice of business fairness and distribution of equal wealth.
		Scarcity	Evaluate the extent of justice in business strategy difficult imitated by competitors, creating superiority competitive.

The analysis of Maqashid Ibn Ashur demonstrates the evident incorporation of SWOT and VRIO. Al Rawaj (Circulation) combines the identification of good aspects and impediments in circulation, whereas VRIO Integration aims to maximize the support of power circulation through source integration. A comparable procedure examines the fundamentals of other practices, such as Al Wuduh, Al Hifz, Al Tsabat, and Al Adl.

Aspect	SWOT integration	VRIO integration
Al Rawaj (circulation)	Identification circulation internal positivity (strength) and potential obstacle circulation (weakness). **Circulation Excellence Alignment (CEA)**	Maximize its source supporting power circulation positive and giving superiority competitive. **Circulation Resource Optimization (CRO)**
Al Wuduh (transparency)	Upgrade transparency in reporting and communication as strengths, identify and overcome weakness transparency. **Transparency Enhancement Strategy (TES)**	Take advantage level of high transparency as a source of tough power imitated. **Transparent Resource Advantage (TRA)**
Al Hifz (preservation)	Identification and conservation potency mark continuity as strength, and respond to threat to continuity as weakness. **Preservation Value Maximization (PVM)**	Consider continuity as mark added and difficult imitated. **Sustainable Preservation Edge (SPE)**

Aspect	SWOT integration	VRIO integration
Al Tsabat (legality)	Take advantage of opportunity for legal and ethical business as strengths, identify and respond to threat law as weakness. **Legal and Ethical Strategic Alignment (LESA)**	Design appropriate strategies and policies with Islamic law as superiority difficult imitated. **Legitimacy Resource Integration (LRI)**
Al Adl (justice)	Maximize justice in business strategy as strengths, identify and improve weaknesses in justice distributive. **Justice Enhancement Initiative (JEI)**	Consider justice as a source of tough power imitated. **Equitable Advantage Framework (EAF)**

The Strategic Maqashid Management Framework (SMMF) integrates the study of outcomes. The principles of Maqashid Ibn Ashur, including Al Rawaj and Al Wuduh, were implemented through various sub-frameworks such as CEA, CRO, TES, TRA, PVM, SPE, LESA, LRI, JEI, and EAF. SMMF provides a detailed perspective on how Maqashid can be merged with SWOT and VRIO analysis to develop a comprehensive framework for strategic management.

Aspect	Draft	SWOT integration	VRIO integration	SMMF
Al Rawaj (circulation)	CEA and CRO	Identification circulation internal positivity (strength) and potential obstacle circulation (weakness).	Maximize its source supporting power circulation positive and giving superiority competitive.	Ensure rotation mark positive and optimal use of internal strength and optimize source power.
Al Wuduh (transparency)	TEST and TRA	Upgrade transparency in reporting and communication as strengths, identify and overcome weakness transparency.	Take advantage level of high transparency as a source of tough power imitated.	Increasing transparency as power, harness level transparency is difficult to imitate.
Al Hifz (preservation)	PVM and SPE	Identification and conservation potency mark continuity as strength, and respond to threat to continuity as weakness.	Consider continuity as mark added and difficult imitated.	Maintain mark continuity as power, creating mark add a difficult one imitated.
Al Tsabat (legality)	LESA and LRI	Take advantage of opportunity for legal and ethical business as strengths, identify and respond to threat law as weakness.	Design appropriate strategies and policies with Islamic law as superiority difficult imitated.	Utilize opportunity legal and ethical business as power, creating superiority difficult imitated.
Al Adl (justice)	JEI and EAF	Maximize its justice in business strategy as strengths, identify and improve weaknesses in justice distributive.	Consider justice as a source of tough power imitated.	Increase justice in business strategy as power, considering justice as a source of tough power imitated.

5 Discussion

This study makes a significant contribution by developing the Strategic Maqashid Management Framework (SMMF), which is a comprehensive framework that applies the ideas of Maqashid Ibn Ashur in the field of strategic management theory. This study incorporates SWOT (Strengths, Weaknesses, Opportunities, and Threats) and VRIO (Value, Rarity, Imitability, and firm) analyses, resulting in a comprehensive methodology that blends the internal and external strengths of the firm with ethical and sharia values.

The strategic management of Islamic companies has been previously examined in the literature, specifically in the book "Islamic Business and Performance Management" by Firdaus and Ahmad (2023), which emphasized the significance of the spiritual dimension. However, this research goes beyond by specifically explaining how Maqashid Ibn Ashur's concepts can be included in SWOT and VRIO analyses, which enhances the basis for making strategic decisions. The SMMF makes a significant contribution by offering a practical framework that can fill the gaps in the existing literature. This framework allows enterprises to effectively employ Maqashid ideas in their strategic decision-making processes. This offers the potential to establish Islamic enterprises that are not only financially viable but also ethically equitable and in compliance with Sharia norms.

The articles "SWOT Analysis of Benchmarking in the Islamic Finance Industry" by Mustafa Omar Mohammed et al. (2021) and "Islamic Financial Management Practices and Business Performance in Small Industrial Companies" by Sirat and Jannang (2022), and "Strategic Plan Analysis Using the Balanced Scorecard Framework at Yogyakarta Islamic Hospital PDHI" by Prihatmoko et al. (2022) provided valuable insights into the incorporation of Islamic values within the context of strategic analysis. Nevertheless, SMMF integrates these components, offering a robust basis for the strategic management of Islamic business. SMMF, or Strategic Management Framework, can be implemented in several industries, ensuring comprehensive and seamless strategic management practices.

This research distinguishes itself from "SWOT Analysis on Research Methodologies and Techniques Adopted in Islamic Economics and Finance" by Irfan Beik and Nursyamsiah (2021) by primarily concentrating on the development of a practical framework (SMMF) that can be effectively utilized by organizations. In contrast, the focus of the research lies more on the research methodology within the realm of Islamic economics and finance. This research conducted by the author contributes by offering a comprehensive and practical perspective on how the concepts of Maqashid Ibn Ashur can be incorporated into strategic management through SMMF, establishing a strong basis for a sustainable and socially equitable Islamic company.

Nevertheless, this research has not yet addressed the practical application of the Strategic Maqashid Management Framework (SMMF) through specific case studies. The limitations of SMMF pose challenges in assessing the efficacy and efficiency of integrating Maqashid concepts with strategic management approaches.

Moreover, this research lacks an effect analysis and specific outcomes derived from the application of SMMF in the chosen study scenarios. Assessing the success or failure of SMMF becomes challenging without a comprehensive grasp of tangible outcomes. A notable limitation of this research is the lack of emphasis on comparing the SMMF approach with conventional strategic management approaches. The lack of comparisons impedes comprehension of the distinct benefits that SMMF may possess in Maqashid integration, rendering it challenging to evaluate its additional worth. Moreover, this research has not explicitly defined the specific difficulties or areas for enhancement that may occur during or after the deployment of SMMF. The limitations of Maqashid ideas pose challenges for corporations in effectively integrating them with strategic management. Another shortcoming is the lack of a comparison of organizational performance prior to and following the introduction of SMMF. Assessing whether any increase or fall in outcomes can be directly due to SMMF implementation becomes challenging. Furthermore, this research has failed to establish a definitive framework or set of measures for accurately assessing the efficacy and efficiency of SMMF. The constraints of SMMF provide challenges for others who wish to implement or assess its application in their corporate context. For optimal results, it is advisable to adopt a more extensive approach for this research, which includes conducting real-life case studies, analyzing the impact, comparing it with traditional methods, and identifying potential obstacles or areas for enhancement that may develop.

6 Conclusion

Based on the study, it can be concluded that the Strategic Maqashid Management Framework (SMMF) was a comprehensive framework that integrates major Maqashid features and appliess them in strategic management theory. This study highlighted the importance of aligning strategic management with the concepts of Maqashid. It demonstrated the usefulness of integrating ethical ideals and Sharia objectives into corporate practices. Recommendation for an organization seeking to implement SMMF, aiming to strike a balance between economic objectives, ethical principles, and sharia compliance.

In order to further research, it is required to do additional exploration connected to the practical implementation of SMMF. This involves conducting concrete case studies to evaluate the impact and success of this framework. Considering the integration of social media marketing strategies in a more efficient manner within the context of regular business operations is a key area of focus for future research. Furthermore, the impact of SMMF implementation and framework changes on development company and environment is examined over an extended period of time in several research fields. Further exploration of the external factors is also essential. Therefore, in conclusion, this confirms the significance of SMMF as a foundation for implementing ethical and appropriate management strategies aligned with the principles of Maqashid.

References

Amalia, R. R., & Mawasti, W. (2022). Social Enterprise Dalam Perspektif Maqashid Syariah: Studi Pada PT Kampung Marketerindo Berdaya. *Realita: Jurnal Penelitian Dan Kebudayaan Islam, 19*(2). https://doi.org/10.30762/realita.v19i2.3516

Assauri, S., & Assauri, F. (2011). *Strategic management: Sustainable competitive advantages*. Fakultas Ekonomi, Universitas Indonesia.

As-Shodiq, M. J. (2020). Strategic Management of Islam University in integrating science and religion (case study at UIN Maulana Malik Ibrahim Malang). *Edukasi, 08*(01), 46–61.

Asyur, M. T. I. (1999). *Maqashid al-Syari ah al-Islamiyah, ed.* Al-Fajr.

Bakhri, S., Saiban, K., & Munir, M. (2022). Peran Lembaga Keuangan Syariah dalam Industri Halal sudut Pandang Maqosid Syariah. *Tasharruf: Journal of Islamic Economics and Business, 3*(1), 11–29. https://doi.org/10.55757/tasharruf.v3i1.88

Beik, I. S., & Nursyamsiah, T. (2021). Swot analysis on research methodologies and techniques Adopted in Islamic economics and finance. In *Teaching and research methods for Islamic economics and finance*. https://doi.org/10.4324/9781003252764-5

Erba, D. M. F., & Nofrianto, N. (2022). Implementation of Maqashid Syariah in sharia business transactions. *AL-FALAH: Journal of Islamic Economics, 7*(1), 125. https://doi.org/10.29240/alfalah.v7i1.3703

Fauzan, H. (2023). Pemikiran Maqashid Syariah Al-Tahir Ibn Asyur. *Al-Mawarid Jurnal Syariah Dan Hukum (JSYH), 5*(1), 101–114. https://doi.org/10.20885/mawarid.vol5.iss1.art7

Firdaus, A., & Ahmad, K. (2023). Islamic business and performance management: The Maslahah-based performance management system. In *Islamic business and performance management: The Maslahah-based performance management system*. Taylor and Francis. https://doi.org/10.4324/9781003390947

Firdaus, M. A. (2018). MAQASHID AL-SYARI'AH: Kajian Mashlahah Pendidikan dalam Konteks UN Sustainable Development Goals. *Journal of Research and Thought on Islamic Education (JRTIE), 1*(1), 73–95. https://doi.org/10.24260/jrtie.v1i1.1068

Fitri, S., Mariko, S., & Nasfi, N. (2021). Potensi transformasi spp konvensional menjadi spp syariah melalui analisa swot (studi pada kelompok spp nagari cubadak). *Jurnal Apresiasi Ekonomi, 9*(3), 301. https://doi.org/10.31846/jae.v9i3.419

Hadi, N. (2019). Maqashid Koperasi Syariah. *I-Economics: A Research Journal on Islamic Economics, 4*(2). https://doi.org/10.19109/ieconomics.v4i2.2562

Hasim, H. F. A., Ariffin, N. M., & Zakaria, M. Z. (2019). Strategic management system framework in Shariah compliant hotel in Malaysia. *International Journal of Advanced Trends in Computer Science and Engineering, 8*(1.6 Special Issue), 416–423. https://doi.org/10.30534/ijatcse/2019/6081.62019

Karim, A. A., & Sahroni, O. (2015). *Maqashid Bisnis Dan Keuangan Islam: Sintesis Fikih Dan Ekonomi*. Rajawali Pers.

Lifia, L., & Rofiq, A. (2023). Maqashid sharia thinking perspective of IBNU Assyur in the development of sharia banking. *Journal of Management and Informatics, 2*(1), 01–17. https://doi.org/10.51903/jmi.v2i1.151

Maharani, J., Diana, I. N., & Rofiq, A. (2022). Pemikiran Ibnu Asyur Tentang Maqashid Syariah Dalam Ekonomi Kontemporer. *Jurnal Ilmiah Ekonomi Islam, 8*(03), 2495–2500. https://doi.org/10.29040/jiei.v8i3.5708

Mohammed, M. O., El Amri, M. C., & Bakr, A. (2021). SWOT analysis of benchmarking in the Islamic finance industry. In *Benchmarking Islamic finance: A framework for evaluating financial products and services* (pp. 56–69). Taylor and Francis. https://doi.org/10.4324/9781003089988-6

Muafi, M. (2021a). Examining the role of human capital and strategic management in Islamic perspective. *International Journal of Business Ecosystem & Strategy*. (2687-2293), *3*(2), 41–50. https://doi.org/10.36096/ijbes.v3i2.255

Muafi, M. (2021b). Examining the role of human capital and strategic Management in Islamic Perspective: A strategic approach for encouraging business sustainability. *International Journal of Business Ecosystem & Strategy, 3*(2), 41–50. https://www.bussecon.com/ojs/index.php/ijbes/article/view/255

Muawaffaq, M. (2021). Maqashid Syariah dalam Perspektif Ibnu Asyur. *Attujjar: Jurnal Ekonomi Syariah, 6*(1), 44–54. https://jurnal.insida.ac.id/index.php/attujjar/article/view/81/183

Muna, M. N., & Hapsari, M. I. (2015). Implementasi Manajemen Strategik Syariah di BMT Amanah Ummah. *Jurnal Ekonomi Syariah Teori Dan Terapan, 2*(12), 1052. https://doi.org/10.20473/vol2iss201512pp1052-1068

Ningsih, B. C. S. (2021). Comparison of Al-Syatibi and Thahir Ibn Asyria's thoughts on Maqashid Shari'ah. *JURNAL ILMIAH MIZANI: Wacana Hukum, Ekonomi, Dan Keagamaan, 8*(1), 11. https://doi.org/10.29300/mzn.v8i1.4685

Poister, T. H., & Streib, G. D. (1999). Strategic Management in the Public Sector: Concepts, models, and processes. *Public Productivity & Management Review, 22*(3), 308. https://doi.org/10.2307/3380706

Pratiwi, F. D., Farhan, A., & Mulyono, R. D. A. P. (2023). Assessing sharia banking commitment through the Shariah Maqashid index. *International Journal of Current Science Research and Review, 6*(4), 2257–2264.

Prihatmoko, C., Kusbaryanto, K., & Hidayah, N. (2022). Strategic plan analysis using the Balanced Scorecard framework at Yogyakarta Islamic Hospital PDHI. *Interdisciplinary Social Studies, 2*(1), 1581–1592. https://doi.org/10.55324/iss.v2i1.312

Priyatno, P. D., Rohim, A. N., & Sari, L. P. (2022). Analisis Kinerja Bank Syariah di Indonesia Berbasis Maqashid Sharia Index. *Jurnal Ilmiah Ekonomi Islam, 8*(3), 2434–2443. https://jurnal.stie-aas.ac.id/index.php/jei/article/view/6581

Salycasanda, N., Febriadi, S. R., & Permana, I. (2022). Tinjauan Nilai-Nilai Maqashid Syari'ah terhadap Penyaluran Zakat Produktif di Lembaga Amil Zakat. *Bandung Conference Series: Sharia Economic Law, 2*(1). https://doi.org/10.29313/bcssel.v2i1.259

Sirat, A. H., & Jannang, A. R. (2022). Islamic financial management practices and business performance in small industrial companies. *Atestasi: Jurnal Ilmiah Akuntansi, 5*(1), 94–108. https://doi.org/10.57178/atestasi.v5i1.25

Sujova, A., & Rajnoha, R. (2012). The management model of strategic change based on process principles. *Procedia–Social and Behavioral Sciences, 62*, 1286. https://doi.org/10.1016/j.sbspro.2012.09.220

Toha, M., & Rozikin, K. (2020). Implementasi Maqasid Al-Shariâ€™ah Dalam Manajemen Strategis Syariah. *JES (Jurnal Ekonomi Syariah), 5*(1). https://doi.org/10.30736/jesa.v5i1.75

Utama, R. W. A., Muhtadi, R., Arifin, N. R., & Mawardi, I. (2019). Tinjauan Maqashid Syariah Dan Fiqh Al-Bi'ah Dalam Green Economy. *Jurnal Ekonomi Islam, 10*, 242–259.

Von Pock, A. (2007). Strategic management in islamic finance. In *Strategic Management in Islamic Finance*. Deutscher Universitats Verlag. https://doi.org/10.1007/978-3-8350-5406-6

Muhammad Majdy Amiruddin, was born in Mangkoso on July 1, 1988. After completing his education from elementary to high school at Pondok Pesantren DDI Mangkoso, he pursued his Bachelor's degree at Al Azhar University in Cairo, Egypt, and later obtained his Master's degree at the Universiti Sains Islam Malaysia. As a lecturer at the State Islamic Institute of Parepare and an educator at Ma'had Ali DDI Mangkoso, Muhammad Majdy Amiruddin has played a crucial role in disseminating knowledge and understanding of Islamic economics. His notable positions as the Head of Publication and Publishing Center, Director of IPN Press, and Chief Editor of *Sao Journal* signify his continuous efforts in advancing academic literature. His scholarly works encompass various articles and books that enrich the understanding of Islamic economics. Furthermore, he has made significant contributions to international journals, adding value to the knowledge base in the field of Islamic economics.

Islamic Business in Malaysia: Scenario and Challenges

Ilhaamie Abdul Ghani Azmi, Hasan Al-Banna Mohamed, and Zarina Kassim

Abstract The Islamic business in Malaysia can be divided into several types, such as large, medium, small, and micro businesses. The majority of Islamic businesses are micro, small, and medium companies. The challenges Muslim entrepreneurs face are categorized into internal and external. The challenges are identified as financial, corruption, interpersonal relationship conflict, emotional challenges, lack of current knowledge, incompetence, risk-taking, changes, competition, less competitive and survivable, crisis, lack of demand, family challenges, customer loyalty, business sustainability, business expansion, entrepreneur attitudes, business blessings, product diversity, partners, requirements vs customer needs, target market, market segment, market placement, and human resource challenges and pricing. To be successful, among necessities are entrepreneurs need to have backup plans such as falling back onto angel capital or venture capital, good planning, basic knowledge in financial management, good interpersonal and time management skills, resilient, persistent, and committed, safeguard the reputation and integrity of the company and update knowledge through training.

1 Introduction

The Islamic business in Malaysia has long been established. However, tribulations and challenges are often a common occurrence that is inevitable in a business. This is because Islamic entrepreneurship is an act of worship and responsibility through the concept of "communal obligation" (*fardhu kifayah*). The Islamic business is

I. Abdul Ghani Azmi
University of Malaya, Kuala Lumpur, Malaysia
e-mail: amieazmi@um.edu.my

H. A.-B. Mohamed (✉) · Z. Kassim
National Defence University of Malaysia, Kuala Lumpur, Malaysia
e-mail: hasanalbanna@upnm.edu.my; zarina@upnm.edu.my

A. Rafiki et al. (eds.), *Strategic Islamic Business and Management*, Contributions to Management Science,
https://doi.org/10.1007/978-3-031-61778-2_11

different from conventional business. This is because Islamic business is based on certain Islamic rules and laws. Hence, this chapter will discuss the Islamic business scenario, especially in Malaysia and the challenges that a company should face.

2 Islamic Business Scenario in Malaysia

2.1 Large Enterprise

Generally, not many large businesses in Malaysia are owned by Muslim entrepreneurs. The top listed companies in Malaysia are Petronas, SimplySiti, Axiata, Ambank, Bank Islam, RHB Bank, MayBank, CIMB, Burger Ramly, Celcom, Felda, Khazanah Nasional, PNB, NAZA, TNB, Sapura, Telekom, Takaful Malaysia, and others.

The richest Muslim entrepreneurs in Malaysia are Tan Sri Syed Mokhtar AlBukhary (BERNAS), Tan Sri Syed Azman Syed Ibrahim (WESTSTAR Axiation), Tan Sri Azman Hashim (AmBank Group), Ahmayudin Ahmad (Westport), Datuk Abdul Karim Abdullah (Serba Dinamik), Datuk Abdul Kadir Sahib (Serba Dinamik), Datuk Kamarudin Meranun (AirAsia), and Tan Sri Mokhzani Mahathir (Petron).

This shows that a business is a *fardhu kifayah* in Islamic teachings. This is because there are not many successful and wealthy Muslim entrepreneurs in Malaysia. In fact, it becomes an advantage for Muslim millionaires as they are able to contribute to the community directly whether in the form of donations, gifts, and waqaf for the future development of Muslims.

2.2 Micro, Small, and Medium Businesses

Small and Medium Enterprises (SMEs) in Malaysia are categorized into service, manufacturing, agriculture, construction, and mining and quarry industries. For the manufacturing sector, SME is defined as a firm that has annual sales not exceeding RM50 million or the number of full-time employees does not exceed 200 people. For the service sector and other sectors, SME is defined as a firm that has annual sales not exceeding RM20 million or the number of full-time employees does not exceed 75 people. To be included in the SME's category, a company needs to be registered with the Malaysian Companies Commission (MCC).

According to the Department of Statistics Malaysia, there were 1,226,494 Small, Micro, and Medium Enterprises (SMEs) by 2021 which accounted for 97.4% of the total organization in Malaysia. The services sector formed 83.8%, with a total of 1,028,403 firms. The construction sector remains the second largest contributor to 8.0% (98,274 firms). Meanwhile, about 5.8% of PMKS (71,612 firms) were involved in the manufacturing sector, followed by 1.9% (23,633 firms) in the agricultural sector, with the remaining 0.4% (4572 firms) in the mining and quarry industry sectors. In 2021, the micro enterprise accounted for 78.6% (964,495 firms) of the total PMKS organization, small-sized firms made up 19.8% (242,540 firms) and the remaining 1.6% (19,459 firms) were medium-sized SME Corporations. In 2022, Malaysia's SMEs made a substantial contribution to the GDP, accounting for 38.4%. This contribution amounted to a value-added of RM580.4 billion, compared to 37.4% (value added: RM520 billion) in the previous year.

The Islamic business in Malaysia is geared toward a business based on *Halal* products and services. This is because, in Islam, other than goods and services, food and beverages must comply with the requirements of *Syara'* which is *Halalan Toyyiban* (free from illegal and harmful things). There are four forms of business which are sole proprietorship, partnership, private limited company, and limited liability partnership.

The *Halal* food demand market has been steadily growing. There was an increase of 4.5% in the export value of the *Halal* market from RM37.7 billion in 2014 to RM39.4 billion in 2015. Halal products and beverages make up a large part of the *Halal* market in Malaysia. The SME company was the largest contributor to the export of *Halal* food products in 2014 involving the average number of SME companies was 475 companies compared to 249 multinational companies. In fact, in 2015, the Bumiputera company contributed a total of RM2.6 billion or 6.6% of the total exports of *Halal* food products in Malaysia (Basaruddin et al., 2019).

2.3 Ministry of Entrepreneur and Cooperatives Development (MECD)

This ministry was first established in 1974 under the name of the Ministry of Public Corporation Coordination (MPCC). Then in 1976, the name was changed to the Ministry of Public Enterprise; 1995 to the Ministry of Entrepreneur Development; 2004 Ministry of Entrepreneur and Cooperative Development (MeCD); 2018 to the Ministry of Entrepreneur Development (MED), and 2020 to the current name, the Ministry of Entrepreneur and Cooperatives Development (MECD).

The Ministry provides online services such as business license application, registration of registered bumiputera building materials vendors, registration of contractor courses, and application for accreditation of social enterprises. Apart from that, it also offers physical services such as market access, access to funding and training programs.

There are 10 agencies under this ministry, namely the Co-operative Commission of Malaysia, the Cooperative Institute of Malaysia, SMECorp, TEKUN, INSKEN, Bank Rakyat, SME Bank, UDA, PERNAS and Amanah Ikhtiar Malaysia (AIM).

2.4 NGOs (Non-Governmental Organizations)

The Malaysian Islamic Chamber of Commerce (MICC), which was established in 1993, is an organization that strives to unite Muslim traders and entrepreneurs for the interests and benefits of the Islamic economy. MICC's vision is to champion business based on the *Quran* and *al-Sunnah*. While the mission of MICC is to make *jihad's* business success by carrying out its responsibilities as a servant to the ummah. MICC also has five core values, which are thinking and cultivating Islamic habits in doing business, strengthening the management system and delivery of MICC services, raising funds through organizing government events and funds, strengthening and diversifying membership, and making MICC an Islamic business leader and intensifying the priorities of the *Halal* and *Shariah*-compliant industries.

3 Challenges of Islamic Entrepreneurs

To become a successful entrepreneur is not an easy and convenient thing. There are many challenges and conjectures that entrepreneurs will face, whether in the form of internal or external challenges:

3.1 Financial

Getting access to finance is one of the biggest challenges for Muslim entrepreneurs. This is because without money or capital, a business is powerless to last long and sustainably forever. Examples of financial challenges are lack of capital, budget preparation, cash flow management, debt, costs and expenses, account management, and so on. These financial challenges are faced at all levels of the business whether it is start-up, growth, maturity, or decline. This is because at each phase of the business, it requires a different financial strategy.

Financial literacy is one of the factors influencing Malaysian entrepreneurs' sustainability performance. Entrepreneurs need to be financially aware, possess clarity on financial matters, and have some finance-based knowledge in order to label them as having been financially literate. Due to low financial literacy among entrepreneurs, financial decisions become ineffective, inefficient management of cash flows is observed, and inefficient planning and management of financial risks occurs. This leads to the fact that SMEs are not able to be used for financing opportunities and invest funds effectively for long-term growth.

Furthermore, entrepreneurs also face problems in terms of getting financial assistance. Usually at the beginning of the business, it is quite difficult to get a bank loan. This matter needs to be solved by having capital from the entrepreneur's own savings, family and friends, angel capital, or venture capital. When a company starts to grow, it is usually quite easy to apply for financial assistance from a bank. However, companies must be careful in taking out bank loans because if the amount is too large, then it is possible that the company cannot afford to pay if the company is weak in managing its cash flow. Therefore, entrepreneurs need to have basic knowledge of financial management. For example, it is possible to perform calculations mentally instead of blindness in calculations. This can lead to the individual being deceived if he is not well versed in cipher and calculation in terms of business.

Customer debt is also a challenge for entrepreneurs. If the company has too much debt, not as far as bank loans but also customer debt, the company may not be able to survive. Therefore, companies need to be good at planning, such as encouraging payments to be made in cash or the entire payment amount. If the customer cannot afford to pay, for example, the bulk price, then it is likely that the company should allow the customer to pay a large part of the price of the goods or services, while the balance is paid in installments.

In terms of business expenses, it is also a challenge that entrepreneurs need to ensure that the cost of expenses is not too high such as the cost of raw materials, labor, rental of office or factory or equipment, water, and electricity. This is intended so that from the customer's payment on the purchase of the product and the use of the service, there is still a profit balance for the company.

3.2 Corruption

The corruption in Malaysia if it is not curbed from the beginning, the business will get more stifled. It is found that many high-caliber Muslims are involved in this activity. This is driven by greed and selfishness. Revealing history from a corruption report by Transparency International in 2005 that reported that Malaysia had fallen 16 steps in 10 years from 23rd place in 1995 to 39th place in 2004. Most recently, a report by Transparency International in 2019 showed an increase in the Corruption Perceptions Index (CPI). The report reported that Malaysia improved its rankings, rising 10 steps to 51 out of 180 countries, with a score of 53 (Transparency International, 2019). This position is the same one that Malaysia achieved in 2014. However, according to BPR, there are still many more unreported cases. In the early days of independence, the problem of corruption was focused on land approvals, forgery of projects, and rural development, but the current era of corruption is becoming more sophisticated and has a variety of patterns.

Thus, corruption is one of the most serious challenges in entrepreneurship. This is because many entrepreneurs take the easy way to make a profit. It does not matter the receiver, the giver, the requester, the abuser of power in corruption, whatever the name is, it is illegal in terms of Islam and should be avoided (Nooh & Nawang, 2022).

As Allah Almighty says in *Surah Al-Baqarah*, verse 188 which means:

> Do not eat up each other's property among yourself in vanity, nor approach with it the authorities to eat up a portion of the property of the people sinfully, while you know (that you are unjust in doing so).

Thus, Muslim entrepreneurs face the challenge of being a trustworthy and truthful person and not being involved in corruption in pursuit of wealth. Therefore, every entrepreneur needs to safeguard the reputation and integrity of the company so as not to be tarnished by the symptoms of corruption. Or then, the public will look down on the company and they no longer believe in the products or services it provides.

3.3 Interpersonal Relationship Conflict

Conflicts can occur between employees, suppliers, customers, and competitors as well as partners. It is found that entrepreneurs spend 20% of the day's time facing and resolving any conflicts. The entrepreneur is likened to a leader in his business. Therefore, it is not easy to become a leader or a leader. It requires interpersonal skills, especially effective communication to ensure that businesses do not face failure and fall out of the market.

In Malaysia, there are many races such as Malays, Chinese, and Indians who need entrepreneurs to have good interpersonal skills. This point is important so that there are no disagreements. This is because different races have different religions, cultures, languages, and dialects as well as different ways of doing business. Lack of understanding and knowledge of this can lead to organizational conflicts. These interpersonal skills are important to gain confidence and trust in business projects from partners as well as assistance from employees as well as support from suppliers and distributors. Hence, a strong network based on the loyalty of all members in the production and distribution channels is essential for the smooth operation and future of the company.

3.4 Emotional Challenges

To succeed, an entrepreneur needs to make many sacrifices such as finances and time. Thus, entrepreneurs will definitely face emotional challenges. In the event of a personal and even organizational conflict, this will add to the emotional burden of an entrepreneur. However, in order to succeed, he needs to be patient and persevering in the face of trials. He should assume that conflicts, losses, and failures in business are a test from Allah SWT so that we can always be patient and grateful.

> Faith is two parts; one part patient and the other part is gratitude
>
> [*Imam* Baihaqi: 9264]

Allah SWT says in Surah *Al-Ankabut* verses 58 and 59, which means:

> And those who believe and do righteous deeds, We will place them in a palace in heaven that flows under it some rivers, they will remain in it. That is the best reward for those who do righteous deeds. Those who are patient and those who submit themselves to their Lord.

Allah SWT says in Surah *Al-Anbiya'*, verse 35 which means:

> And We test you with distress and pleasure as a trial, and to Us you will all be returned.

Furthermore, the statement of Allah SWT in *Surah Al-Baqarah* verse 155, means:

> And indeed We will test you, with a little fear, hunger, lack of wealth, soul and fruit. And give good tidings to those who are patient.

From Syuhaibin, the Prophet Muhammad SWT said:

> It is miraculous that the affairs of the believers, all their affairs are good, and there is nothing like that but believers alone. If the pleasure he will be grateful for, then it is good for him. If hardship is behaved, then it is good for him.
>
> [*Imam* Muslim: 2999]

3.5 Lack of Current Knowledge

The next challenge is that there are a handful of entrepreneurs who have a passion for success quickly but are too lazy to strive. For example, not having the latest knowledge and knowledge on technology and the development of current issues. Thus, it can lead to failure in the business of an individual. While the future progress and development of entrepreneurship significantly depend on innovative technologies. Moreover, technology literacy could be considered as a crucial factor impacting the competitiveness of entrepreneurs in doing business during this digital age. Therefore, successful Muslim entrepreneurs need to be good at managing their businesses and spend time with experts whether field experts or business experts for in-depth discussions and specific training for the improvement of the current knowledge. For example, the use of social media for marketing.

There are entrepreneurs who think that paid marketing is better than free marketing such as marketing through social media. This misunderstanding can cause the company to incur high marketing expenses, however, the profit result is not so. Therefore, companies need to diversify the medium or channel for marketing to offset the cost of marketing so that it is not too high to burden the company.

Focusing on a product or service alone will also put the business in danger. However, all entrepreneurs can likely afford to do so due to financial constraints. If entrepreneurs choose excess cash, the more they invest in other businesses with less cost and loss implications such as high-demand products or services.

3.6 Incompetence

Several other factors can prevent Bumiputera SME entrepreneurs from implementing the development of *Halal* products such as lack of human resources, time constraints, and lack of management system and facilities. In addition, premise factors, technology, and lack of experience in terms of marketing business development are among the constraints of SME entrepreneurs in implementing *Halal* product development (Basaruddin et al., 2019). This has resulted in 70% of new businesses failing due to weaknesses in management, especially financial cash flow management.

Efficiency in communication is also important in the negotiation process. Accordingly, misunderstandings can be eliminated. Cross-cultural understanding is also important to ensure coordination of information and no misunderstandings. This is especially true of many races in Malaysia. Next, if Muslim entrepreneurs want to expand their business internationally. This is important to ensure that Muslim entrepreneurs can move quickly in their businesses and achieve the success they want.

This shows that Muslim entrepreneurs need to follow the training to deepen any knowledge and improve their skills. The background of the entrepreneur's parents and business experience and interests can also help Muslim entrepreneurs in facing any challenges in their business.

3.7 Risk-Taking

Another challenge that a Muslim entrepreneur often faces is risk-taking. Muslim entrepreneurs face a dilemma in risk-taking whether the risk will be profitable or otherwise. Thus, a successful Muslim entrepreneur is a good person in assessing and researching all the risks he faces as well as converting the risks into opportunities that bring profit to his business. Examples of risks include competition and demand clashes as well as the expansion of companies both locally and internationally. They are also determined in taking risks and do not stop halfway.

3.8 Technology and Digitalization

The world is constantly changing and every day there will be changes. In business, a Muslim entrepreneur will face the challenge of change, especially in terms of technology and financial management. Technology and digitalization are essential to business sustainability, especially innovation. Innovation is important in business. Without creativity and innovation, the business will be on the notch of the old, while competitors are already moving forward. Thus, in order to succeed, a Muslim entrepreneur needs to face and adapt to the changes. For example, e-commerce and digital marketing. Due to the ever-changing marketing trends over time, customer tendencies and interests have created fierce competition and become a challenge that every entrepreneur should focus on. Marketing now needs to be more creative in order to address the challenge.

3.9 Competition

Competition challenges are common in business. This is because a person does not actually trade alone in the market. There are other businesses available in the current market. Therefore, entrepreneurs should always ensure that there is a difference between other businesses, not only in terms of company name but also in terms of products or services even if it is a little bit in terms of quality. In order to ensure that the products or services of the business are not replicated by competitors, it is appropriate for entrepreneurs to invest in patenting their own products or services. Let no trader take dirty tactics such as the use of shamans or magic to bring down rival companies. This is cursed and prohibited in Islam. Hence, to ensure that entrepreneurs obtain the blessings and success of the world and the hereafter, it is only natural for them to ensure that their business is in accordance with the ruling of *Syara'* set by the teachings of Islam.

Aside from that, there are a handful of Muslim entrepreneurs who are in a state of dilemma in order to organize a marketing communication strategy based on an Islamic perspective. This is because the opponent to honesty and trust in marketing is fraud and betrayal in deceiving consumers who are seen as more easily profitable. In addition, honesty in marketing carried out on social media is very important to be practiced to make Muslim entrepreneurs' products remain long competitive in the market. This honesty covers all aspects, especially the method of display and notification of the content of the content or material of the product sold. Product descriptions should be thorough and no information should be hidden from users. The testimonials or "reviews" presented are also not mere fabrications. Price display is also very important because it shows the transparency of sellers and their willingness to compete with other entrepreneurs in the market.

Branding is also one of the competition. Sometimes companies do not want to survive due to poor branding strategies. Even if their product or service is good, however, the company does not make a good investment in branding, then the company can be buried.

3.10 Less Competitive and Survivable

In the face of challenges, entrepreneurs can easily shake their faith and stance, especially when faced with failures and losses. This may be due to the lack of skills, experience, and piety. They are also less critical, creative, and innovative. Hence,

Muslim entrepreneurs need to be resilient and persistent in the face of competitive challenges. This is in accordance with the Quranic evidence which means:

فَإِنَّ مَعَ ٱلْعُسْرِ يُسْرًا ﴿٥﴾

So then (establish your belief) that every hardship is accompanied by ease.

إِنَّ مَعَ ٱلْعُسْرِ يُسْرًا ﴿٦﴾

(Once again), it is affirmed that every hardship is accompanied by ease.

3.11 Crisis

Muslim entrepreneurs face crisis challenges such as the COVID-19 pandemic, SAR, floods, and so on. The COVID-19 pandemic, for example, has affected businesses run by SMEs as the pandemic has put pressure on management and has led to failures in the business. Small companies are the most affected enterprises due to the lower level of availability, they face constraints in terms of resources and have a relatively weak market position and a higher dependence only on local governments. In addition, SME enterprises usually suffer financial losses, reduced sales volume, inability to qualify contracts, cash flow problems, and a reduction in staff in the business which has put emotional strain on SME entrepreneurs (Rahman et al., 2021).

According to Malaysia Productivity Corporation (MPC) (2020), a survey of the impact of COVID-19 on productivity of companies found that more than 50% of respondents had experienced a decrease in productivity levels ranging from 25% to 100%. Meng and Mui (2021) asserted that over 70% of the business experiencing impact in earnings and facing losses due to sanctions movement on individuals and goods during the MCO. There are companies that have to lay off employees, directing employees to leave without pay or reducing wages to reduce the impact of losses on the company so that the company can still be maintained. This causes another negative effect on the workforce which is the increase in rates of unemployment, loss of assets, and accumulated debt.

However, in Malaysia, there are various aid channels for Muslim entrepreneurs such as assistance from governments and NGOs. In the aspect of zakat assistance to asnaf small traders and hawkers, among the state zakat institutions that implement special assistance to these groups are the states of Johor, Kelantan, Melaka, Negeri Sembilan, Pahang, Perak, Perlis, Penang, Selangor, Terengganu, and Federal Territory.

3.12 Lack of Demand

These challenges are usually faced by entrepreneurs in the introductory and declining phases of a product or service. In the early stages of business, the community does not yet know and recognize a new product or service, while at the downward stage, it is common for the community to put aside the product or service. This may be due to the fact that the company underperforms marketing activities such as advertising and sales promotion compared to competitors or the quality or quality of the product or service is not given serious attention and pressure by the company and causes less favorability compared to competitors. Furthermore, it is possible that the products or services offered by the company do not keep pace with technological changes and customer demand that causes demand to decline.

Therefore, companies need to improve the quality of products or services without increasing production costs. They need to strive for quality recognition such as ISO from SIRIM to increase customers' confidence in their products. In addition, entrepreneurs need to be creative and innovative in improving their output to suit customers' needs.

3.13 Family Challenges

Entrepreneurs also have trouble balancing time between business and family. To do business requires entrepreneurs not only to sacrifice in terms of ringgit, and energy but also time with family. This is especially true mostly before and also in the early stages of the business which requires the close attention of the entrepreneur to ensure the success of the business. Therefore, entrepreneurs need to be able to balance the time between the two. Make sure not to lose both because both are equally important.

3.14 Customer Loyalty

Customer loyalty is also a challenge and an issue for entrepreneurs. This is because if the customer is not satisfied, the customer will run away from the company by purchasing the product or using the services of another company. Therefore, customer satisfaction is important to ensure that customers are loyal to the company's products. In this regard, companies need to adopt various marketing techniques such as advertising, promotional sales, and publicity to ensure that customers remain with the company and do not purchase goods or use the services of other companies. Customer loyalty is essential to the sustainability of a business. Without customer satisfaction and loyalty, a company can fall out of the market and can easily cause a company to be shut down.

3.15 Business Sustainability

Ensuring sustainable business is also a challenge for all businesses. Therefore, entrepreneurs need to be careful and careful in planning and implementing. Any implementation is necessary for monitoring. If the objectives of any implementation are achieved, the entrepreneur must take corrective and improvement actions. Business sustainability is important to ensure that the business does not fall out of the market. Hence, the commitment of entrepreneurs is required for the continuity of the business in a market.

3.16 Business Expansion

Not only failure is a challenge for entrepreneurs, but also expansion. This is because expansion requires financial commitment, energy, and time. Therefore, entrepreneurs need to do careful planning for this purpose not only for the resilience and sustainability of the company. In short, business expansion needs to serve as a business vision. At the same time, companies need to be competitive and move forward. Thus, an entrepreneur cannot deny this from the beginning of the installation of the intention to do business. They need to put this as their aspiration and direction in doing business.

3.17 Entrepreneurial Attitude

The attitude of the entrepreneur himself is also one of the challenges faced by any entrepreneur. Usually, when the business starts to be accepted and the company starts to experience success, the entrepreneur starts to change his/her attitude from cautious to comfortable and careless to take action to do so. They start spending company money down the wrong path by buying hours and luxury cars or not being careful in doing expenses. It should be invested in the right path either for the sustainability or expansion of the company such as zakat payments, money storage, investments, and so on as well as employee welfare such as education payments and network expansion with industry.

Some even start to be arrogant and become more arrogant when they succeed. They need to remember that their success is a Divine blessing and does not last forever. It is also the result of the help and sacrifice of family, friends, employees, and partners. Without faith and piety, Muslim entrepreneurs can simply be buried. Therefore, entrepreneurs need to recognize and appreciate the help of those involved and not just spend the company's money on the wrong path.

4 Business Blessings

In pursuit of world success, many Muslim entrepreneurs have lost midway through the way in defending their faith and religious beliefs. For example, performing prayers. In the bustle of pursuing the timing of tender delivery and execution of tasks and ensuring in a timely manner, some do not have time to fulfill their responsibilities as a Muslim. This suggests they are not planning properly so much that these obligations are ignored.

Fraud is also often practiced such as in scales, specifications of goods and consultations. It shows that piety is not in the entrepreneur. This is because they are more concerned with the worldly than the success of the Hereafter. So, how can we get a blessing in business?

Doubts in business are also among the elements in the difficulty of obtaining blessings in business. For example, the subtleties in investments such as interest earned. Therefore, as a progressive Muslim entrepreneur, it is necessary to seek information and ensure that there is no element of doubt in business.

In fact, there are many aspects that Muslim entrepreneurs need to take care of. This is so because many things cannot be seen with the naked eye. However, as a Muslim entrepreneur, it is necessary to take precautions so as not to miss out on ensuring that the goods sold and services delivered are *Halalan Toyyiban*. Every matter should be investigated and ensured to comply with *Syara'*.

4.1 Product Diversity

The diversity of products or services is also a challenge that an entrepreneur will face in ensuring that his business is successful. It is due to the fact that a homogeneous product or service is easy to get out of the market if the entrepreneur is incompetent in using marketing and business strategies. Thus, ensuring that there is a diversity of products or services requires creative and innovative entrepreneurs. To be a creative and innovative entrepreneur requires the knowledge and knowledge and attitude of entrepreneurs who love to learn and observe. Observation is necessary to ensure that the product or service is different from other organizations. This is a bit difficult if an entrepreneur does not like to observe another business and does not like the important details of the product or service differentiation. Differentiation is not necessary, little is enough to distinguish from other businesses. Some customers are capable and have the power to buy products or use different services from other businesses.

4.2 Partners

Sometimes, some partners overreact from the mold making the original objectives of the business unattainable and the business unable to survive. As the business begins to progress, the partner's faith is challenged and the eyes begin to look to the world. Lust wants to quickly succeed in causing missteps by investing in the wrong place. Too high trust placed in partners can lead to businesses rolling up and facing bankruptcy cases for not being able to pay off bank debts. Therefore, Muslim entrepreneurs need to be careful in choosing partners. Entrepreneurs need to recognize their hearts before choosing them as partners. The chosen partner must be someone with high faith and piety. Once selected, they need to be treated like family members so that friendship can be strengthened and they have no intention of betraying their own partners.

4.3 Requirements Versus Customer Needs

Many entrepreneurs are not clear between the needs and wants of customers. For them, the customer's needs or wishes are the same. This can obscure entrepreneurs in choosing target customers and doing market segmentation and positioning. Mistakes in choosing the customer's needs as the basis of the target market can reduce the potential of the market. This is so because basically, customers have needs. Whereas, their passions have wills. At the same time, they also face financial constraints. Thus, usually they will emphasize the fulfillment of their needs rather than their wishes.

Muslim entrepreneurs are also often confused with the theory of conventional needs such as Maslow, Herzberg, and so on. Instead, the needs of Muslim customers are ignored. The needs of Muslim customers are in accordance with religion, life, intellect, generation of descent, and property, or in other words *Shariah Maqasid*. Therefore, it is the basis for the subsequent Muslim customer proficiency.

4.4 Target Market

In starting a business, one of the challenging things is the target market. This is so because, at the moment, entrepreneurs are not clear who their customers are. Most entrepreneurs usually shock themselves with their ideas and products or services. However, they are not sure who will use their product or service. If they are not clear about the needs and wants of customers, their business will not bring much profit. This is because they are not sure who among them will become loyal customers and bring repeat purchases to the entrepreneur. With the latest technology such as the use of social media or digital marketing, this can be overcome by doing a neater and neater marketing program. In short, the needs and wants of customers can be clearly known.

4.5 Market Segmentation

In choosing customers who will bring profit and business sustainability, most entrepreneurs are also confused about which one is diamond and which one is glass. Which of the customers will bring profit and which not. Entrepreneurs should be aware of rising costs such as production, shipping, and other costs. Therefore, entrepreneurs should not rush to choose customers who will be targeted for their products or services. Entrepreneurs also need to be relieved in sacrificing less profitable customers such as simple users (medium users) or less (light users). On the other hand, entrepreneurs need to focus on frequent customers (heavy users) using their products or services.

4.6 Market Placement

Once the target customer has been carefully selected and reviewed, the next thing that is a challenge for entrepreneurs is the placement of the market. This is because entrepreneurs need to ensure that customers can differentiate their products or services from rival organizations. If the client cannot remember and identify the brand of the entrepreneur's product or service, this indicates the presence of a problem. Most likely the customer will not choose to purchase and use the entrepreneur's product or service. On the other hand, they are more passionate and admire the products or services of rival companies. Therefore, entrepreneurs need to be aware and see what is the difference in their products or services that will be the competitive advantage of the company. It is important that they can do the market placement best.

4.7 Human Resources

Human resources or talent are also a problem for business companies. The practice and process of hiring and selecting employees seem simple. However, the problem is human capital. There are many qualified and experienced human capital in the market. However, human capital that can be trusted and can do its job and work well is a problem and a challenge for Muslim entrepreneurs. Therefore, Muslim entrepreneurs need to be careful in choosing and hiring their employees.

In the meantime, entrepreneurs also need to remember the future of their human capital. This is because all of them not only want to earn a salary but also they need career development. This is because as they age, their family members grow. Hence, they need extra money to fund their family's needs such as education and so on.

4.8 Pricing

The placement of product prices or prices is also a challenge for Muslim entrepreneurs. This is because errors in price placements can cause losses to the company. Cheap prices or high prices can both bring losses to the company. If the price is too low, the cost of production and delivery is not covered causing losses to the company. On the other hand, if the price is too high, the customer does not want to buy unless the customer is rich or can afford it. Therefore, Muslim entrepreneurs should be careful in choosing and placing any price on their products or services. In-depth studies need to be done to ensure that profits are obtained rather than losses. In the meantime, price wars may occur between competitors. It is not wrong if the entrepreneurs want to increase the price of the goods sold, but there needs to be improvements. The improvements that can be done can be in the form of selling products in several packages, as well as additions in terms of functionality, color, and design. Therefore, Muslim entrepreneurs need to be careful about taking action. Mistakes in taking response can cause losses to the company, especially if the price is lowered.

5 Summary

In conclusion, Islamic business can be divided into large, medium, small, and micro businesses and each type of business must be based on Islamic principles. Small and Medium Businesses (SMEs) is categorized into services, manufacturing, agriculture, construction, and mining and quarry industries. Registration under the Companies Commission of Malaysia (CCM) is an obligation in the SME category. In general, large, medium, small, and micro businesses are the division of Islamic business in Malaysia. Each of these business divisions has parameters that need to be followed for the determination of distribution such as annual sales and full-time employees. In addition, there are those involved in facilitating Islamic businesses such as the Ministry of Entrepreneur and Cooperative Development (MECD), and those who unite Islamic entrepreneurship traders such as the Islamic Trade Hall of Malaysia (ITHM).

Every businessman is sure to face various challenges and conjectures before becoming a successful Islamic businessman. The main challenges faced by businessmen are financial problems such as the problem of getting financial assistance, the problem of paying off the debt, the cost of expenses, and so on. Next, the challenge of corruption resulting from greed and selfishness. In addition, interpersonal relationship conflict is also one of the challenges in the Islamic business and it can occur between employees, suppliers, partners, customers, and competitors.

Furthermore, other challenges include emotions, lack of knowledge and the latest knowledge, incompetent entrepreneurs, lack of risk-taking, inability to adapt to change, problems of lack of competitiveness, and weak face failure. In addition, the crisis that has endured by Muslim businessmen, less demand for products from customers, the level of customer loyalty, and the sustainability of the business are other challenges that can affect the business. Finally, business development, business blessings, product diversification, target and market segmentation, and human resources are also the main challenges faced by businessmen.

References

Basaruddin, S. K., Omar, S. N. Z., Mohamad, M., & Zainal, H. (2019). Cabaran Usahawan Asnaf Dalam Perlaksanaan Pensijilan Halal Malaysia. *International Journal of Zakat and Islamic Philanthropy, 1*(2), 23–30.

Meng, K. H., & Mui, H. K. D. (2021). SMEs and Sustainability Achieving Sustainable Business Growth in the New Normal. Handbook of Research on Sustaining SMEs and Entrepreneurial Innovation in the Post-Covid-19 Era. (pp.331-351). Publisher: IGI Global

Nooh, M. N., & Nawang, W. R. W. (2022). *Usahawan Muslim: Isu dan Cabaran.* Seminar Pembudayaan & Pendidikan Keusahawanan Remaja. Accessed Jan 11, 2024, from https://adoc.pub/seminar-pembudayaan-pendidikan-keusahawanan-remaja.html

Profil PMKS pada Tahun. (2016–2022). *SMECorp Malaysia*: Accessed Jan 11, 2024, fromhttps://www.smecorp.gov.my/index.php/my/polisi/2020-02-11-08-01-24/profil-dan-kepentingan-kepada-ekonomi

Rahman, R. A., Noor, H. M., & Ramli, A. M. (2021). Cabaran dan Kelangsungan Pengusaha Makanan Kecil dalam Pandemik COVID-19. *Malaysian Journal of Social Sciences and Humanities (MJSSH), 6*(12), 13–23.

Transparency International Global Corruption Barometer. (2019). Accessed Feb 10, 2024, from https://www.transparency.org.my/pages/what-we-do/indexes/transparency-international-malaysia-corruption-perceptions-index-2019

Ilhaamie Ab. Ghani Azmi is an Associate Professor at the Department of Syariah and Management, Academy of Islamic Studies, University Malaya, Kuala Lumpur. She obtained her PhD at USM, MBA from UKM and BBA, IIUM. Her current research interests are human resource management, entrepreneurship, service quality, and Islamic management. To date, she has authored and co-authored books, book chapters, journal articles, and conference papers. She has presented her papers in UAE, Singapore, Indonesia, Hong Kong, Turkey, Cambodia, Australia, Korea, Thailand, Spain, Italy, and the UK. Four of her papers have won best paper award at the international conferences. Her research teams also won diamond, silver, and bronze medals in research poster competitions locally and internationally. She and her team were the finalists for "Anugerah Harta Intelek Negara" and the National Innovation Award. Due to her passion for research, she is registered with Global Agency Research and the International Association of Computer Science and Information Technology for a lifetime membership.

Hasan Al-Banna Mohamed is a Professor in the Department of Management, Faculty of Management and Defence Studies, National Defence University of Malaysia (NDUM), Kuala Lumpur. He obtained his PhD from Malaya University in the Islamic Quality Management in 2015. In addition to the academic post, he experienced working as the Director for the Division of Quality Assurance and Data Management, UPNM from 2010 until 2022. After that, he was appointed as a Senior Director for the Project Management and Strategic Planning Division, in the Assistant Vice Chancellor's office (Project Management and Strategic Development), NDUM. His areas of expertise and research are in Islamic Management, Quality Management, Human Resource Development and HRM, Islamic Studies.

Zarina Kassim is currently a Senior Lecturer at the Faculty of Defence Studies and Management, National Defence University of Malaysia (NDUM). She earned a Bachelor of Economics (honor) in 2010 and a Master's in Economics in 2013 from the Faculty of Economics and Management, the National University of Malaysia (UKM) and also completed a Doctor of Philosophy (PhD) in Economics Education from the Faculty of Education at the same university in 2019. Her research interests include Economics Development, Economic and Social Well-being, Defence Economics, and Economics Education. She is an ambitious person that interested in professional development and research, ongoing learning and growth, who likes to communicate and work with new ideas and people.

The Role of Religiosity and Red Flag on Financial Statement Fraud

Ratno Agriyanto, Musahadi, Fitra Istianah Turahman, Farah Amalia, Rahman El-Junusi, W. Warno, Dessy Noor Farida, and Mohammad Irfan

Abstract Uncovering the facts of factors that influence financial statement fraud is the purpose of this study. The role of religiosity and detection of red flags from external auditors is a concern to uncover financial statement fraud. This study used a quantitative approach. Purposive sampling method was used and produced 85 research samples. Research shows evidence that red flags and religiosity affect the tendency of financial statement fraud. This research makes an important contribution that companies must provide a good atmosphere for the implementation of employee worship. Providing sufficient time opportunities and good places of worship is the solution to increase religiosity. Indications of red flags in employee's behavior can be used as early detection to prevent fraud.

1 Introduction

Large or small companies are required to make financial statements at the end of the period to account for the company's performance. Financial statements become a form of output from the financial accounting process and the input is in the form of transactions and evidence. Financial statements become important information for investors and creditors. The company strives to display the best possible financial statements both in good and in bad conditions. The company's poor financial condition is the company's biggest scourge. Good packaging of information will be the main point in making rational decisions in investing, credit, and other similar decisions.

R. Agriyanto (✉) · Musahadi · F. I. Turahman · F. Amalia · R. El-Junusi · W. Warno · D. N. Farida
Universitas Islam Negeri Walisongo Semarang, Semarang, Indonesia
e-mail: ratnoagriyanto@walisongo.ac.id

M. Irfan
Department of Management Studies, CMR Institute of Technology, Bangalore, India

A. Rafiki et al. (eds.), *Strategic Islamic Business and Management*, Contributions to Management Science,
https://doi.org/10.1007/978-3-031-61778-2_12

Some companies justify all means so that financial statements appear to experience profits or cover losses. This manipulative or dishonest act is called fraud. Priantara (2013) defines fraud as a deliberate act used as a technique of fraud, lying, or dishonesty in order to rob or eliminate property, money, or property of others due to actions or fatal effects of the actions themselves. The Association of Certified Fraud Examiners (ACFE, 2022) divides three forms of fraud called Fraud Tree. Fraud Tree is Corruption, Asset Misappropriation (misuse of assets/wealth), and Financial Statement Fraud. Based on the survey results (ACFE, 2022), financial statement fraud data is one of the three least common, but most detrimental forms of fraud.

At the top level, employees who engage in asset misappropriation, such as 86% of cases are stolen or employer resources are misused. This scheme is classified as a cause of loss with the lowest median around USD 100,000/case. Second, violations in the form of corruption cases such as bribery, extortion, and conflicts of interest are in the medium median in terms of frequency and loss. Fifty percent points to the level of case schemes occur, so the impact of losses ranges from USD 150,000/case. In contrast to misappropriation of assets, financial statement fraud schemes where the perpetrator intentionally causes a misstatement or material omission of the organization's financial statements is the least common category with the 9% scheme, but the most detrimental up to USD 593,000/case.

Cheating is certainly inseparable from the reason for the perpetrator. The red flags method which is a point for auditors in indicating fraudulent financial statement behavior is important to note. Auditors as independent parties are considered capable of intermediating interested parties in ensuring the fairness of financial statements (Huda et al., 2021). Red flags are a series of circumstances that are beyond activity in general (Coenen, 2008). Red flags function as an early warning signal for possible fraud. These indicators or conditions require a more detailed investigation (Werastuti, 2017). These warnings can help auditors in more accurate investigations in detecting fraud.

Groveman (1995) suggests that audit failure in detecting financial misstatements is an inaccurate reaction to various warning signals from the audit team. Red flags in SAS 99—Consideration of Fraud in a Financial Statement Audit are essential (Ramos, 2004). Auditors are required to make the risk assessment of misstatement more specific. Efforts to overcome fraud certainly vary. Assessment of individuals is more emphasized because the assessment of characteristics on one object will be more focused. Research conducted by Albrecht et al. (1986) found that red flags are considered effective by audit partners to be used in detecting fraud through personal character while corporate character is not effective in detecting fraud.

Religiosity is theoretically a unity of elements so that individuals do not just adhere but can be called religious. Empirically, individual moral standards are influenced by religiosity which contributes to shaping one's idealism (Muhaimin., 2021). Religiosity is binding oneself to a sacred supernatural power (Muhaimin & Mudzakir, 2005). Religiosity has to do with spirituality. Najoan (2020) suggests that the meaning of religiosity is the rebinding of divine values that are spiritual. Religiosity is not spirituality but is the cause of the formation of one's spirituality.

A person's motivation for cheating has to do with spirituality. This benchmark is because aspects of life in the world of work, business, leadership, and religion are related to spirituality. Spirituality affects a person's psychology (Inayah et al., 2018). Religiosity embedded in the spirituality of an auditee can be a controlling factor so as not to commit bad actions. Research by Apsari and Suhartini (2021) shows that religiosity has a negative influence on the level of tendency to commit fraud. Religiosity in individuals becomes the basis that determines individual attitudes to trigger an action, such as not cheating.

Red flags and religiosity have involvement in a person based on the Theory of Planned Behavior (TPB). Ajzen (1991) explains that the formation of a person's behavior is due to intention. Attitude toward the behavior, subjective norms, and perceived behavioral control influence intentions. Prevention through understanding the existence of red flags alarms that have the potential to affect the three components of the SDGs allows it to more effectively anticipate so that fraud does not occur. The nature of religiosity in a person can be a barrier when someone intends to commit fraud because the mindset of these actions is prohibited in religion (Apsari & Suhartini, 2021).

Knowing that red flags and religiosity affect the level of propensity for financial statement fraud is the goal of this study. The originality of the study is to present a variable of religiosity based on the auditor's perception in explaining the tendency to commit financial statement fraud. This research is expected to be an alternative solution to explain the problem of financial statement fraud.

2 Literature Review and Hypothesis Development

2.1 Theory of Planned Behavior (TPB)

The Theory of Reasoned Action was developed into the Theory of Planned Behavior (TPB). Ajzen (1991) explains that the formation of a person's behavior is due to intention. Attitude toward the behavior, subjective norms, and perceived behavioral control influence intentions.

Attitude toward the behavior is something that is formed due to the level of a person in giving an assessment or evaluation of the positive or negative impact of a behavior he does. The determination is based on a person's belief that manipulating financial statements includes unethical behavior, then someone will not intend to commit fraud.

Subjective norm is something that arises because of social impact. Someone who lives with the presence or absence of pressure accompanied by positive or negative responses will see a difference in determining a course of action. An environment that supports someone under the pretext of rationality for cheating financial statements will stimulate someone to realize it, and vice versa.

Perceived behavioral control is something related to the position of control that a person feels. Control of one's behavior arises due to easy or difficult actions to be carried out by considering various perceptions. An obstacle that comes from oneself and the environment can be a control. A person's intention to commit financial statement abnormalities tends to be controlled when there are no loopholes or opportunities for fraud.

The SDGs estimate religiosity and red flags have involvement in a person. Prevention through understanding the existence of red flags alarms that have the potential to affect the three components of the SDGs allows it to more effectively anticipate so that fraud does not occur. The existence of religiosity in a person can be a barrier when someone intends to commit fraud because the mindset of these actions is prohibited in religion (Apsari & Suhartini, 2021).

2.2 *Red Flags*

Red flags are commonly applied to various audit literature in order to signal red flags. Clues to the presence of these danger signs can be marked through an unusual condition, so further investigation is needed. Coenen (2008) defines red flags as follows: "A red flag is a set of circumstances that are beyond activity in general. This is intended as a signal that may need further investigation. Keep in mind that red flags are not to indicate guilt or innocence, but can indicate possible signs of fraud."

Ramos' (2004) SAS 99—Assessing the risk resulting from misstatement as a result of fraudulent flows is a special consideration for auditors in assessing fraud in Financial Statement Audit. A person cannot be justified as guilty or not absolutely through red flags, but red flags act as a warning alarm when fraud has, is, or will occur. Based on data by ACFE (2022), from 85% of all scams will display at least one red flag of eight key leads.

There are eight red flag indicators that are most commonly seen, namely: (1) Behavior pursuing wants rather than needs. They seem to live beyond their means. More concerned with lifestyle interests by buying consumptive goods such as luxury homes, cars, electronic goods, and so on; (2) The second indicator is financial difficulties. As a result of fulfilling desires, the debt usually accumulates from more than one source; (3) have an unreasonable relationship with a customer or supplier. A very close relationship is an indication that there is a transactional relationship. This incident needs to be watched out by the company; (4) Not wanting to share tasks with fellow colleagues is also an indicator of red flags. Self-inflicted tasks indicate that something is kept secret. So this is a control problem; (5) Irritability, irritability, and suspicion with fellow co-workers are indicators that there are confidential things. The fear of being disturbed by his existence in a matter becomes a trigger for suspicion and anger; (6) The occurrence of intimidation or bullying, incidents of intimidation or bullying usually occur because there is a threat to the comfort that has been obtained; (7) Family problems or divorce are indicators of red flags. A happy family is a person's main capital to be calm at work. Broken families have the potential to interfere with work activities. Broken families also reflect a lack of ability to manage disagreements within the family. Differences of opinion certainly also occur in corporate linkups; and (8) dishonest or unfair attitudes toward purpose or success are indicators of red flags. Red flags were identified in at least 76% of each case.

Accounting scandals on financial statement fraud prove that audit failures lead to a loss gap. An auditor's unresponsiveness to red flags alarms is one factor. Red flags need special attention again, considering the results of research on the importance of the red flags method as an effort to detect fraud has again become a special concern since the Enron case. The American Institute of Certified Public Accountants revealed that before the case that brought down Enron, there were actually clear red flags. The claim that red flags are effectively used as an effort to detect fraud also agrees with several other researchers. A similar case occurred at Global Crossing, Worldcom in the United States, which was able to cause a big stir in the world of capital markets. In the manufacturing sector in Indonesia, such as PT Kimia Farma Tbk, PT Indofarma Tbk, and PT Garuda Indonesia (Persero) Tbk, and others.

2.3 *Religiosity*

The term religion is derived from the word *religie* (Dutch), or *religion* (English), which was instilled into Indonesian by Westerners (Dutch and English) as a strategy for spreading Catholicism and Christianity when colonizing Indonesia. The origin of Latin word *relegere* or *relegare* is used to mean the word "religion." *Relegare* has the basic meaning of "being careful," and adhering to the norms or rules strictly. *Relegare* also means "to bind," with the intention of binding oneself to a sacred supernatural power. Individuals who are referred to as *being religious* and not just claiming to have religion (*having religious*) are manifestations of this element.

Empiris and Lpd (2021) explain religiosity as a belief that exists in a person related to the religion adopted by implementing the values contained in it for the life lived. Someone with a high level of religiosity will affect a person's consideration in behavior. Conformity taught in religion is able to anticipate a person to behave negatively. Considering these conditions, religiosity has a relationship with spirituality. Najoan (2020) asserts that a form of rebinding divine values of a spiritual nature is the meaning of religiosity. Religiosity is not spirituality but is the cause of the formation of one's spirituality. This certainly requires actors to manage it in the form of leaders. Simply put, leaders can start from themselves in managing good or bad things that race on spiritual values. Junusi et al. add that Spiritual Global Leadership (SGL) will energize in guiding, empowering, and giving a broader global vision of improving organizational performance. More responsibility in the work of organizations is based on spiritual values that they believe will be a good thing for the future. Strak and Glock (1970) hold the view that there are five dimensions of religiosity, including belief, practice, knowledge, experience, and consequential.

2.3.1 The Belief Dimension

The dimension of belief means that the views of religious people are related to theology by justifying religious teachings. The various instruments of belief that each religion maintains are expected so that its adherents can ratify them. The dimension of belief becomes doctrinaire that must be obeyed by adherents of their religion, such as beliefs about the existence of God, Angels, books, Prophets and Apostles, doomsday, heaven, hell, and others that are dogmatic in nature.

2.3.2 Religious Practice

Religious practice includes worship and devotion activities carried out by individuals as a form of religious commitment. Religious practice has two important levels. At the first level, rituals are related to a series of religious ceremonial instruments or rites and other sacred practices in the hope that they can be carried out by their adherents. At the second level of devotion which has a parallel concept to ritual, but refers more to individual piety rather than the public.

2.3.3 The Experience Dimension

All religions have certain expectations is a dimension of experience considering. Each individual experiences different spiritual sensations. Society will involve communication with the divine essence, although there is a diversity of experiences that gives rise to varying degrees as a sign of the individual's religiosity. The appreciation felt by believers can be in the form of fear of sinning, feeling close to God and others.

2.3.4 The Knowledge Dimension

The dimension of knowledge relates to an expectation for religious people to have at least a little information/knowledge taught in the religion adhered to. Such knowledge can be the basis of believed faith, a series of rites, scriptures, and others. This dimension is continuous with the dimension of belief because the ability of individuals to know about a belief they hold is a prerequisite.

2.3.5 The Consequential Dimension

The consequential dimension of religious commitment is different from the other four dimensions. This dimension identifies the effects of religious beliefs, practices, experiences, and knowledge on an individual's daily life. Religion plays a decisive role in much of how adherents should think and act in daily life, but religious experience depends on the individual adhering to a religion. The attitudes and behaviors carried out by individuals in accordance with religious guidance will reflect their level of religiosity.

2.4 *Financial Statement Fraud*

According to the Association of Certified Fraud Examiners (ACFE, 2022) defines Financial Statement Fraud is intentional behavior that causes misstatement of numbers in financial reporting. Such behavior may take the form of creating notes for false expenses. Creating fictitious trips or buying food that did not actually make the purchase. Two forms of financial fraud that can be done are net income understatement and net income overstatement. Such fraud can be carried out through the components of playing the difference in transaction time, by recording, presentation of nominal that does not match the facts, deliberate errors in asset valuation, fictitious income, fictitious costs, concealed liabilities, and expenses.

Wells said several modes that often occur in financial statement fraud, such as: (1) manipulation in financial records and transaction/supporting documents; (2) intentionally omitting traces of information, accounts, transactions, and other events to be presented in financial statements; (3) implementation or deliberate misconduct, measurement, reporting, recognition and disclosure of business transactions and economic events through incorrect policies, accounting principles, and procedures; (4) intentional omission of information in the form of accounting policies and principles used for making financial statements. Based on the foundation of Islam, it is certainly inseparable from the underlying propositions, one of which is in surah Al-Mutafifin verses 1–3:

٣٢ الَّذِيْنَ اِذَا اكْتَالُوْا عَلَى النَّاسِ يَسْتَوْفُوْنَ وَاِذَا كَالُوْهُمْ اَوْ وَّزَنُوْهُمْ يُخْسِرُوْنَۗ ١وَيْلٌ لِّلْمُطَفِّفِيْنَ

> It means: (1) It is a great accident for those who cheat, (2) Those who when they receive a measure from others they ask to be fulfilled, (3) And when they measure or weigh for others, they reduce. (Al-Mutafifin: 1-3)

Based on the Tafsir of the Ministry of Religious Affairs of the Republic of Indonesia, paragraph 1 explains that those who cheat in measuring and weighing will get tremendous punishment and humiliation on the Day of Judgment. The terrible threat has been conveyed by Allah Almighty to those who measured and weighed fraudulently at the time of buying and selling in places that were in Mecca and Medina at that time.

It is narrated that a man named Abu Juhainah was in Medina. He had a small and large kind of dose. He used a large amount when buying dates or wheat from farmers, but he used a small measure when he sold it to others. This action shows greed in a person by seeking his own benefit, even though it harms others.

In verses 2 and 3, an explanation from Allah SWT is related to the deeds of those who will inhabit hell. They are people who do not want to lose by wanting a full dose or scale when buying something while when selling, they will reduce the dose or scale.

Actions done by people by reducing the measure or scale will get a great sin. The reason is that the act is considered to have eaten other people's property without the sincerity of the legal owner. Allah forbids such acts. Allah said:

وَلَا تَأْكُلُوْٓا اَمْوَالَكُمْ بَيْنَكُمْ بِالْبَاطِلِ

It means*:* And do not eat of treasures among you in a vanity way. And eat not of treasures among you in a vanity way. (Al-Baqarah: 188)

This intended measure includes all scales and measures commonly used in buying and selling and their relation to the reduction of the property rights of others. The majority of various reductions found to harm others, including reducing liters of gasoline, filling gas cylinders that are not according to standards, and so on. Other dose reductions that are very detrimental and dangerous are such as manipulation of financial statements by manipulating nominal. Perpetrators reduce or increase the nominal in order to obtain their own benefits, regardless of the impact of being able to harm others and even the state.

This verse admonishes people to stay away from the prohibitions of Allah Almighty in the form of actions that harm others with the threat of extraordinary punishment in the world and in the hereafter. A harmonious verse commands man to meet or perfect the scales as in the word of Allah:

وَاَوْفُوا الْكَيْلَ اِذَا كِلْتُمْ وَزِنُوْا بِالْقِسْطَاسِ الْمُسْتَقِيْمِۗ ذٰلِكَ خَيْرٌ وَّاَحْسَنُ تَأْوِيْلًا ٣٥

It means: And perfect the measure when you measure, and weigh it with the correct scales. That's the main thing (for you) and the better the result. (Al-Isra': 35).

3 Hypothesis Development

3.1 The Relationship Between Red Flags and Financial Statement Fraud

The red flags method is a method used by auditors in detecting indications of fraud. The results with this method are not absolutely the same among auditors, but red flags can be used as a detection tool for the early stages of fraud. The red flags method for auditors does not have specific guidelines that are most effective in detecting financial statement fraud, so all indicator assumptions for auditors are important. Indications of fraud arise because of the encouragement of someone's intentions. The background of a person's life is indicated in the 8 key warning signs on red flags that guide the level of attitude toward behavior, subjective norms, and control of one's behavior to do so. This shows that the tendency of financial statement fraud is influenced by red flags.

High red flags will result in a high tendency for financial statement fraud. According to Tedjakusuma (2012), red flags can support auditors to focus more on assessing fraud risk. Rustiarini et al. (2016) showed that for auditors, red flags indicators are effectively used to detect fraud. The formulation of the first hypothesis is based on previous research, and the logic of the relationship is described as follows:

H1. Red flags positively affect the tendency of financial statement fraud.

3.2 The Relationship Between Religiosity and Financial Statement Fraud

Religiosity is a comprehensive unity of elements, so that it is able to shape a person's personality not just claiming to embrace religion. Religiosity in a person is considered as a control of the intention not to commit fraud (stay away from prohibitions) and do financial reporting reasonably (strengthen confidence in carrying out the commands of Allah SWT). Someone who has high religiosity will not be able to commit financial statement fraud. Understanding and belief in everything that is done is always monitored and will be accounted for in the future. Barnet et al. in Giovano et al. (2020) stated that religiosity is a tool that regulates one's behavior according to standards. Someone who has high religiosity will have a low impact on committing fraud. The formulation of the second hypothesis is based on previous research, and the logic of the relationship is described as follows:

H2. Religiosity negatively affects the tendency of financial statement fraud.

4 Method

A quantitative approach was used in this study. Quantitative research explains the factors that influence financial statement fraud, namely red flags and religiosity. The study population is auditors working in Public Accounting Firms. Eighty auditors were used as samples. They are those who work in Public Accounting Firms with at least 1 year of work experience. Progression analysis is used to process research data. Hypothesis testing used t-test with a significance value of 5%.

Red flags are early warning signals that function in alerting auditors to the possibility of fraud. These indicators or conditions require a more detailed investigation (Werastuti, 2017). Red flags are considered an early warning alarm that is useful as a notifier of the signal of ignorance, so as to minimize the risk of fraud not being detected. The warning will help further investigation so that the audit results are more in-depth and accurate. The instrument used in measuring red flags is to identify eight key warning signs (ACFE, 2022). Red flag measurements indicators such as a luxurious lifestyle beyond financial means, difficult finances, too close relationships between employees and customers/vendors, suspicious, irritability, bullying or threatening, broken homes, and fraudulent attitudes in achieving goals.

Religiosity is a person's belief in religion that is adopted through various religious practices in his life. A person who has high religiosity will well consider his actions. This is because the religious teachings adopted will fortify or bind someone in anticipation of negative actions (Empiris & Lpd, 2021). The instruments used in measuring religiosity consist of five dimensions (Strak & Glock, 1970). These measurement indicators include dimensions of religious belief, practice of observing religious law, dimensions of religious experience, dimensions of knowledge about religion, and dimensions of consequences of worship.

Financial statement fraud is a form of fraud that misrepresents material financial statements committed by management so that it has an impact on investors and creditors. In 2022, based on the results of the ACFE survey related to fraud, financial statement fraud data is one of the three least common but most detrimental forms of fraud. The instruments used in measuring financial statement fraud consist of 5 indicators (ACFE, 2022). These measurement indicators include differences in recording time with economic events, fictitious income and expenses, concealment of liabilities and expenses, improper valuation of assets, and improper disclosure of nominal financial statements.

5 Result and Discussion

5.1 Descriptive Statistical Analysis

The variables used in this study include red flags and religiosity variables to determine the influence of financial statement fraud tendencies. The results of this test are shown in Table 1 by considering the levels of the assessment indicators (Table 2).

Based on Table 1, it can be explained that descriptive red flags have a mean value of 25.07 (medium category). The variable religiosity has a mean value of 43.14 (high category). The variable financial statement fraud has a mean value of 21.77 in the low category (Table 3).

Table 1 Descriptive statistical analysis

	N	Minimum	Maximum	Mean	Std. Deviation
Red flags	71	10	40	25.07	**8.400**
Religiosity	71	20	50	43.14	**6.386**
Financial statement fraud	71	10	50	21.77	**11.184**
Valid N (listwise)	**71**				

Source: Data processed, 2023

Table 2 Assessment indicators

Mean	Category
10–23,3	Low
23,4–36,7	Medium
36,8–50	High

Sumber: Data processed, 2023

Table 3 Validity test

Variable	Item	R count	R table	Remark
Red flags	1	0,723	0,2335	**Valid**
	2	0,856	0,2335	**Valid**
	3	0,853	0,2335	**Valid**
	4	0,848	0,2335	**Valid**
	5	0,834	0,2335	**Valid**
	6	0,809	0,2335	**Valid**
	7	0,828	0,2335	**Valid**
	8	0,740	0,2335	**Valid**
	9	0,830	0,2335	**Valid**
	10	0,617	0,2335	**Valid**
Religiosity	1	0,823	0,2335	**Valid**
	2	0,860	0,2335	**Valid**
	3	0,860	0,2335	**Valid**
	4	0,868	0,2335	**Valid**
	5	0,889	0,2335	**Valid**
	6	0,892	0,2335	**Valid**
	7	0,887	0,2335	**Valid**
	8	0,839	0,2335	**Valid**
	9	0,891	0,2335	**Valid**
	10	0,893	0,2335	**Valid**
Financial statement fraud	1	0,877	0,2335	**Valid**
	2	0,870	0,2335	**Valid**
	3	0,883	0,2335	**Valid**
	4	0,911	0,2335	**Valid**
	5	0,914	0,2335	**Valid**
	6	0,894	0,2335	**Valid**
	7	0,891	0,2335	**Valid**
	8	0,856	0,2335	**Valid**
	9	0,899	0,2335	**Valid**
	10	**0,920**	**0,2335**	**Valid**

Source: Data processed, 2023

5.1.1 Validity Test

The validity test is used as a measure of the validity of the questionnaire, with the following results:

The results of observation in the r table get the value of sample (N) = 71, amounting to 0.2335. In the validity test, it is produced that all instruments produce r count positive and r count > r table. This shows that all instruments in this study can be said to be valid.

5.1.2 Reliability Test

Reliability tests are performed in order to assess the consistency of the instrument. A research instrument is said to be reliable when it gets a Cronbach Alpha score of >0.07. Table 4 shows the results of the 3 variables used in this study after going through reliability tests.

The results of the reliability test showed the value of Cronbach's Alpha in the variables red flags of 0.935, religiosity of 0.964, and financial statement fraud of 0.971 with each instrument of this study reliable because it has a value of Cronbach's Alpha >0.70. This shows that each statement item produces consistent data. If the statement is given to the respondent again, it will get a relatively similar answer to the previous one (Fig. 1).

5.2 Hypothesis Testing

The figure 1 show hypothesis testing (Fig. 1).

The significance value for red flags for financial statement fraud is 0.000 < 0.01 and the calculated t value is 6.157 > t Table 1.995, so it can be concluded that the first hypothesis is accepted. This shows that red flags have a significant positive effect on financial statement fraud. The significance value of religiosity to financial statement fraud is 0.000 < 0.01 and the value of t count −4.045 < t table −1.995, so it can be concluded that religiosity is accepted. This shows that religiosity has a significant negative effect on financial statement fraud. This also means that the higher the religiosity, the smaller the financial statement fraud.

Table 4 Reliability test

Variable	Cronbach's Alpha	Number of items
Red flags	0.935	**10**
Religiosity	0.964	**10**
Financial statement fraud	**0.971**	**10**

Source: Data processed, 2023

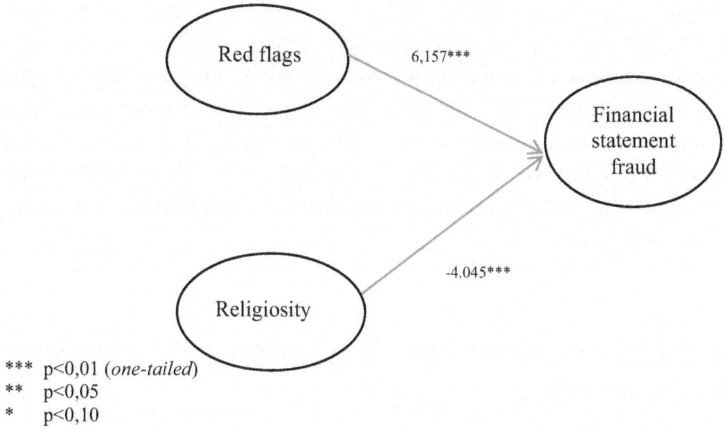

Fig. 1 Full model results of Hypothesis Testing. *** $p < 0{,}01$ (*one-tailed*) ** $p < 0{,}05$ * $p < 0{,}10$

6 Discussion

This study aims to analyze how red flags and religiosity affect the tendency of financial statement fraud through explanatory quantitative methods with the distribution of questionnaires. This is because fraud in financial statements is the most detrimental fraud although fraud schemes include categories that rarely occur compared to corruption and misuse of assets. Referring to the report (ACFE, 2022), red flags have an influence on someone triggered to commit financial statement fraud. For auditees who are exposed to red flags indicators, it is undeniable that in general, they will look for ways/solutions even though by justifying all means. The red flag indicators show that there is a tendency for financial statement fraud. However, considering the pressure felt and an adequate opportunity to modify financial statements for personal benefit rationalized by the auditee, then it is considered reasonable. This mindset needs to be addressed and the auditor's perspective is needed in this study as a foundation for determining auditees with integrity as managers and makers of financial statements. In addition to red flags, religiosity in the auditee becomes a system control related to the motivation or encouragement of fraud, so it is necessary to consider the auditor's perspective, because in the process of assessing, the auditor's financial statements not only correct the nominal but also consider the auditee's background information.

As a result of financial statement fraud, based on the report of ACFE (2022), the loss rate reached $593,000, even though the percentage of the scheme was only 9%. The very high loss compared to corruption of $ 150,000 with a percentage of 50% scheme and asset misappropriation of $ 100,000 with a percentage of 56% scheme needs special attention. If this is not overcome or minimized, it may increase. The increase will certainly cause negative impacts that can spread to all aspects of life, especially unhealthy economic growth.

The results of this statistical test study use sample data that can be processed as much as 71 or 83.5% of the data that should have been 85 auditors (respondents) in total spreading across 17 active Public Accounting Firms that have establishment permits from the Ministry of Finance in the scope of Semarang. The tendency of financial statement fraud is analyzed through explanatory quantitative methods with red flags and religiosity variables. These variables show that there is a relationship in influencing the tendency of financial statement fraud positively or negatively and partially and stimulatively.

6.1 Red Flags Positively Affect the Tendency of Financial Statement Fraud

The results of the study provide evidence that significance value for X1 to Y is $0.000 < 0.05$ and the calculated t value is $6.157 > t$ Table 1.995, so it can be concluded that H1 is accepted. This shows that red flags have a significant positive effect on Financial Statement Fraud, so it can be concluded that H1 is accepted. This means that when an auditee has a high level of red flags, the potential tendency to commit financial statement fraud will be high. Similarly, if an auditee has a low level of red flags, then the potential tendency to commit financial statement fraud will be low. When referring to the theory of planned behavior proposed by Ajzen (1991), it can be said that behavior is formed because of intention. Due to problems in the auditee's life included in the 8 key warning signs red flags, such as living beyond means, financial difficulties, having unusually close relationships with customer vendors, excessive control problems or unwillingness to divide tasks, irritability, suspicious, unusual defensiveness, experiencing bullying and intimidation, experiencing confusion or family problems, and having a wheeler-dealer attitude can shape behavior to intend to do something to be able to solve the problems faced. The existence of 8 key warning signs triggers someone to commit fraud by rationalizing the action, especially if there is an adequate opportunity.

This research is in line with research conducted by Rustiarini et al. (2016) that red flags have a positive effect in detecting financial fraud, although the study is derived from a survey of 84 auditor perspectives, there is still a minority of auditors who realize the importance of using fraud indicators in detecting fraud levels. The assessment results also show that each red flag indicator has a difference in effectiveness in the fraud detection process. The most effective component of red flags according to auditors' perceptions is the opportunity with the highest means of 2.29. This can be in the form of bank accounts or subsidiaries/branches operating showing uncertainty, dominance in management without clear control, and so on for personal gain.

6.2 Religiosity Negatively Affects the Tendency of Financial Statement Fraud

The results of the analysis show that the significance value for X2 to Y is $0.000 < 0.05$ and the calculated t value is $-4.045 < t$ table -1.995, so it can be concluded that H2 is accepted. This shows that religiosity has a significant negative effect on Y, so it can be concluded that H2 is acceptable. This means that when an auditee has a high level of religiosity, the potential tendency to commit financial statement fraud will be low. In contrast, if an auditee has a low level of religiosity, then the potential tendency to commit financial statement fraud will be low. When referring to the theory of planned behavior proposed by Ajzen (1991), it can be said that behavior is formed because of intention. The religious nature of an auditee is what is needed in intention control when in conditions indicated by 8 key warning sign red flags. The growth of religiosity in auditors is not only through religious self-recognition but with the implementation of behavior in accordance with the religious values adhered to. Understanding and awareness of the limits or rules that in religion bind individuals not to deviate from the provisions.

This research is in line with research conducted by Giovano et al. (2020) showing that religiosity has a negative influence on the tendency of accounting fraud. The results of the study justify that someone with a high level of religiosity, will be low in committing fraud.

7 Conclusion

Based on the results of research on the analysis of the influence of red flags and religiosity derived from various perceptions of auditors working in Semarang scope of public accounting on the tendency of financial statement fraud carried out by auditees in Semarang, it can be concluded that red flags partially have a positive influence on the tendency of financial statement fraud in Semarang with a significant value of $0.000 < 0.05$ and a calculated t value of $6.157 > t$ table of 1.995. Therefore, the higher the red flags indication on the auditee, the higher the tendency of financial statement fraud committed by the auditee, and vice versa.

Partial religiosity has a negative influence on the tendency of financial statement fraud in Semarang with a significant value of $0.000 < 0.05$ and a calculated t value of $-4.045 < t$ table—1.995. This concludes that the higher the religiosity of the auditee, the lower the tendency of financial statement fraud committed by the auditee, and vice versa.

Red flags and religiosity simultaneously or together have an influence on the tendency of financial statement fraud in Semarang with a significant value of $0.000 < 0.05$ and a calculated F value of $45.303 > F$ Table 3.130. This in fact has a continuous influence role based on the high or low red flags and religiosity, thus affecting the level of financial statement fraud tendencies in Semarang.

The limitation of the study lies in the auditor's perception of data with the timing of the early year picking. We know that at the beginning of the year auditors are busy doing work. To get maximum results, data collection should be carried out evenly, for example at the beginning, middle, and end of the year. Based on the results of the research, the suggestions that can be given are: (1). For companies, the impact of financial statement fraud is certainly very detrimental. Therefore, the selection of internal accountants or managing financial statements needs to be demanded in terms of performance so that it is clearly not just a family relationship or people without considering their integrity. It is expected that individuals who occupy positions can be trusted without easily looking at the privilege; (2). For internal accountants, they must present financial statements in accordance with actual facts. This effort is to minimize increasing losses that have a negative impact on companies and economic growth. Increased religiosity can change bad intentions when red flags are indicated; (3). For auditors, it is also necessary to pay attention to personal information that can be the basis for additional information in conducting audits to detect fraud, not just content in financial statements. Personal information is important to increase the validity of the opinion to be given.

Attachment Questionnaire

Red Flags

No.	Statement	Answer				
		1	2	3	4	4
1.	A thing that does not become a problem in auditing, if the auditor ignores the *background* of the auditee's life.					
2.	A thing that is not a problem for the emergence of the *financial statement fraud* scheme, if the auditee is a person with a lifestyle beyond his financial ability.					
3.	A thing that is not a problem for the emergence of the *financial statement fraud* scheme, if the auditee is a person who is under pressure or lack of financial.					
4.	A thing that is not a problem for the emergence of the *financial statement fraud* scheme, if the auditee is a person who has an unusual close relationship with vendors and customers.					
5.	A thing that is not a problem for the emergence of the *financial statement fraud* scheme, if the auditee is a person who has excessive control problems or reluctance to divide tasks.					
6.	A thing that is not a problem for the emergence of *financial statement fraud* schemes, if the auditee is an unusual irritable, suspicious or defensive person.					

No.	Statement	Answer				
		1	2	3	4	4
7.	A thing that is not a problem for the emergence of the *financial statement fraud* scheme, if the auditee is a victim or perpetrator of bullying or intimidation internally or externally.					
8.	A thing that is not a problem for the emergence of the *financial statement fraud* scheme, if the auditee is a person who has a divorce or has family problems.					
9.	A thing that is not a problem for the emergence of *financial statement fraud* schemes, if the auditee is a person who has a "*Wheeler-dealer*" attitude that involves clever or immoral behavior.					
10.	Audit failure is solely due to an auditor's mistaken opinion.					

Religiosity

No.	Statement	Answer				
		1	2	3	4	5
1.	The auditee needs to believe that god exists and always supervise every activity carried out by his servant.					
2.	Religion becomes a religious element for auditees as a guideline for life.					
3.	Auditees need to understand the permissible and prohibited limitations of their religion.					
4.	Auditees who are pious will stay away from God's prohibitions, including conducting *financial statement fraud.*					
5.	Auditees with integrity will feel remorse and sin when committing bad or prohibited acts by religion, such as committing *financial statement fraud.*					
6.	Auditees need to understand and believe that every action will get a reward from god.					
7.	Auditees who are able to understand well the laws regarding fraud in their religion have better behavior, subjective norms and behavioral control in performance.					
8	A knowledgeable auditee will prioritize adab to minimize *financial statement fraud actions.*					
9.	Auditees with integrity will firmly reject invitations that violate norms, such as conducting *financial statement fraud.*					
10.	The auditee must always maintain the mandate given.					

Financial Statement Fraud

No.	Statement	Answer				
		1	2	3	4	5
1.	It is natural for the auditee, if for a certain purpose the costs recorded are greater than they should be.					
2.	It is natural for the auditee, if the recording of proof of transaction is carried out without authorization from the authorized party.					
3.	Something natural for the auditee, if the budget user includes other needs that are not in accordance with the budget.					
4.	It is normal for auditees to use blank receipts for purchases.					
5.	It is natural for the auditee, if an office equipment and equipment purchased is not in accordance with the specifications that must be purchased.					
6.	It is natural for the auditee, if the transaction time is not in accordance with the proof of transaction.					
7.	It is natural for the auditee, if it is found that there are expenses and income without supporting documents.					
8.	It is natural for the auditee to distribute the remaining budget to employees as a bonus.					
9.	It is natural for the auditee, if for certain purposes the company increases profits in order to attract investors / creditors.					
10.	It is natural for the auditee, if it rationalizes unilateral profiteeship for its services without reporting in the financial statements.					

References

ACFE. (2022). Occupational fraud 2022: A report to the nations. *Association of certified fraud examiners*, 1–96.

Ajzen, I. (1991). The theory of planned behavior. *Organizational Behavior and Human Decision Processes, 50*(2), 179–211. https://doi.org/10.1016/0749-5978(91)90020-T

Albrecht, S., Howe, K., & Romney, M. (1986). Red-flagging management: A validation. *Advances in Accounting*.

Apsari, A. K., & Suhartini, D. (2021). Religiosity as moderating of accounting student academic fraud with a hexagon theory approach. *Accounting and Finance Studies, 1*(3), 212–231. https://doi.org/10.47153/afs13.1512021

Coenen, T. (2008). Red flags of fraud. In *Essentials of corporate fraud* (pp. 47–69). Wiley. https://doi.org/10.1002/9781118386156.ch3

Empiris, S., & Lpd, P. (2021, April). *Hita Akuntansi dan Keuangan Universitas Hindu Indonesia Edisi*, 297–312.

Giovano, A., Wibowo, A. S., & Yanuarisa, Y. (2020). Pengaruh love of money dan Religiusitas Terhadap Kecenderungan fraud accounting Dana Desa dengan Gender sebagai Variabel Moderasi Pada Desa di Kecamatan Katingan Tengah. *Balance, 12*(6), 11–24.

Groveman, H. (1995). How auditors can detect financial statements misstatement. *Journal of Accountancy*.

Huda, C., Agriyanto, R., Lestari, H. S., & Pangayow, B. (2021). Financial distress as a moderating variable of the influence of audit opinion and public accounting firm size on voluntary auditor switching. *Journal of Islamic Accounting and Finance Research, 3*(2), 155–176. https://doi.org/10.21580/jiafr.2021.3.2.8609

Inayah, N., Agriyanto, R., & Warno, W. (2018). The role of spirituality in the behavior of sharia bank mobile banking: Evidence from Indonesia. *Walisongo: Jurnal Penelitian Sosial Keagamaan, 26*(1), 197–224.

Muhaimin. (2021). Pengaruh love of money dan Religiusitas terhadap fraud accounting Anggaran Dana Desa pada Kecamatan Sinjai Tengah. *Journal of Management, 4*(2), 121–133. https://doi.org/10.37531/yum.v11.75

Muhaimin, Mujib A., & Mudzakir, J. (2005). Kawasan dan Wawasan Studi Islam.

Najoan, D. (2020). Memahami Hubungan Religiusitas Dan Spiritualitas Di Era Milenial. *Educatio Christi., 1*(1), 64–74.

Priantara, D. (2013). *Fraud auditing and investigation*. Mitra Wacana Media.

Ramos, M. J. (2004). Fraud detection in a GAAS. *Audit Revised Edition., 99*.

Rustiarini, N. W., Suryandari, A., & Nova, S. (2016). Red flags and fraud prevention. *Buletin Ekonomi Moneter Dan Perbankan, 19*(2), 177–206.

Strak, R., & Glock, C. Y. (1970). *American piety: The nature of religious commitment*. University Of California Press.

Tedjakusuma, F. N. (2012). Pentingnya Red Flag Bagi Auditor Independen untuk Mendeteksi Kecurangan dalam Laporan Keuangan. *Berkala Ilmiah Mahasiswa Akuntansi, 1*(3), 47–52.

Werastuti, D. N. S. (2017). Pengaruh Kerakteristik dan Kompetensi Auditor Pada Persepsi Auditor Atas Efektivitas Red flags Untuk Mendeteksi Kecurangan Pelaporan Keuangan. *Indonesian Journal of Accounting and Governance, 1*(1).

Ratno Agriyanto, M.Si, Akt, CA, CPA is a lecturer and he is still active as the central administrator of the Sharia Accounting Compartment–Indonesian Institute of Accountants.

Musahadi is a professor and he is an expert in the field of Islamic studies.

Fitra Istianah Turahman is a Sharia accounting student at Walisongo State Islamic University.

Farah Amalia, MM is a lecturer and expert in behavioral finance.

Rahman El-Junusi, SE, MM is a lecturer and expert in the field of unity of science and integration of Islamic science.

W Warno, SE, M.Si, SAS is a lecturer and also active as editor-in-chief of Islamic accounting and finance research journals.

Dessy Noor Farida, M.Si, Ak, CA is a lecturer and expert in finance, management.

Mohammad Irfan, MBA, M.Com, MA, PhD is a lecturer and expert in the field of Islamic Finance, Banking and Capital Market.

Exploring Hurdles in Accessing Islamic Microfinance for Sustainable Poverty Reduction in Nigeria

Joy E. Akahome and Sidney Ogodo

Abstract This study investigates the challenges hindering underprivileged Nigerians from accessing Islamic microfinance loans. Employing qualitative methods multiple case study approach, the research analyzes the experiences of three individuals who faced such challenges. The objective is to identify and analyze primary barriers encountered by Nigerians seeking financial services from Islamic Microfinance Institutions (IMFIs). The identified challenges include Convenient Access for the Unbanked Community, Limited Awareness, Stringent Eligibility Criteria, and High Cost of Financing. The narrative of the case study findings sequentially presents each business owner's account, incorporating excerpts from their interviews. The study concludes by emphasizing the pivotal role that Islamic microfinance institutions can play in sustainable poverty alleviation efforts.

1 Introduction

The exploration of poverty and its reduction is a longstanding concern across Africa and globally (Adeleke & Alabede, 2021; Ishak, 2020; Madise, 2019; Salako & Azeez, 2018). Therefore, in several developing nations, microfinance is widely regarded as a pivotal tool for alleviating poverty (Dogarawa & Uthman, 2019). The prevalence of this assertion suggests that when financial services are offered to people who are impoverished, it can empower them to engage in microenterprises, amass savings, augment their incomes, stabilize consumption patterns, reduce

J. E. Akahome (✉)
Department of Entrepreneurship and Marketing, Faculty of Management Sciences, Federal University, Otuoke, Bayelsa State, Nigeria
e-mail: akahomejj@fuotuoke.edu.ng

S. Ogodo
Department of Business Administration and Marketing, School of Management Sciences, Babcock University, Ilisan, Ogun state, Nigeria
e-mail: ogodo0096@pg.babcock.edu.ng

A. Rafiki et al. (eds.), *Strategic Islamic Business and Management*, Contributions to Management Science,
https://doi.org/10.1007/978-3-031-61778-2_13

financial risks, and ultimately break free from the cycle of poverty. It is evident that scholarly discussions have progressed beyond merely criticizing the "idealistic approach to microfinance," this acknowledges that microfinance falls short of the miraculous solution it was once heralded to be (Dogarawa & Uthman, 2019; Dogarawa, 2012, 2013). However, the international development community has remained hopeful about increasing the awareness of microfinance as a potential tool for alleviating poverty, which asserts its sustainability to address some financial exclusion among the poor people. This statement rests on the agreement that microfinance has demonstrated considerable success in furnishing sustainable financial services to marginalized populations who would otherwise be excluded from mainstream financial systems (Egboro, 2015; Gumel et al., 2014; Nwanyanwu, 2011).

Recently, there has been a growing international consensus to prioritize poverty alleviation in sustainable development goals. Governments have committed themselves to formulate and implement some anti-poverty policies with achievable objectives for substantial poverty reduction (Salako & Azeez, 2018). Bilateral or multilateral organizations, governments, and non-governmental organizations (NGOs) underscore the crucial role of the latter group in global poverty alleviation endeavors. Many NGOs have challenged governments' historical inadequacies in addressing wealth distribution and resource allocation, which are root causes of poverty (Okpara, 2010). In addition, Nigeria has witnessed various efforts by successive administrations to combat poverty through microfinance initiatives. Numerous microfinance platforms, including traditional and Islamic microfinance, have been established to counter the poverty threat. Despite these efforts, poverty persists as a significant concern for the government, private sector, and NGOs due to widespread financial services exclusion among many Nigerians.

Despite the considerable number of microfinance banks in Nigeria, which exceeds 800, their contribution to the country's economy remains less than 1%, compared to the banking sector's 10%, as reported by the National Bureau of Statistics (NBS) (2020). Unfortunately, the scarcity of microfinance facilities and limited financial inclusion have worsened the poverty rate in the country. Moreover, over 60% of Nigeria's Small and Medium Enterprises (SMEs) face financial exclusion from mainstream financial opportunities due to conventional microfinance banks' stringent requirements, as highlighted by the National Bureau of Statistics (2019).

However, the daily establishments of small-scale businesses remain a significant need for micro-credit facilities, a service that Islamic microfinance banks are well-positioned to provide especially in the northern part of Nigeria. Despite the presence of over 800 Microfinance banks in Nigeria, less than 10 of them are Islamic microfinance banks. Additionally, the establishment of Islamic microfinance institution (IMFI) in Nigeria has been unproductive. This situation has led to financial exclusion, lack of awareness, high operational costs (due to the Central Bank of Nigeria's huge capital requirements), inadequate human resources, and a dearth of Islamic banking-based financial technology, among other challenges, hindering their progress (Dogarawa, 2012; Gumel et al., 2014; Onakoya & Onakoya, 2014). Additionally, some individuals still harbor distrust toward microfinance bank industries as a whole due to its financial burdens and challenges caused by existing conventional microfinance banks.

The banks operate based on the principles of the Islamic faith concerning commercial dealings. This framework extends to permissible investments according to Shariah guidelines. The foundational principles guiding the non-interest bank are sourced from both primary and secondary Shariah texts. A prominent difference between conventional and Islamic banking lies in the practices and principles deemed impermissible under Shariah, as highlighted by Dogarawa (2013). Islamic banking distinctly prohibits practices such as Riba (usury) and speculative transactions, setting it apart from conventional banking practices. While the introduction of Islamic banking into both local and international financial markets has garnered recognition, it still encounters numerous obstacles before establishing itself as a viable institution capable of meeting the demands of the global financial landscape. Its initial success can be attributed partly to the existence of a niche market for Islamic funds and partly to its ethical and intrinsic values. To sustain and enhance this system, Islamic banks must address all impediments hindering their growth and sustainability (Onakoya & Onakoya, 2013).

In countries like Nigeria, particularly in cities like Abuja, Kaduna, Nassarawa, Kano, where the conventional banking system dominates, Islamic banks face huge challenges. The northern region of Nigeria faces various socio-economic challenges, including high unemployment rates, inadequate infrastructure, and limited economic opportunities. These factors contribute to the persistence of poverty and make it difficult for Taj and Jaiz Banks to implement sustainable poverty alleviation programs effectively.

Therefore, existing literature often lacks an in-depth exploration of the diverse perspectives from customers' experiences in accessing the Islamic microfinance loans in the northern part of Nigeria. This raises the research questions:

- What are the experiences of customers with the application process and loan approval faced by these individuals in seeking financial services from Islamic Microfinance Institutions (IMFIs) in Nigeria?
- How do customers perceive the negative impact of Islamic microfinance services in addressing their financial needs and contributing to poverty reduction efforts?

The objective of the study is to identify and analyze the primary barriers and challenges encountered by individuals in Nigeria who seek financial services from Islamic Microfinance Institutions (IMFIs).

Addressing these issues of awareness and understanding is crucial for the successful integration and acceptance of Islamic banking within Nigerian society.

This book chapter will use a qualitative case study of three cases in northern parts of Nigeria.

2 Microfinance and Poverty Alleviation

Microfinance implies the provision of financial services to individuals at the grassroots level, particularly those in the lowest rung of the economic ladder and without access to mainstream financial institutions, notably commercial banks.

Microfinance institutions are defined as organs whose primary task is the delivery of microfinance services (CBN, 2017). These institutions were established to provide financial inclusion for millions of rural and urban dwellers who are marginalized and underserved by traditional banking establishments.

Despite various government-initiated schemes in Nigeria, such as the Rural Banking Programme in 1977 and the Agricultural Credit Guarantee Scheme, their impact on improving the economic status of rural populations has been limited. The emergence of entities like the Nigerian Agricultural and Cooperative Bank Limited and the People's Bank of Nigeria in 2000 aimed to enhance support to the agricultural sector. Nevertheless, these initiatives fell short of significantly improving the economic conditions of rural populations. Over the past three decades, there has been elevated interest in making microfinance accessible in Nigeria. Different and diverse organizations have actively contributed to the development of microfinance services. These interventions have made microfinance in Nigeria evolve from initial skepticism to playing a central role in development discussions about poverty reduction.

2.1 Microfinance and Poverty Reduction Strategies

Microfinance serves as a crucial strategy in addressing poverty, with its effectiveness hinging on the enabling environment. While United Nations agencies offer technical assistance and models for poverty reduction through microfinance, capital provision typically falls under government purview, often supported by bilateral donors and international financing institutions (United Nations Report, 1995). In Nigeria, government-sponsored microfinance programs operate within the formal sector, while informal schemes like savings clubs/pools, "Esusu," Ajo, and money lenders play vital roles. Esusu, a revolving loan scheme, is prevalent in Nigeria and has contributed significantly to poverty alleviation, especially among market women. Formal microfinance models involve established financial institutions, but challenges exist in understanding the needs of the poor and fostering collaboration between formal and informal institutions. Efforts by the World Bank Group, the International Monetary Fund, and international donors have supported poverty reduction strategies, emphasizing participatory approaches. The Central Bank of Nigeria has launched various schemes to enhance access to finance for the poor and low-income groups.

2.2 Islamic Finance Principles and Practices

Islam is characterized by its encompassing values and aspirations, catering to both the spiritual and material needs of individuals. It offers a holistic framework for human life, encompassing social and economic dimensions. Sharia, which translates to "clear path," constitutes the body of Islamic laws that guide various aspects of life, including finance and banking.

Contrary to the principles of Islamic finance, contemporary banking systems often involve elements that are inconsistent with Islamic teachings. In this context, it becomes imperative to explore the core principles of Islamic finance and banking, shedding light on their distinctiveness.

One fundamental principle of Islamic finance is Profit and Loss Sharing. This principle advocates for equitable sharing of profits and losses among partners based on their respective contributions to a venture. Unlike conventional banking systems that guarantee fixed returns, Islamic finance fosters a partnership approach, ensuring that participants engage as partners rather than mere creditors in business endeavors.

Shared Risk Islamic banking promotes the concept of shared risk in economic transactions. This principle entails that when two or more parties engage in a transaction, the risk is distributed among them based on Islamic banking principles. By dividing and dispersing the burden of risk among multiple parties, economic activity within the state is enhanced.

Riba Riba, or the prohibition of interest, is a cornerstone principle in Islamic finance. It encompasses several key aspects:

- Wealth accrues returns without commensurate risk or effort.
- Irrespective of the outcome of economic activities, borrowers are obligated to repay the loan amount along with interest to the lender.
- Islamic finance principles deem it unjust to exploit the financial difficulties of others.

Gharar According to Islamic finance principles, Muslims are prohibited from engaging in transactions characterized by ambiguity and uncertainty. Both parties involved in a transaction must exercise proper control over the business, and complete information must be transparently shared to ensure equitable sharing of profits and losses.

Gambling In Islam, the accumulation of wealth through unethical means or engagement in gambling activities is strictly prohibited. This prohibition extends to conventional insurance products, as they are deemed akin to gambling. Conversely, Islamic banking operates on the principles of Takaful, which emphasize mutual responsibility and the sharing of risks among participants.

No Investment in Prohibited Industries: Islamic principles prohibit investment in industries that pose harm to society or violate social responsibilities. These industries include:

- Pornography
- Prostitution
- Alcohol
- Pork
- Drugs

Individuals are forbidden from investing in such industries or participating in mutual funds that support their growth and sustainability.

Zakat Zakat, a form of property tax mandated by Islam, facilitates the equitable distribution of wealth. Islamic banking adheres to this principle by deducting a fair amount of Zakat from the accounts of Muslims during the holy month of Ramadan. Islamic banks actively promote social responsibility by allocating these funds to support the needs of the less fortunate in society.

2.3 Existing Studies on Islamic Microfinance in Nigeria and Other Contexts

Kulik and Molinari have explored the impact of technology on the financial sustainability of microfinance institutions (MFIs), which emphasizes the role of technology in streamlining operations and enhancing outreach. Also, Ghani et al. argued that robust technological infrastructure was instrumental to the growth of microfinance institutions over the past decade. Additionally, research in Indonesia proposed innovative models for the long-term viability of Islamic microfinance institutions. Hassan and Mollah examined the role of Islamic microfinance and concluded that MFIs have made appreciable contributions to poverty alleviation among women through provision of resources and enhancement of revenue growth. Studies in Bangladesh, Oman, and Nigeria have highlighted the challenges and growth of Islamic microfinance, emphasizing the alignment of Islamic financing models with fundamental human needs. These studies underscore the positive impact of technology on microfinance sustainability while addressing challenges hindering its full utilization. Overall, research emphasizes the significant role of Islamic microfinance in poverty alleviation and sustainable development.

Hassan and Mollah delved into the role of Islamic microfinance in small-scale solutions for poverty alleviation. Their study highlighted the significant contribution of resources and revenue growth among women to the development of MFIs.

In a study by Dutta and Banerjee, the focus was on evaluating the impact of microfinance on sustainable entrepreneurial creativity among female borrowers in Bangladesh. Data analysis, employing descriptive methods, indicated that microfinance interventions lead to reduced poverty consumption and improved living conditions for participants.

3 Research Methodology

The method of choice for this research is the case study, as it aligns well with a detailed interpretive analysis of the dynamics within a single research setting (Eisenhardt, 1989; Walsham, 1995; Stake, 1995). By employing an inductive process that draws upon a set of practices, activities, and/or perceptions within a specific situation, the case study aims to generate a comprehensive understanding of that setting to address specific research questions (Stake, 1995). Thus, the case study was selected as the appropriate method for this study.

Utilizing a case study allows researchers to explore deeply into the particular context, operations, and outcomes of Islamic Microfinance Institutions (IMFIs) operating in Nigeria. This in-depth understanding is essential for thoroughly examining the role of IMFIs in poverty reduction efforts, particularly considering Nigeria's socio-economic and cultural context significantly influences the operations and impact of Islamic Microfinance Institutions. Conducting a case study within the Nigerian context provides valuable insights into the distinctive challenges, opportunities, and dynamics shaping poverty reduction efforts through Islamic microfinance.

In 2019, TAJBank Limited was established and granted a license by the Central Bank of Nigeria, the country's national banking regulatory authority, to function as a regional bank. Operations commenced on December 2, 2019, in Abuja, with subsequent expansion to other locations. By December 12, 2019, TAJBank had inaugurated its second branch in Kano, part of its strategic expansion plan across urban centers in Nigeria. The institution attained significant milestones, including the acquisition of a National License from the Central Bank of Nigeria in August 2022. Additionally, on August 24, 2020, TAJBank extended its reach by opening a branch in Sokoto State. Furthermore, on February 14, 2023, TAJBank successfully listed its N10 billion Mudarabah Sukuk issuance on the Nigerian Exchange Limited (NGX).

Jaiz Bank is another Islamic bank that was established in 2003 under the name Jaiz International Plc. Subsequently, on November 11, 2011, Jaiz International obtained a license from the Central Bank of Nigeria, the national banking regulatory authority, permitting it to function as a regional bank. Operations under the new name, Jaiz Bank Plc, officially commenced on January 06, 2012, with offices and branches set up in Abuja, Kaduna, and Kano.

Jaiz and Taj Bank were chosen as the research settings in December 2023. These banks are notable Islamic financial institutions in Nigeria, playing significant roles in the Islamic finance sector. Investigating these banks offers insights into the unique challenges and opportunities faced by Islamic Microfinance Institutions (IMFIs) in Nigeria, given the country's substantial Muslim population.

Data collection involved structured interviews with clients and beneficiaries of Jaiz and Taj Bank's microfinance services, aiming to capture their experiences, perceptions, and socio-economic outcomes resulting from access to financial products and services. Three in-person interviews, lasting from 30 to 55 min, were conducted in December 2023. The interview questions sought to understand the needs, aspirations, and challenges of these customers, crucial for evaluating the effectiveness of poverty reduction interventions. Additionally, two subsequent interviews were conducted with entrepreneurs and small business owners between December and January 2023 to provide further context to the interview data. One of the entrepreneurs was available for a follow-up interview of approximately 60 min, conducted in December 2023, to delve deeper into emerging themes.

The interview data were then coded and analyzed based on thematic categories derived from the literature review and themes that emerged from the initial analysis of the interviews. The occurrence and variance of codes were analyzed to identify common themes among the interview participants. It is important to acknowledge that the study's generalizability is limited due to the small number of entrepreneurs included. Nonetheless, besides providing insights into the actions and perspectives of entrepreneurs in a single setting, the study serves as a foundation for scholars gathering data from business owners in future contexts, offering a grounded perspective on larger issues within the emerging field of OER scholarship.

3.1 Overview of Study Site and Participants

The three participants for this research who are entrepreneurs and small business owners have complained about access to finance which has posed a significant challenge for small and medium-scale enterprises (SMEs) across various sectors. SMEs often encounter steep qualification thresholds, high interest rates, and stringent requirements for collateral, financial track records, and other conditions when seeking funds.

While traditional bank financing remains crucial for SMEs, there is growing concern that credit constraints may become the norm for SMEs and entrepreneurs. Therefore, it is imperative to diversify the range of financing instruments available to SMEs and entrepreneurs to support their role in investment, growth, innovation, and job creation.

Convenient Access for the Unbanked Community According to the Enhancing Financial Innovation & Access (EFInA) report on financial exclusion in Nigeria for the year 2020, only 67.5 million adults, which is 64% of the total 105.5 million adult population, were financially included. However, the report also indicates that 36% of Nigerian adults, approximately 38 million individuals, remained financially excluded by the end of 2020 (Adeleke & Alabede, 2021). Despite improvements in financial inclusion over the past decade, Nigeria has not met the targets set out in the National Financial Inclusion Strategy for 2020. The strategy aimed to provide formal financial services to 70% of Nigerians, but the actual figure stood at 51% (National Bureau of Statistics, 2020).

The advent of digital financial services and agent banking presents opportunities to accelerate progress toward financial inclusion, particularly in Northern Nigeria. Consequently, the Central Bank of Nigeria (CBN) has taken proactive steps to promote financial inclusion in the country. Under the leadership of the CBN Governor, initiatives like the establishment of E-Naira have been launched to incorporate millions of unbanked individuals into the formal financial system.

Limited Awareness Numerous entrepreneurs in Nigeria, particularly those in rural areas, may lack awareness regarding the availability and advantages of Islamic microfinance bank loans (Bello & Abubakar, 2014). This dearth of knowledge concerning Sharia-compliant financial options can impede entrepreneurs from exploring these financial services.

The lack of awareness regarding Islamic Financial Literacy (IFL) remains a significant issue within the global Islamic financial sector, especially in Africa. Many average Muslims in regions like Nigeria struggle to differentiate between Islamic and conventional banking systems. Unfortunately, Nigeria's understanding of Islamic banking, its procedures, products, and principles, is still at an early stage.

Stringent Eligibility Criteria Islamic microfinance banks often enforce rigorous eligibility standards for loan applicants, which may encompass criteria related to collateral, credit history, and business documentation (Adepoju & Oyesanya, 2014). Meeting these prerequisites can pose challenges for entrepreneurs, particularly those hailing from poor rural communities with limited assets or formal credit backgrounds (Al-Ameen, 2016).

High Cost of Financing Islamic microfinance loans may involve neck-throat costs compared to conventional loans due to the absence of interest-based financing (Enyia & Inyang, 2018). Entrepreneurs may encounter difficulties in covering the higher profit rates or fees associated with Islamic microfinance products, especially if their businesses operate on slender profit margins.

The forthcoming presentation delineates the narrative of the case study findings, sequentially presenting each business owner's account by incorporating excerpts from their interview responses.

A. *Fatima Daudu*

Once upon a time in Abuja, Nigeria, there lived a woman named Fatima Daudu who dreamed of expanding her small catering business. Fatima has been running her catering service for 5 years, specializing in delicious traditional Nigerian dishes for weddings, parties, and corporate events. Despite her passion and dedication, Fatima struggled to meet the increasing demand for her services due to limited resources and equipment.

As Fatima explored avenues for financing her business expansion, she learned about Islamic microfinance banks offering loans tailored to entrepreneurs like herself. Excited by the prospect of accessing Sharia-compliant financing, Fatima decided to apply for a loan to purchase additional kitchen equipment, hire more staff, and promote her catering services.

However, Fatima soon encountered several challenges in her quest for an Islamic microfinance loan. Firstly, she discovered that the eligibility criteria set by the microfinance bank were stringent, requiring collateral and a solid credit history. Being a small business owner with limited assets and no formal credit history, Fatima found it difficult to meet these requirements.

> I've worked hard to build my business, but without access to funds, I fear it will all go to waste. It's frustrating to see my dreams slipping away because of bureaucratic hurdles.
>
> I wish the Islamic microfinance institution would be more flexible and understanding of entrepreneurs like me who are trying to make a difference in our communities. Instead, I feel like I'm constantly fighting an uphill battle.

Furthermore, Fatima faced challenges in navigating the loan application process, which seemed complex and time consuming. She struggled to gather the necessary documentation and complete the paperwork required by the microfinance bank.

As Fatima continued her efforts to secure financing, she also realized that Islamic microfinance loans often came with higher costs compared to conventional loans due to the absence of interest-based financing. This posed a significant financial burden for Fatima, whose business operated on narrow profit margins.

Despite her determination and resilience, Fatima's journey to access a loan from Islamic microfinance proved to be arduous. She encountered numerous obstacles along the way, highlighting the challenges that many Nigerian small business owners face in accessing financing for their entrepreneurial ventures.

Nevertheless, Fatima remained hopeful and determined to overcome these challenges. She sought assistance from business mentors, attended financial literacy workshops, and explored alternative financing options to realize her dream of expanding her catering business and serving more customers across Lagos. With perseverance and tenacity, Fatima believed that she would eventually succeed in securing the funding needed to take her business to new heights.

The Impact of the Challenges on Fatima

Financial Strain: Fatima will experience increased financial strain due to the challenges in accessing Islamic microfinance services. This could result in limited access to capital for their businesses or personal needs, that hinders her ability to invest in growth opportunities or cope with financial emergencies.

Limited Economic Opportunities: The barriers to accessing Islamic microfinance services may restrict Fatima, Aisha, and Yusuf's ability to seize economic opportunities and improve their livelihoods. Without access to affordable financing, they may struggle to expand their businesses, acquire assets, or pursue education and training opportunities that could lead to higher income levels.

B. *Aisha Bello*

In the bustling city of Kano, Nigeria, lived Aisha, a resilient woman and a mother of six children. Aisha had always dreamt of starting her own small tailoring business to support her family and provide a better future for her children. With her exceptional sewing skills and unwavering determination, she decided to turn her dream into reality.

Aisha embarked on her entrepreneurial journey, setting up a modest tailoring shop in her neighborhood. Despite facing numerous challenges as a single mother, she poured her heart and soul into her work, stitching beautiful traditional garments for weddings, ceremonies, and everyday wear.

As her tailoring business grew, Aisha realized the need to expand her shop, invest in more sewing machines, and hire additional tailors to meet the increasing demand for her services. Eager to take her business to the next level, she explored various financing options and learned about Islamic microfinance loans offered by a local microfinance bank.

Filled with hope and determination, Aisha approached the microfinance bank to inquire about the possibility of securing a loan to support her business expansion. However, she soon discovered that the eligibility criteria set by the bank were stringent and challenging to meet.

The bank required collateral and a strong credit history, neither of which Aisha possessed. As a single mother with six children, she had dedicated all her time and resources to supporting her family and building her business, leaving little room to establish a formal credit history or accumulate significant assets.

She expressed frustration toward the application of a loan.

In her own words

> The lack of transparency in their loan approval process leaves me feeling helpless and discouraged.
>
> It's disheartening to see how little support they offer to entrepreneurs like me who are trying to make a difference in our community.
>
> I can't believe how long it's taking to process my loan application. It's like they don't understand the urgency of my financial needs.

Despite her impeccable work ethic and proven track record as a skilled tailor, Aisha found herself ineligible for the Islamic microfinance loan due to the stringent requirements set by the bank. She felt disheartened and frustrated, realizing that her dream of expanding her tailoring business might remain unattainable without access to the necessary financing.

Nevertheless, Aisha refused to give up hope. She sought guidance from local business advisors, participated in financial literacy workshops, and explored alternative avenues for funding her business expansion. With determination and resilience, she remained committed to overcoming the obstacles in her path and providing a brighter future for herself and her children through her entrepreneurial endeavors.

Despite the challenges posed by the stringent eligibility criteria of the Islamic microfinance loan, Aisha remained determined to pursue her dreams and build a better life for herself and her family, one stitch at a time.

C. *Yusuf Kazim*

In the serene countryside of Kaduna State, Nigeria, lived Yusuf Kazim, a hard-working farmer and a devoted father of six children. Yusuf had spent most of his life tending to his small farm, cultivating crops, and raising livestock to provide for his family and contribute to his community's food supply.

With a passion for agriculture and a desire to expand his farm, Yusuf dreamed of investing in modern irrigation systems, purchasing high-quality seeds and fertilizers, and acquiring additional land to increase his crop yield. He believed that by expanding his farming operations, he could improve his family's livelihood and contribute to the local economy.

Eager to turn his vision into reality, Yusuf explored various financing options and learned about Islamic microfinance loans offered by a nearby microfinance institution. Excited by the prospect of accessing Sharia-compliant financing, he visited the microfinance bank to inquire about the possibility of securing a loan to support his farm expansion plans.

However, Yusuf soon encountered challenges as he navigated the loan application process. The microfinance institution required applicants to meet stringent eligibility criteria, including providing collateral, demonstrating a strong credit history, and presenting detailed business plans.

As a small-scale farmer with limited assets and no formal credit history, Yusuf found it challenging to meet the bank's requirements. Despite his years of experience and proven track record as a successful farmer, he felt disheartened by the realization that he might not qualify for the Islamic microfinance loan.

Yusuf's dreams of expanding his farm and providing a better future for his family seemed out of reach due to the stringent eligibility criteria set by the microfinance institution. He felt frustrated and discouraged, knowing that access to financing was essential for realizing his aspirations and improving his family's livelihood.

Nevertheless, Yusuf refused to give up hope. He sought advice from local agricultural experts, attended workshops on financial management and business planning, and explored alternative avenues for funding his farm expansion.

With determination and perseverance, Yusuf remained committed to his goal of expanding his farm and providing a better life for his family. Despite the challenges posed by the stringent eligibility criteria of the Islamic microfinance loan, he continued to work tirelessly to achieve his dreams and make a positive impact on his community through his agricultural endeavors. Yusuf expressed a mix of determination, frustration, and disappointment as he navigated the process of accessing the Islamic microfinance loan. In his own words:

> It's frustrating to see others getting financial assistance while I struggle to even get my application considered. It feels like the system is stacked against people like me.
>
> I've been trying to expand my business for months now, but without access to financing, I'm stuck in the same place. It's disheartening.

he conveyed his deep desire to expand his farming operations and provide a better life for his family. However, he also expressed frustration at the stringent eligibility criteria set by the microfinance institution, which seemed to present insurmountable barriers to his aspirations.

Yusuf described feeling overwhelmed by the complexities of the loan application process and the challenges of meeting the institution's requirements for collateral, credit history, and detailed business plans. Despite his years of experience and proven track record as a successful farmer, he felt disheartened by the realization that he might not qualify for the loan due to his limited resources and lack of formal credit history.

However, Yusuf's determination to overcome these challenges shone through in his words. He spoke of seeking advice from local agricultural experts, attending workshops on financial management, and exploring alternative avenues for funding his farm expansion. Despite the setbacks, Yusuf remained committed to his goal of expanding his farm and providing a better future for his family.

Yusuf's expression reflected the resilience and perseverance of countless small-scale entrepreneurs facing similar challenges in accessing financing for their businesses. His words highlighted the need for greater accessibility and inclusivity in financial services, particularly for rural entrepreneurs with limited resources and formal credit histories.

4 Conclusion

The challenges of financial strain, limited economic opportunities, and stringent eligibility criteria present significant obstacles for entrepreneurs seeking financial assistance from Islamic microfinance banks. These institutions often enforce rigorous eligibility standards, including requirements related to collateral, credit history, and business documentation. For entrepreneurs, especially those from poor rural communities with limited assets or formal credit backgrounds, meeting these prerequisites can be exceptionally challenging.

Moreover, the high cost of financing further worsens the difficulties faced by entrepreneurs who seek funding from Islamic microfinance banks. The cost associated with Islamic financing products and services can place a considerable financial burden on entrepreneurs, which makes it difficult for them to access the capital needed to start or expand their businesses.

In conclusion, the challenges mentioned have described the importance of addressing structural barriers within Islamic microfinance institutions to foster greater financial inclusion and support entrepreneurship among marginalized communities. Efforts to streamline eligibility criteria, reduce the cost of financing, and expand economic opportunities can help empower entrepreneurs and catalyze sustainable economic development within Islamic finance ecosystems. In addition to this, there should be collaboration between policymakers, financial institutions, and community stakeholders to create an enabling environment that promotes equitable access to financial services and facilitates entrepreneurial success.

In addressing the conclusion with regard to providing sustainable poverty alleviation, it is important to recognize the pivotal role that Islamic microfinance institutions can play in this endeavor. Sustainable poverty alleviation requires not only addressing immediate financial needs but also fostering long-term economic empowerment and resilience within marginalized communities.

To achieve sustainable poverty alleviation through Islamic microfinance, several key strategies can be implemented:

There should be promotion of financial inclusion where Islamic microfinance institutions prioritize the expansion of access to financial services for poor populations, especially those that reside in rural or remote areas. By reaching out to marginalized communities and offering tailored financial products and services, these institutions will be able to empower individuals to improve their livelihoods and build financial freedom.

The improvement of financial literacy and capacity building should be a deliberate effort to alleviate poverty. It will also necessitate financial education and skill development. Islamic microfinance institutions can invest in workshops that provide financial literacy programs and entrepreneurship training to equip customers with the knowledge and skills that are needed to manage their finances effectively and run businesses successfully.

To provide the facilitations of good social investment where Islamic microfinance serves as a vehicle for social investment, it can direct funds toward productive and socially beneficial projects that will contribute to poverty reduction and community development. By incorporating financing with ethical and sustainable principles, Islamic microfinance institutions can catalyze positive social change and address pressing societal needs.

It should support the development of microenterprises. Islamic microfinance should focus on supporting microenterprises and small businesses, particularly those operated by women and those with physical disabilities. The provision of access to capital, technical assistance, and market linkages, can empower entrepreneurs to generate income, create employment opportunities, and contribute to local economic growth.

Strategies for creating collaboration and partnerships should be encouraged because Sustainable poverty alleviation requires collaboration and partnerships between Islamic microfinance institutions, governments, civil society organizations, and other stakeholders. By working together toward common goals, these stakeholders can leverage their resources, expertise, and networks to maximize impact and address systemic barriers to poverty reduction.

By adopting a unified approach that integrates financial inclusion, capacity building, social investment, and enterprise development, Islamic microfinance institutions will contribute significantly to sustainable poverty alleviation efforts. By empowering individuals, strengthening communities, and fostering economic freedom. These institutions can help create a more inclusive and equitable society where all individuals have the opportunity to thrive.

References

Adeleke, R., & Alabede, O. (2021). Understanding the patterns and correlates of financial inclusion in Nigeria. *GeoJournal, 87*, 2405. https://doi.org/10.1007/s10708-021-10378-6

Adepoju, R. I., & Oyesanya, O. S. (2014). A critical analysis of poverty alleviation strategies of Al-Hayat relief foundation in Nigeria. *Journal of Islamic Finance, 3*(2), 49–59.

Al-Ameen, A. (2016). Implementing Islamic microfinance in Nigeria: A matter of equity and social justice. *Journal of Sustainable Development Law and Policy (The), 7*(2), 211.

Bello, A., & Abubakar, M. (2014). Challenges and solutions to Islamic banking system in a pluralistic-secular country like Nigeria. *Mediterranean Journal of Social Sciences, 5*(6), 25–34. https://doi.org/10.5901/mjss.2014.v5n6p25

Central Bank of Nigeria (CBN). (2017). *Guidelines for regulation and supervision of non-interest (Islamic) Microfinance banks in Nigeria.* 5, 1 January–June 2022 47Adnan Opeyemi Salaudeen, Habeebullah Zakariyah.

Dogarawa, A. B. (2012). Framework for an effective Islamic microfinance in northern Nigeria. *SSRN Electronic Journal.* https://doi.org/10.2139/ssrn.1624748

Dogarawa, A. B. (2013). Critique of the non-interest banking framework in Nigeria. *ISRA International Journal of Islamic Finance, 5*(1), 115–134. https://doi.org/10.12816/0002760

Dogarawa, A. B., & Uthman, M. B. (2019). Regulatory and operational challenges of Islamic microfinance in Nigeria. *SSRN Electronic Journal, 1–15.* https://doi.org/10.2139/ssrn.3492057

Egboro, E. (2015). Creation of microfinance banks in Nigeria: What is their Main object? *British Journal of Economics, Management & Trade, 7*(3), 158–174. https://doi.org/10.9734/bjemt/2015/15107

Eisenhardt, K. M. (1989). Building theories from case study research. *The Academy of Management Review, 14*(4), 532–550-38.

Enyia, J. O., & Inyang, P. E. (2018). The regulatory and operational framework of microfinance banks on consumers in Nigeria: An appraisal. *International Journal of Managerial Studies and Research, 6*(7). https://doi.org/10.20431/2349-

Gumel, G. B., Saad, N. M. D., & Kassim, S. H. (2014). Assessing the impact of Islamic microfinance on poverty alleviation in northern Nigeria. *Journal of Islamic Economics, Banking and Finance, 10*(4), 37–49.

Ishak, N. S. (2020). *Demand for Takaful and Microtakaful in Malaysia.* University of Sussex. Retrieved from http://sro.sussex.ac.UK/ This 48 el Barka: Journal of Islamic Economics and Business Challenges Hindering Islamic Microfinance Banks' Sustainable….

Madise, S. (2019). *The concept of financial inclusion.* https://doi.org/10.1007/978-3-030-13831-8_6.

National Bureau of Statistics (NBS). (2019). Selected banking sector data: Sectorial breakdown of credit, ePayment channels. *Deposits & Domestic Credit Breakdown and Staff Strength, Q3.*

National Bureau of Statistics (NBS). (2020, March). *Annual abstract of statistics.* Annual Abstract of Statistics, 1–69. Retrieved from https://nigerianstat.gov.ng/

Nwanyanwu, O. (2011). Micro finance in Nigeria: Problems and prospects. *African Research Review, 5*(2). https://doi.org/10.4314/afrrev.v5i2.67307

Okpara, G. C. (2010). Microfinance banks and poverty alleviation in Nigeria. *Journal of Sustainable Development in Africa, 12*(6), 177–191.

Onakoya, A. B., & Onakoya, A. O. (2013). Islamic microfinance as a poverty alleviation tool: Expectations from Ogun state, Nigeria. *Scholarly Journal of Business Administration, 3*(2), 36–43.

Onakoya, A. B., & Onakoya, A. O. (2014). Alternative financial inclusion strategy: Islamic expectations from Ogun sub–National Government, Nigeria. *Singaporean Journal of Business Economics and Management Studies, 2*(11), 169–180. https://doi.org/10.12816/0006791

Salako, T., & Azeez, A. (2018). Semi-formal Islamic microfinance institutions in Nigeria: The experience of Al-Hayat relief foundation. *Journal of Contemporary Issues and Thought, 8*, 41–54. https://doi.org/10.37134/jcit.vol8.5.2018

Stake, R. (1995). *The art of case research.* Sage Publications.

Walsham, G. (1995). Interpretive case studies in IS research: Nature and method. *European Journal of Information Systems, 4*(74–81), 39.

Joy Eghonghon Akahome is a Senior Lecturer at the Federal University Otuoke, Bayelsa State, Nigeria and a Contract staff at The University of Economics and Human Sciences, Warsaw, Poland. She has an extensive expertise in Entrepreneurship and Marketing. She is a member of the American Marketing Association, Chicago, USA. Her contribution to the upcoming entrepreneurship book explores the investigation of the prevalence and nature of gender bias in family businesses in Nigeria. As a chapter contribution, Dr. Akahome's research focuses on Entrepreneurship and Marketing. She has previously achieved a Coimbra Group Scholarship at Jagiellonian University, Krakow, Poland. Her eagerness for future endeavors envisions goals or projects in advancing entrepreneurial knowledge.

Sidney Ogodo is currently studying for a Master of Philosophy (M.Phil) in strategic marketing management at the Babcock University, Ilisan, Ogun state; in preparation for a Doctor of Philosophy (PhD) in strategic marketing management. He has an extensive industry background in consumer marketing with a decade of cognate experience in brand management in a multinational company.

He also worked with the National Institute of Marketing of Nigeria (NIMN)—the statutory body responsible for regulating and controlling the practice of marketing in Nigeria, and rose to the position of Registrar/Secretary to the governing Council of the Institute.

Sustainability of Islamic Social Banking: A Systematic Literature Review

Junaidah Abu Seman, Syahidawati Shahwan, Nurul Aini Muhamed, Nur Ainna Ramli, and Aimi Fadzirul Kamarubahrin

Abstract Islamic banking should transcend conventional commercial banking, extending its reach to all segments of society through diverse product and service offerings, distribution channels, and sustainable initiatives. While Islamic banks are assessed based on financial performance metrics, they face challenges in quantifying their social impact. This is partly due to the absence of a clear conceptual framework defining Islamic social banking and its overarching themes. This paper seeks to systematically review and consolidate the sustainability aspects of Islamic social banking. A systematic review was identified literatures by electronic open-access database searches (Scopus, ScienceDirect, and Emerald Insights). The study applied specific inclusion criteria, including publication dates from 2018 to 2022, country, methodology, model/theory, and dimension. It performed a narrative synthesis and systematic analysis in accordance with the Preferred Reporting Items for Systematic Review and Meta-Analyses (PRISMA) search methodology. Initially, 599 studies were identified from various sources, but after applying the selection criteria, 15 relevant studies were retained. Recommendations include conducting more qualitative studies, adopting a standardized systematic review method for research synthesis in Islamic social banking, and employing complementary searching techniques such as citation tracking, reference searching, and expert consultation. Overall, there is a scarcity of research in Islamic social banking, and this study contributes to addressing this gap.

J. Abu Seman(✉) · S. Shahwan · N. A. Muhamed · N. A. Ramli · A. F. Kamarubahrin
Faculty of Economics and Muamalat, Universiti Sains Islam Malaysia,
Nilai, Negeri Sembilan, Malaysia
e-mail: junaidah@usim.edu.my; syahidawati@usim.edu.my; nurulaini@usim.edu.my; nurainna.ramli@usim.edu.my; aimi_fadzirul@raudah.usim.edu.my

A. Rafiki et al. (eds.), *Strategic Islamic Business and Management*, Contributions to Management Science,
https://doi.org/10.1007/978-3-031-61778-2_14

1 Introduction

Social banks can be referred to as financial institutions offering financial products and services for individuals and institutions that promote social, environmental, and sustainability benefits (Mohd Nor, 2022; Nasir & Abu Seman, 2022; Hamidi & Worthington, 2018). Social banking embraces a bottom-line strategy as its operational model, rooted in communities, bolstering the tangible economy, and nurturing the evolution of inventive business models. A mutually beneficial relationship with the client would be developed to understand economic operations and risk mitigation better and demonstrate social banking's long-term, self-sustaining, transparent, inclusive, and resilient to external disruption characteristics (Weber, 2014). These traits represent the long-term, autonomous, transparent, inclusive, and resilient to external dispute options nature of social banking (Weber, 2014). It is claimed that social banking, or development banking, facilitates access to financing (Hamidi et al., 2019).

Meanwhile, an Islamic social bank must hold the characteristics of a bank such as a commercial entity and a financial intermediary that collects deposits (investments) and provides financing according to Shariah (Nasir & Abu Seman, 2022). As a commercial entity, the bank must target financings that generate profit. However, as a social bank, it must provide financing only if a project has a positive social impact. In the annual report, the bank must report on the social impact it has achieved during the period. To serve the commercial needs and social causes, the sources of capital and funds of the bank will be three types: (i) Capital and funds expecting a profit. This will be from the shareholders, depositors, and investors who wish a profit. Depending on the bank's model and market condition; (ii) Capital and fund that do not require any profit, but require repayment of the investments; (iii) The capital and fund providers will not require any profit or return of money. Sadaqah and permanent waqf will be primary principles (Ahmed, 2019).

According to Zafar et al. (2022), Islamic banks of Pakistan must work on a better Corporate Social Responsibility (CSR) policy, it would enhance the reputation of Islamic banks. On the other hand, it would also increase their capabilities to cope with future reputation damage and negative news, thus it would protect and secure the financial results. Furthermore, investors in Islamic banks of Pakistan should consider the social and ethical commitments of these institutions when making investment decisions. In line with the study to enhance social achievements in Islamic bank, Dusuki (2008) suggests the CSR approach. Furthermore, CSR could amplify Islamic banks' responsiveness to environmental effects, advance social well-being, and cultivate constructive endeavors. As an alternative, Raimi et al. (2014) recommend combining CSR with waqf (Islamic endowments) and zakat (Islamic compulsory charity) systems.

Using two different simulation cases, Shahimi et al. (2013) argue that cash waqf contributes to alleviating poverty by up to 50% in Malaysia. More profoundly, however, other studies propose the establishment of Islamic social bank as a response to the apparent failure of current Islamic bank in delivering its purported social outcomes (Asutay, 2012; Mohd Nor, 2016). The Islamic banks are entrusted with assisting the community in lifting out of poverty, preserving it, and leading to a more prosperous standard of living (Hamidi et al., 2019). The offerings and services by Islamic banks should reach people who do not have financial access, such as micro-businesses. The fundamental rationale behind the Islamic social banking concept is supported by compelling reasons. Islamic scholars are increasingly focusing on integrating Islamic social banking practices into Islamic banks. Ongoing discussions reflect the need to further strengthen the current understanding of Islamic social banking's alignment with Shariah regulatory requirements and underlying ethical principles.

According to Hamidi et al. (2019), Islamic social banking practices can maximize benefits and improve overall societal well-being. Utilizing the triple bottom-line approach is anticipated to strengthen Islamic social banks as it aligns with the broadened social obligations of Islamic banks. This approach emphasizes consideration for all stakeholders, thereby enhancing the accountability of Islamic social bank. With the establishment of Islamic social bank, it would facilitate more accessible funding for project groups seeking funds, thereby fostering a more responsible approach to development. Islamic banks could integrate certain practices and standards from social banking as well. For example, financial inclusion takes the lead as an enabler of other developmental goals in the 2030 SDG. SDG 17 highlights about enhancing implementation and promoting higher savings for investment and consumption that can improve growth, which leads to the critical role of promoting greater financial inclusion (UNCDF, 2024). With the conflicting objectives of social banking and financial inclusion (Tyle 2013), Islamic banks should take the lead in social banking to accomplish Maqasid al-Shariah (Nasir & Abu Seman, 2022).

The literature on Islamic social banking has been increasingly expanding (see, for example, Mohd Nor, 2022; Nasir & Abu Seman, 2022; Mirjalili et al., 2020; Hamidi et al., 2019; Hamidi & Worthington, 2018; Mohd Nor, 2016). However, the study is limited because it focuses on the current practices of Islamic social banking in general, given that it is still a new industry (Mohd Nor, 2022; Nasir & Abu Seman, 2022; Mirjalili et al., 2020). In spite of the fact that ISB is not a new topic in research, it lacks empirical studies, particularly on its sustainability. The challenges underscore the importance of solid evidence to guide best practices, elucidating the impact on the nation's economic growth through improved performance and enhanced financial access. In light of this necessity, this systematic literature review endeavors to assess the sustainability of Islamic social banking through a comprehensive analysis. The study aims to identify, review, and consolidate existing literature to understand recurring patterns and the current state-of-the-art research in this important topic. Through a systematic literature review, the researchers aim to discern various factors contributing to the variability in study outcomes. After the introduction, the next section covers methodology of the study employ, followed by findings and discussion part. The final section provides the conclusion for the study.

2 Methodology

The researchers adopted the procedures of Muhamed et al. (2022) and Khairi et al. (2021). It serves as a mechanism designed to generate a methodical summary of evidence within a specific domain, diverging from the conventional narrative review approach. The intended inclusion criteria were limited to the following characteristics shown in Table 1:

A systematic review attempts to collate all empirical evidence that fits pre-specified eligibility criteria to answer a specific research question (Desrosiers et al., 2020). It employs clear and systematic techniques chosen to reduce bias, thereby yielding dependable findings upon which conclusions and decisions can be based. According to Bichler et al. (2022), systematic literature review can be defined as the statistical analysis of the results of individual studies, to integrate them. According to Linares-Espinós et al. (2018), a systematic review is characterized by the application of statistical methods to amalgamate a collection of distinct yet comparable experiments. Siddaway et al. (2019) defined the systematic review as an analysis of the "empirical analysis," which attempts to explain the differences in results between studies. A systematic literature review is a simultaneous analysis of a set of studies addressing the same question, to obtain the information that none of these studies taken singly could provide and explain the differences in the results of these studies (Pati & Lorusso, 2018). The aim was to mitigate the expenses associated with experimental studies, which frequently yielded disparate outcomes. Rapidly, this approach proliferated into other research domains, including environmental studies, marketing, and the social sciences.

Based on electronic open-access database searches (Emerald Insights, Scopus, and ScienceDirect), this study listed 599 studies that can form the basis of a

Table 1 Summary of inclusion criteria

Item	Description
Annual	This item described the "year" of the articles in this study.
Region	The item described the "region" studied in the articles.
Approach	The item described the "methodology" adopted in the articles.
Concept	The model described the model/theory used in previous studies.
Factors	Factors recognized specific studies related to the present study.

Source: Muhamed et al. (2022) & Khairi et al. (2021)

systematic literature review. The selection of studies was based on keywords derived from the titles of articles found in the database: "Islamic," "Social," and "Banking." The researcher selected online databases based on their familiarity with platforms commonly used by scholars studying the sustainability of Islamic social banking and the accessibility of these databases. Khairi et al. (2021) recommended searching multiple databases to ensure comprehensive coverage and avoid biased reviews. An open-access electronic repository was utilized to access full-text articles, resulting in the examination of over 599 studies. This underscores the importance, as suggested by Khairi et al. (2021), of identifying relevant online databases to streamline the search process. After completing the initial search, the researcher continued to identify relevant literature in the secondary search phase. Any suitable papers found were included in the list of studies eligible for synthesis.

The analysis used data from 599 manuscripts published within the past 5 years. The search terms in electronic databases included specific phrases such as "Islamic," "Social," and "Banking." Results from both sections were manually organized, with a focus on Islamic social banking. Figure 1 illustrates the process and outcomes of the inclusion phase. Three inclusion criteria were established for this systematic literature review. The first criterion required the inclusion of only empirical studies. The second criterion involved studies providing adequate statistics for calculating effect sizes, based on mean, standard deviation data, or population proportion data. The third criterion limited studies to those presenting statistics relevant to Islamic social banking.

The comprehensive search conducted over a span of 5 years, from 2018 to 2022, across seven electronic databases resulted in 599 search findings, with only 15 of them being scrutinized for this study, as noted by Khairi et al. (2021). Additionally,

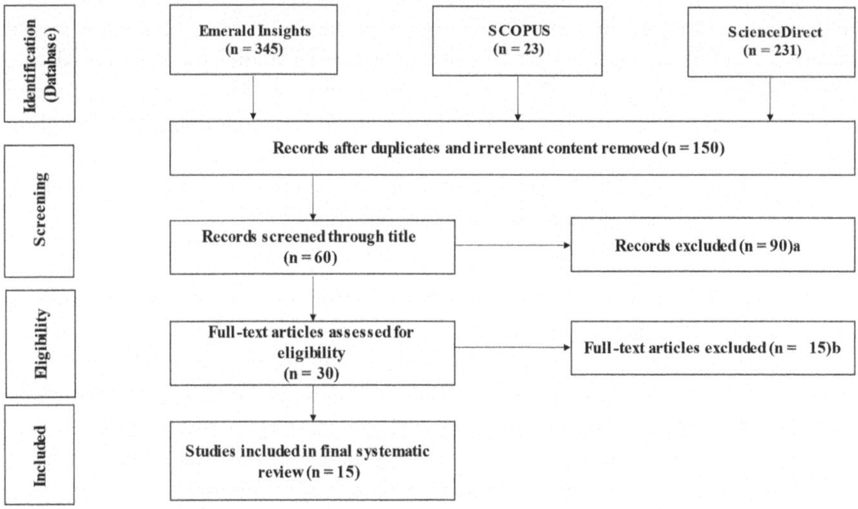

Fig. 1 The flowchart depicting the article search strategy in accordance with PRISMA guidelines. *Source: Authors*

the references of the reviewed manuscripts were analyzed to identify further relevant studies. Each manuscript's relevance was assessed based on its title. However, not all of the reviewed manuscripts were relevant to this study. Consequently, out of the 599 eligible findings, only 15 manuscripts provided data appropriate for inclusion in the systematic literature review.

From various perspectives on Islamic social banking (Nasir & Abu Seman, 2022; Mirjalili et al., 2020; Hamidi et al., 2019; Hamidi & Worthington, 2018; Mohd Nor, 2016, 2022), the selection process for this meta-analytic study unfolded in two stages. Initially, a frequency search was conducted among the gathered manuscripts to locate highly cited works on Islamic social banking. The number of papers mentioning specific factors indicated the level of attention these factors received in the literature, justifying their inclusion as variables in the systematic literature review. During the data collection phase, the second step of selection occurred. Variables that lacked adequate data for computing effect sizes were omitted from the study. These exclusions involved studies containing unclear terms (like "Islamic social banking"), infrequent mentions in the literature, and cases where data regarding means, standard deviations, proportions, or sample sizes were absent.

3 Findings and Discussion

The findings are examined based on outcome categories, interventions used, and individual studies. Summarizing quantitative data was difficult due to the diverse nature of the studies analyzed. These studies varied greatly in terms of their outcomes, locations, and databases. Therefore, conducting a quantitative summary was not feasible or appropriate for the reasons mentioned. The review identified four main themes, outlined in Table 2, offering a thorough analysis of Islamic social banking by scholars over the past 5 years. A total of 15 studies focusing on Islamic social banking were identified in this systematic review. This suggests a limited scholarly discourse on Islamic social banking, indicating a shortage of research in this area.

The examination of articles published within the realm of Islamic social banking from 2018 to 2022 has yielded the following insights:

1. There is a lack of research in the area of Islamic social banking.
2. The majority of the research has been carried out in Malaysia (4 studies), Indonesia (4 studies), Pakistan (3 studies), the United Kingdom (1 study), Uganda (1 study), Morocco (1 study), and an unidentified country (1 study). The notable volume of research in these countries can be attributed to various factors. Islamic banking is a popular topic for exploration and experiences rapid growth in these regions. Consequently, studies focusing on the scope or domain of Islamic social banking have become popular topics for publication.
3. Based on the systematic literature review, 15 relevant articles were analyzed. Out of these, 10 studies employed a qualitative methodology involving document

Table 2 Characteristics of studies involved in the sustainability of Islamic social banking

Authors	Year	Country	Title	Dimension
Zafar, Sulaiman & Nawaz	2022	Pakistan	Does corporate social responsibility yield financial returns in Islamic banking?	Corporate Social Responsibility (CSR)—Financial Performance—Islamic Banking (IB)
Mohd Nor	2022	Malaysia	Islamic social bank: A sustainable banking approach	Islamic Social Bank (ISB)
Darmalaksana	2022	Indonesia	The need for social theology to strengthen the social functions of Islamic banking in Indonesia	Islamic Bank (IB)—Social Finance
Muflih	2021	Indonesia	The link between corporate social responsibility and customer loyalty: Empirical evidence from the Islamic banking industry	Corporate Social Responsibility (CSR)—Islamic Banks (IB)
Hamidi & Worthington	2021	Indonesia	How social is Islamic banking?	Islamic Social Bank (ISB)—Sustainable Development Goals (SDGs)
Lui, Zainuldin, Wahidudin & Foo	2021	Malaysia	Corporate Social Responsibility Disclosures (CSRDs) in the banking industry: A study of conventional banks and Islamic banks in Malaysia	Islamic Bank (IB)—Corporate Social Responsibility (CSR)
Zafar & Sulaiman	2020	Pakistan	Measuring CSR in Islamic banking: What matters?	Islamic Bank (IB)—Corporate Social Responsibility (CSR)
Ur Rehman, Zahid, Rahman, Asif, Alharthi, Irfan & Glowacz	2020	Pakistan	Do CSR disclosures improve financial performance? A perspective of the Islamic banking industry in Pakistan	Islamic Bank (IB)—Corporate Social Responsibility (CSR)
Tahiri Jouti	2019	Morocco	An integrated approach for building sustainable Islamic social finance ecosystems	Islamic Social Finance (ISF)—Sustainability
Putri, Fitriyanti, Sulistiana, Fahria & Hidayati	2019	Indonesia	The effect of Islamic social reporting index on Islamic banking financial performance in Indonesia	Islamic Social Reporting (ISR)—Islamic Banking (IB)
Lekpek	2019	–	CSR in Islamic banking: Theory and practice	Islamic Bank (IB)—Corporate Social Responsibility (CSR)

(continued)

Table 2 (continued)

Authors	Year	Country	Title	Dimension
Amran, Mohamed, Yusuf & Rozzani	2019	Malaysia	Financial and social performances of Islamic microfinance service providers with mobile banking	Islamic Microfinance—Social
Platonova, Asutay, Dixon & Mohammad	2018	United Kingdom	The impact of CSR disclosure on financial performance: Evidence from the GCC Islamic banking sector	Islamic Bank (IB)—Corporate Social Responsibility (CSR)
Shahabuddin, Alamer, Sukor & Naseem	2018	Malaysia	Weighted CSR index for measuring Islamic Banking's social performance: How to develop it?	Islamic Bank (IB)—Corporate Social Responsibility (CSR)
Hamidi & Worthington	2018	Indonesia	Islamic social banking: The way forward perbankan sosial Islam: Langkah kehadapan?	Islamic Social Bank (ISB)

reviews of annual reports and interviews. The analysis is based on content analysis. Meanwhile, the other 5 articles utilized a quantitative methodology in their studies.

4. A quantitative study was applied for rest of the articles found by conducting a survey questionnaire.
5. Due to the limited research on Islamic social banking, future studies in this field may need to expand and explore other areas, including Islamic finance and other related research topics.

(Source: Authors' compilation from literature)

This systematic review seeks to comprehensively analyze and evaluates the sustainability of Islamic social banking. In the 15 studies, the dimension focusing on Islamic bank—CSR (Muflih, 2021; Lui et al., 2021; Zafar & Sulaiman, 2020; Ur Rehman et al., 2020; Lekpek, 2019; Platonova et al., 2018), Islamic social bank (Mohd Nor, 2022; Hamidi & Worthington, 2018), CSR—financial performance—Islamic banking (Zafar et al., 2022), Islamic bank—social finance (Darmalaksana, 2022), Islamic social bank—sustainable development goals (Hamidi & Worthington, 2021), Islamic social finance—sustainability (Tahiri Jouti, 2019), Islamic social reporting—Islamic banking (Putri et al., 2019) and Islamic microfinance—social (Amran et al., 2019).

Although the results of the studies did not cover all aspects due to limited literature, this review found there is a need for exploring sustainability of Islamic social banking since most study are only focusing on Islamic bank—corporate social responsibility segment and their views on sustainability of Islamic social banking are limited to only certain such as corporate social responsibility and sustainable development goals (Muflih, 2021; Lui et al., 2021; Hamidi & Worthington, 2021; Zafar & Sulaiman, 2020; Ur Rehman et al., 2020; Lekpek, 2019; Platonova et al., 2018).

In addition, there is no established framework or guidelines available for Islamic social banking (Hamidi & Worthington, 2021). The applicable framework or guidelines for Islamic social banking are important to ensure misconduct of funds. The current practices, unfortunately, are not sufficient to fully support the newly structured Islamic social banking (ISB) model due to differences in the scope of certain practices compared to the broader purposes of ISB. While a few studies explored the scope of ISB, they did not do so in a specific or comprehensive manner (Mohd Nor, 2022; Muflih, 2021; Lui et al., 2021; Zafar & Sulaiman, 2020; Ur Rehman et al., 2020; Lekpek, 2019; Hamidi & Worthington, 2018; Platonova et al., 2018).

Another notable discovery from this review is the consistent upward trajectory observed in Malaysia and Indonesia regarding the advancement and promotion of Islamic social banking studies. It shows Malaysia and Indonesia are the countries that encourage Islamic social banking in order to serve commercial needs and social causes (Mohd Nor, 2022; Darmalaksana, 2022; Muflih, 2021; Hamidi & Worthington, 2021; Lui et al., 2021; Putri et al., 2019; Amran et al., 2019; Shahabuddin et al., 2018; Hamidi & Worthington, 2018).

A social bank provides financial services to individuals and organizations that yield social, environmental, or sustainability benefits. These banks prioritize long-term relationships with clients and possess a thorough understanding of their economic activities and associated risks (Weber, 2014). Despite being theoretically committed to social responsibility (Asutay & Harningtyas, 2015; Haniffa & Hudaib, 2007), Islamic social banking appears to lack concrete implementation strategies. Some argue that Islamic social banking has minimal impact on disadvantaged individuals (Asutay, 2007, 2008; Mohd Nor, 2016), while others contend that it prioritizes profit-seeking similar to conventional banks (Abdul-Baki & Uthman, 2017; Abdul-Rahman et al., 2014).

Although the outcomes are not definitive, this study may contribute to addressing scholars' inquiries. For instance, scholars often emphasize the necessity of a credible database and trustworthy information sources. The results of this study suggest that they could provide a basis for future research exploring the sustainability of Islamic social banking on a global scale. Moreover, it underscores the importance of sustainability in Islamic social banking, given its substantial positive societal impact. Furthermore, this systematic review highlights the innovative aspects of sustainability within Islamic social banking, particularly in Malaysia, and suggests the importance for regulatory bodies to enhance and develop effective frameworks or guidelines.

4 Conclusion

In terms of social objectives, Islamic scholars hold divergent views on the role of social objectives in Islamic banking. Some believe that Islamic banks should function as regular businesses as long as they adhere to Shariah principles (Lewis & Algaoud, 2001; Satkunasegaran, 2003). Conversely, the majority of Islamic scholars argue that Islamic banking should strive to promote an economic system that fosters a just, fair, and balanced society (Ahmad, 2000; Naqvi, 2016; Siddiqui, 2001). They advocate for economic interactions within society to enhance overall public happiness, reinforce social bonds, and uphold societal values (Tripp, 2006). Similarly, Al-Zuhayli (2003) asserted that the primary objective of Islamic financial institutions is not profit-making but rather the advancement of social goals such as socio-economic development and poverty alleviation.

To implement such values, Islamic banks should include the important tenets of compliance in their daily activities, and place an emphasis not solely on financial profit, but also on justice and societal obligations (Taqi-Usmani, 2002). Furthermore, Islamic banking is expected to be more socially responsible than conventional banks (Asutay & Harningtyas, 2015; Haniffa & Hudaib, 2007). As a result, Kamel (1997) argued that Islamic goals should include objectives relating to economic development, the creation of value added, more exports and fewer imports, job creation, the rehabilitation of the incapacitated, and the training of others. To date, this strong view of the social responsibility of Islamic bank has found legislative support in just a few countries, including Indonesia and Jordan (Hamidi & Worthington, 2017).

In summary, this study establishes a foundation for future research in this area, marking the first documented effort to conduct a systematic literature review aimed at understanding the different factors contributing to variations in study findings. The key contributions of this paper include the use of a powerful new tool that is readily available, as well as its comprehensive analysis of extensive data, allowing for a deeper understanding of the gaps in the sustainability of Islamic social banking. Additionally, the study follows the Preferred Reporting Items for Systematic Reviews and Meta-Analyses (PRISMA) guidelines, which are widely accepted as the standard for reporting evidence in systematic reviews and meta-analyses. The benefits of adhering to PRISMA include demonstrating the review's quality, enabling readers to evaluate its strengths and weaknesses, facilitating the replication of review methods, and structuring and formatting the review using PRISMA headings.

Pooling opinions from a large dataset has addressed the ambiguity in the literature concerning the sustainability of Islamic social banking. The findings of this study hold great significance as they establish foundational knowledge on the subject. Importantly, this study is pioneering and effectively addresses the research gap in understanding this subject matter.

Despite its contributions, this study has some limitations worth mentioning. Firstly, the study aimed to conduct a systematic literature review on the sustainability of Islamic social banking from a broader perspective. However, there may be variations in the practices of Islamic social banking. Therefore, further investigations should explore other aspects of Islamic social banking practices. Secondly, the focus of this study is on the sustainability of Islamic social banking, which may not be as prominent subject as others. Studies on more popular subjects may yield different results. Given that the sustainability of Islamic social banking is not universally recognized across all databases, stakeholders need to understand the importance and necessity of such studies. The findings of this study can be beneficial for industry professionals, practitioners, academics, and entrepreneurs to purposefully explore the significance of sustainability in Islamic social banking for industry development.

Acknowledgments This research was funded by the Ministry of Higher Education (MOHE) under the Fundamental Research Grant Scheme (FRGS): FRGS/1/2021/SS01/USIM/02/5. Faculty of Economics and Muamalat, Universiti Sains Islam Malaysia.

References

Abdul-Baki, Z., & Uthman, A. B. (2017). Exploring the "social failures" of Islamic banks: A historical dialectics analysis. *Journal of Islamic Accounting and Business Research, 8*(3), 250–271.

Abdul-Rahman, A., Latif, R. A., Muda, R., & Abdullah, M. A. (2014). Failure and potential of profit-loss sharing contracts: A perspective of new institutional, *economic (NIE) theory. Pacific-Basin Finance Journal, 28*, 136–151.

Ahmad, K. (2000). Islamic finance and banking: The challenge and prospects. *Review of Islamic Economics*, 57–82.

Ahmed, M. U. (2019). *Islamic social bank*. LinkedIn. Accessed Nov 2, 2019, from https://www.linkedin.com/pulse/islamic-social-bank-mezbah-uddin-ahmed

Al-Zuhayli, W. (2003). *Al-Fiqh Al-Islami Wa Adillatuh (Islamic Jurispudence and its proofs)*. (M. A. El-Gamal, Trans. (Vol. 5). Dar al-Fikr al-Mouse.

Asutay, M., & Harningtyas, A. F. (2015). Developing maqasid al-shari'ah index to evaluate social performance of islamic banks: A conceptual and empirical attempt. *International Journal of Islamic Economics and Finance Studies, 1*(1), 5–64.

Asutay, M. (2012). Conceptualising and locating the social failure of Islamic finance: Aspirations of Islamic moral economy vs the realities of Islamic finance. *Asian and African Area Studies, 11*(2), 93–113.

Asutay, M. (2008). Islamic banking and finance: Social failure. *New Horizon, No. 169*, 1–3. https://doi.org/10.2139/ssrn.1735674

Asutay, M. (2007). Conceptualisation of the second-best solution in overcoming the social failure of islamic finance: Examining the overpowering of homoislamicus by homoeconomicus. *IIUM Journal in Economics and Management, 15*(2), 167–195.

Amran, A. M., Mohamed, I. S., Yusuf, S. N. S., & Rozzani, N. (2019). Financial and social performances of Islamic microfinance service provider with Mobile banking. *International Journal of Financial Research, 10*(5), 181–190. https://doi.org/10.5430/ijfr.v10n5p181

Bichler, B. F., Petry, T., Kallmuenzer, A., & Peters, M. (2022). Get on task: A pragmatic tutorial on planning and conducting a systematic literature review. In F. Okumus, S. M. Rasoolimanesh, & S. Jahani (Eds.), *Contemporary research methods in hospitality and tourism* (pp. 39–53). Emerald Publishing Limited. https://doi.org/10.1108/978-1-80117-546-320221004

Darmalaksana, W. (2022). The need for social theology to strengthen the social functions of Islamic banking in Indonesia. *HTS Teologiese Studies/Theological Studies, 78*(1), a7342.

Desrosiers, A., Betancourt, T., Kergoat, Y., Servilli, C., Say, L., & Kobeissi, L. (2020). A systematic review of sexual and reproductive health interventions for young people in humanitarian and lower-and-middle-income country settings. *BMC Public Health, 20*(666), 1–21. https://doi.org/10.1186/s12889-020-08818-y

Dusuki, A. W. (2008). What does Islam say about corporate social responsibility. *Review of Islamic Economics, 12*(1), 5–28.

Hamidi, L., & Worthington, A. C. (2021). How social is Islamic banking? *Society and Business Review, 16*(1), 51–70. https://doi.org/10.1108/SBR-03-2020-0036

Hamidi, M. L., & Worthington, A. C. (2018). Islamic social banking: The way forward. *Jurnal Ekonomi Malaysia, 52*(1), 179–190. https://doi.org/10.17576/JEM-2018-5201-15

Hamidi, M. L., & Worthington, A. C. (2017). Islamic banking plus social banking equals Islamic social banking: An equation in the making. In K. Jerzy (Ed.), *Banking: Services, opportunities and risks*. Nova Science Publishers, Inc.

Hamidi, M. L., Worthington, A. C., West, T., & Ismal, R. (2019). The prospects for Islamic social banking in Indonesia. *Journal of Islamic Monetary Economics and Finance*, 5(2), 237–262.

Haniffa, R., & Hudaib, M. (2007). Exploring the ethical identity of Islamic banks via communication in annual reports. *Journal of Business Ethics, 76*(1), 97–116.

Khairi, K. F., Laili, N. H., & Kamarubahrin, A. F. (2021). Takaful scheme for mental health disorders: A systematic literature review. *Al-Uqud: Journal of Islamic Economics, 5*(1), 29–42. https://doi.org/10.26740/al-uqud.v5n1.p29-42

Kamel, S. (1997). *Development of Islamic banking activity, problems and prospects* (Vol. 12). IDB Prize Winners' Lecture Series.

Lekpek, A. (2019). Corporate social responsibility in Islamic banking: Theory and practice. *Sociologija, 61*(1), 32–54. https://doi.org/10.2298/SOC1901032A

Lewis, M. K., & Algaoud, L. M. (2001). *Islamic banking*. Islamic Edward Elgar.

Linares-Espinós, E., Hernández, V., Domínguez-Escrig, J. L., Fernández-Pello, S., Hevia, V., Mayor, J., Padilla-Fernández, B., & Ribal, M. J. (2018). Methodology of a systematic review. Metodología de una revisión sistemática. *Actas Urologicas Espanolas, 42*(8), 499–506. https://doi.org/10.1016/j.acuro.2018.01.010

Lui, T. K., Zainuldin, M. H., Wahidudin, A. N., & Foo, C. C. (2021). Corporate social responsibility disclosures (CSRDs) in the banking industry: A study of conventional banks and Islamic banks in Malaysia. *International Journal of Bank Marketing, 39*(4), 541–570. https://doi.org/10.1108/IJBM-04-2020-0192

Mirjalili, S. H., Soroosh, A., Baan, H., & Botshekan, M. H. (2020). Islamic social banking platform (case of Resalat Islamic Bank). *Journal of Money and Economy, 15*(1), 35–54. https://doi.org/10.29252/jme.15.1.35

Mohd Nor, S. (2022). Islamic social Bank: A sustainable banking approach. In *Institutional Islamic economics and finance* (pp. 65–79). Routledge.

Mohd Nor, S. (2016). Islamic social bank: An adaptation of Islamic banking? *Jurnal Pengurusan, 46*, 43–52.

Muflih, M. (2021). The link between corporate social responsibility and customer loyalty: Empirical evidence from the Islamic banking industry. *Journal of Retailing and Consumer Services, 61*, 102558. https://doi.org/10.1016/j.jretconser.2021.102558

Muhamed, N. A., Nor, F. M., Nawai, N., Shahwan, S., Abu Seman, J., Abdullah, M., Nooh, M. N., & Kamarubahrin, A. F. (2022). Financial literacy module for micro businesses entrepreneur: Systematic literature review. *International Journal of Academic Research in Accounting Finance and Management Sciences, 12*(2), 537–548. https://doi.org/10.6007/IJARAFMS/v12-i2/14119

Nasir, N. S. M., & Seman, J. A. (2022). Sustainability and Islamic social banking: A review of literature. In *Islamic banking, accounting and finance international conference-the 10th iBAF 2022* (pp. 509–519). Faculty of Economics and Muamalat, Universiti Sains Islam Malaysia.

Naqvi, S. N. H. (2016). *Perspectives on morality and human Well-being: A contribution to Islamic Economics*. Kube Publishing Ltd..
Pati, D., & Lorusso, L. N. (2018). How to write a systematic review of the literature. *HERD, 11*(1), 15–30. https://doi.org/10.1177/1937586717747384
Platonova, E., Asutay, M., Dixon, R., & Mohammad, S. (2018). The impact of corporate social responsibility disclosure on financial performance: Evidence from the GCC Islamic banking sector. *Journal of Business Ethics, 151*, 451–471.
Putri, A. K., Fitriyanti, E., Sulistiana, I., Fahria, I., & Hidayati. (2019). The effect of Islamic social reporting index on Islamic banking financial performance in Indonesia. *Humanities & Social Sciences Reviews, 7*(4), 609–616. https://doi.org/10.18510/hssr.2019.7481
Raimi, L., Patel, A., & Adelopo, I. (2014). Corporate social responsibility, waqf system and zakat system as faith-based model for poverty reduction. *World Journal of Entrepreneurship, Management and Sustainable Development, 10*(3), 228–242.
Satkunasegaran, E. (2003). Corporate governance and the protection of customers of Islamic banks. *Paper presented at the Proceedings of the International Islamic Banking Conference*.
Shahimi, S., Mohd Marzuki, M., & Embong, Z. (2013). Potential of cash waqf for poverty alleviation in Malaysia: A system dynamics approach. *Jurnal Ekonomi Malaysia, 47*(2), 149–163.
Shahabuddin, A., Alamer, A. R. A., Sukor, M. E. A., & Naseem, N. A. M. (2018). Weighted corporate social responsibility index for measuring Islamic Banking's social performance: How to develop it? *International Journal of Economics and Management, 12*(1), 233–246.
Siddaway, A. P., Wood, A. M., & Hedges, L. V. (2019). How to do a systematic review: A best practice guide for conducting and reporting narrative reviews, meta-analyses, and meta-syntheses. *Annual Review of Psychology, 70*, 747–770. https://doi.org/10.1146/annurev-psych-010418-102803
Siddiqui, S. H. (2001). Islamic banking: True modes of financing. *New Horizon, 109*, 15–20.
Tahiri Jouti, A. (2019). An integrated approach for building sustainable Islamic social finance ecosystems. *ISRA International Journal of Islamic Finance, 11*(2), 246–266. https://doi.org/10.1108/IJIF-10-2018-0118
Taqi-Usmani, M. (2002). *An introduction to Islamic finance (Arab and Islamic Laws)*. CQ.
Tripp, C. (2006). *Islam and the moral economy: The challenge of capitalism*. Cambridge University Press.
Tyle, I. (2013). *Revolting subjects: Social abjection and resistance in neoliberal Britain*. London, UK: Zed Books.
UNCDF. (2024). *Financial Inclusion and the SDGs*. https://www.uncdf.org/financial-inclusion-and-the-sdgs?ref=hackernoon.com.
Ur Rehman, Z., Zahid, M., Rahman, H. U., Asif, M., Alharthi, M., Irfan, M., & Glowacz, A. (2020). Do corporate social responsibility disclosures improve financial performance? A perspective of the Islamic banking industry in Pakistan. *Sustainability, 12*(8), 3302. https://doi.org/10.3390/su12083302
Weber, O. (2014). Social banking: Concept, definitions and practice. *Global Social Policy, 14*(2), 265–267.
Zafar, M. B., Sulaiman, A. A., & Nawaz, M. (2022). Does corporate social responsibility yield financial returns in Islamic banking? *Social Responsibility Journal, 18*(7), 1285–1310. https://doi.org/10.1108/SRJ-04-2020-0160
Zafar, M. B., & Sulaiman, A. A. (2020). Measuring corporate social responsibility in Islamic banking: What matters? *International Journal of Islamic and Middle Eastern Finance and Management, 13*(3), 357–388. https://doi.org/10.1108/IMEFM-05-2019-0227

Junaidah Abu Seman is a Senior Lecturer at the Faculty of Economics and Muamalat, Universiti Sains Islam Malaysia (USIM). She earned her Doctor of Philosophy degree in Finance from the University of Salford (2017). She has extensively written on Islamic banking, finance, takaful, and business management across seminar papers, journal articles, and case studies. Also, she serves as an Evaluator for the FAA Programme Accreditation (FPA) under the Finance Accreditation Agency (FAA). Furthermore, she is a lifetime member of the Malaysian Muamalat Association (MMA) and a member of the International Council of Islamic Finance Educators (ICIFE). Her research interests encompass inclusive finance, Islamic banking and finance, and takaful.

Syahidawati Shahwan is an Associate Professor at the Faculty of Economics and Muamalat, Universiti Sains Islam Malaysia (USIM). She holds a Doctor of Philosophy degree in Islamic Banking and Finance from the International Institute of Islamic Banking and Finance, International Islamic University Malaysia (IIiBF) in 2017. She specializes in Shariah and Muamalat, Islamic Finance, and Shariah Governance research. She is involved in creating Islamic finance products and educational materials, notably the award-winning Muamalat Interactive Game (MIG) for basic Islamic finance training. She provides training to industry professionals through MIG Training and Consultancy and AMANIE Academy.

Nurul Aini Muhamed is an Associate Professor at the Faculty of Economics and Muamalat, Universiti Sains Islam Malaysia (USIM), and a research fellow at the Institute of Halal Research and Management (IHRAM). She earned a PhD in Business and Management from the University of South Australia in 2009. She is editor "Halal Governance and Management: Malaysian and ASEAN Countries," and authored "Malaysian Debt from the Perspectives of Economics and Islam." Her research focuses on Islamic Transaction Law, Islamic Banking and Finance, Halal Compliance, and Corporate Governance.

Nur Ainna Ramli is an Associate Professor at the Faculty of Economics and Muamalat, Universiti Sains Islam Malaysia (USIM). She has established herself as a prominent figure in academia through her extensive contributions to teaching, research, and scholarly activities. She continues to inspire and empower future generations of scholars, leaving an indelible mark on the academic landscape.

Aimi Fadzirul Kamarubahrin is an Associate Researcher at the Faculty of Economics and Muamalat, Universiti Sains Islam Malaysia (USIM). He holds a Doctor of Philosophy degree in Economics and Muamalat, majoring in Marketing and specializing in Consumer Behavior from USIM. He has served as a research fellow at various public and private institutions, contributing as a member, advisor, strategic decision maker, and data analyst. His expertise spans consumer behavior, research methods, marketing, Islamic finance, non-profit Islamic organizations, and halal management.

Professionalism and Competency of Amil Zakat in Malaysia and Indonesia: Strategies for Sustainability

Azman Ab Rahman, Ahmad Rafiki, Hery Syahrial, and Fitriani Tobing

Abstract This chapter aims to elaborate and discuss on professionalism and competency of amil in Malaysia and Indonesia. The study adopts a content analysis technique where various sources are referred including research database named Scopus, reports, official documents, and research articles. The chapter also explains on the research (active countries and affiliations in publishing research documents) of zakat management, professionalism, and competency of amil zakat in Malaysia and Indonesia. There are challenges related to the amil zakat where expectations are high and the impact of zakat on the economy is expected. Various keywords associated with zakat management which are importantly to be discussed by practitioners, academicians, and other stakeholders, in getting positive solutions to the countries.

1 Introduction

The management of zakat faces several challenges, including transparency, accessibility, fairness, and compatibility due to cultural, geographical, and political contexts. These difficulties require significant attention to ensure effective and efficient zakat management (Akhyar Adnan and Barizah Abu Bakar (2009). Among the issues impeding the implementation and management of zakat are the understanding and awareness of people about zakat, rigid interpretations, and the crisis of confidence in zakat agencies (Owoyemi, 2020). Additionally, low public awareness and the lack of human resources have been identified as challenges in zakat management (Mohammed et al., 2021). Strategies to address these challenges include the establishment of standard operating procedures, increasing zakat education, and improving the quality of governance and human resources in zakat institutions

A. A. Rahman
Universiti Sains Islam Malaysia, Nilai, Negeri Sembilan, Malaysia

A. Rafiki (✉) · H. Syahrial · F. Tobing
Universitas Medan Area, Medan, Indonesia

A. Rafiki et al. (eds.), *Strategic Islamic Business and Management*, Contributions to Management Science,
https://doi.org/10.1007/978-3-031-61778-2_15

(Mawardi et al., 2023). Furthermore, there is a growing interest in digital management techniques for zakat, reflecting efforts to modernize its administration and enhance its effectiveness (Hudaefi et al., 2022; Bin-Nashwan et al., 2020b). These challenges highlight the need for continuous improvement and innovation in the management of zakat to ensure its impact on social welfare and economic prosperity.

The research has also highlighted the significance of zakat as a mechanism for promoting social solidarity, addressing economic inequality, and fostering a sense of communal responsibility. Furthermore, studies have examined the management and institutional arrangements of zakat systems, including the collection, distribution, and eligibility criteria for recipients. The research emphasizes the multifaceted nature of zakat, encompassing its religious, social, and economic dimensions, and its potential to bring about positive change in society (Al-Bawwab, 2023; Alshater et al., 2021).

2 Literature Review

2.1 Amil Zakat

Amil in zakat are all parties who take action related to the collection, storage, safeguarding, recording, and distribution or distribution of zakat assets. They are appointed by the government and obtain permission from it or are selected by authorized government agencies or by the Islamic community to collect and distribute and other tasks related to zakat, such as raising awareness or educating the public about zakat law, explaining the characteristics of property owners who are subject to the obligation to pay. The zakat management institutions and committees that exist today are contemporary forms of institutions authorized to administer zakat as stipulated in Islamic law. Therefore, officers (amil) who work in these institutions must fulfill the specified requirements.

2.2 Development of Zakat

The development of zakat has evolved over time and has become a global phenomenon, with the potential to generate sustainable income and contribute to the eradication of poverty among Muslims worldwide (Sarif et al., 2024). Over time, the collection and distribution of zakat have become more organized, with governments and non-governmental organizations establishing institutions to facilitate the process (Mutamimah et al., 2021). The development of zakat has been influenced by the United Nations Sustainable Development Goals (SDGs), which aim to eradicate poverty, reduce inequalities, and promote peace and environmental protection. Zakat aligns strongly with these goals, as it focuses on helping the poor and needy,

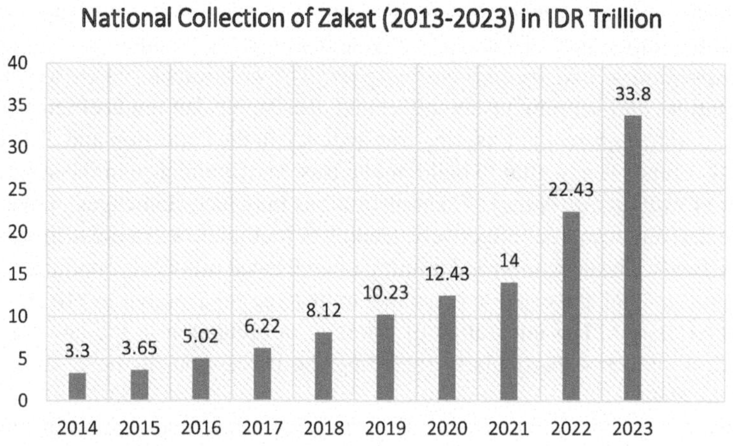

Fig. 1 National collection of Zakat (2013–2023). Source: Badan Amil Zakat Nasional (2023)

Table 1 Zakat collection in Malaysia 2015–2022

	2015	2016	2017	2018	2019	2020	2021	2022
Zakat collection (in million)	2483.38	2846.56	2923.1	2854.82	3175.72	3365.55	3164.52	4042.69

Source: PPZ-MAIWP Malaysia (2023)

refugees, and displaced people, among other causes. The following is the national zakat collection data in Indonesia for 2013–2023 and Malaysia in 2015–2022 (Fig. 1 and Table 1):

The zakat system currently faces several challenges that need to be addressed to ensure its effectiveness in alleviating poverty and supporting the needy. The challenges of zakat management in Muslim developing countries are numerous and complex. Some of the key challenges include (Mawardi et al., 2023; Kamal et al., 2024; Ardani & Pujiyono, 2021; Bin-Nashwan et al., 2020a):

- *Lack of integration with government programs:* There is a need for zakat programs to be integrated with government programs, particularly those aimed at poverty reduction, to maximize their impact.
- *Low public awareness:* Increasing public awareness about zakat is crucial for its successful implementation and acceptance by the community.
- *Low reporting:* Many zakat institutions do not report their zakat management activities, which makes it difficult to track and optimize the collection and distribution of zakat funds.
- *Regulatory issues:* The current Zakat Act does not yet integrate zakat as part of state finances, which needs to be addressed to facilitate its use as a source of state revenue.

- *Socialization and community engagement:* The community must be educated about the importance of zakat and its role in poverty alleviation.
- *Instrumentation and institutional support:* The government needs to establish institutions and instruments to collect and distribute zakat funds efficiently.
- *Trust and transparency:* Ensuring transparency in the collection and distribution of zakat funds is essential to build public trust and confidence in the system.
- *Lack of staff and resources:* Zakat institutions may face challenges in managing their staff and resources effectively, leading to inefficiencies in the distribution of zakat funds. The education and training of amil zakat officers is crucial for ensuring professional and capable personnel to manage zakat activities effectively.
- *Political issues:* The integration of zakat as an obligation in the law may raise political concerns, particularly regarding the perception of the country as an Islamic state.
- *Corruption and mismanagement:* Ensuring that zakat funds are not misused or misappropriated is a significant challenge, as it requires strong governance and accountability mechanisms.
- *Technological limitations:* The use of technology to improve the efficiency of zakat distribution and monitoring is limited, which can lead to issues in tracking recipient information and the distribution process.
- *Governance of zakat institutions:* Ensuring the transparency, accountability, and effectiveness of zakat institutions is crucial for maintaining public trust and confidence in the zakat system.

To address these challenges, it is essential to strengthen the governance of zakat institutions, improve transparency and accountability, and increase public awareness and education about the importance of zakat in supporting the needy and alleviating poverty. Moreover, the development of zakat has been driven by its potential to address social issues and contribute to the SDGs. It is a significant form of philanthropy that has been adopted globally and has the potential to make a significant impact on poverty, inequality, and other social issues.

3 Methods

This study adopts a content analysis based on past research and formal documents related to professionalism and competency in managing zakat funds in Malaysia and Indonesia. It is also includes a few items of bibliometric approach that manifests interconnections among the articles whereby the data was retrieved from Scopus, a reputable and comprehensive database. The primary reason for choosing Scopus is its availability, nearly 60% larger than the Web of Science (WoS) database (Zhao & Strotmann, 2015).

4 Discussion

Table 2 lists some interpretations of the research on zakat in 10 years. Out of 568 articles retrieved from the Scopus database, 20 of them have the highest number of cited documents, which indicates the most referred documents by the readers/ researcher. The highest cited document (66 citations) is an article written by Syed M.H. et al. in 2020 emphasized the artificial intelligence on zakat. Followed by an article written (cited 52 times) by Ali I. and Hatta Z.A. in 2014 focused on zakat which is used as an instrument for poverty reduction. The least cited documents (26 citations) are an article written by Sohag K. et al. in 2015 that concerned on zakat system in alleviating rural poverty, an article written by Wulandari P. and Kassim S. in 2016 that discussed issues and challenges in financing the poor, and an article written by Ismail C.Z. et al. in 2015 that explained on administration and management of waqf land (Table 2).

As shown in Fig. 2, both Malaysia and Indonesia lead the list as the most productive countries that produced articles on zakat compared to, i.e., Saudi Arabia, the United Kingdom, and Pakistan. This indicates that both countries are pursuing more aggressive research on the topic of zakat. Additionally, more graduates and young researchers are now putting an interest to study on zakat management. This means that zakat becomes an interesting topic to be discussed by many groups of people regardless of differences in religion.

As shown below Table, Universiti Utara Malaysia, International Islamic University Malaysia and Universiti Teknologi MARA as the top three affiliations in the list that actively doing research on zakat in 10 years. There are two Indonesian Universities (Universitas Airlangga and Universitas Islam Negeri Syarif Hidayatullah) include as the ones that seen the important role of zakat in many aspects particularly social economy which need to be enhanced by various stakeholders (Fig. 3).

From Table 3, the 20 top keywords are related to the research on zakat management which covers many aspects and factors. This shows that zakat is considered an important instrument in social economy. Both countries Indonesia and Malaysia actively published articles on zakat, which became the focus of this study. Other keywords like poverty, zakat institutions, waqf, and Islamic banks were among the most adopted keywords in zakat studies (Table 3).

1. *Amil's Role and Practice of Professional Integrity*

Integrity is an important foundation in forming an amil's personality. Even al-Nawawi's basic view of integrity must start from a person's just personality. Integrity is one of the determinant factors for measuring the level of community support for a service. As an individual who, in managing public funds, amil should prioritize the public interest above his own needs by acting in accordance with moral, ethical, and legal principles and carrying out tasks efficiently and with quality. However, according to Yamaludin et al. (2023), amil's integrity in Malaysia is still at a problematic level and always gets a lot of attention. Through research in previous

Table 2 20 highest citation article on Zakat in 10 years

No	Authors	Title	Year	Source title	Cited by
1	Syed M.H. et al.	An artificial intelligence and NLP-based Islamic FinTech model combining zakat…	2020	Intl. J. of Economics and Business Administration	66
2	Ali I.; Hatta Z.A.	Zakat as a poverty reduction mechanism among the Muslim community: …	2014	Asian Social Work and Policy Review	52
3	Javaid S.; Al-Malkawi H.-A.N.	Corporate social responsibility and financial performance in Saudi Arabia: …	2018	Managerial Finance	49
4	Rosli M.R.B. et al.	Distribution management of zakat fund: Recommended proposal…	2018	Intl. J. of Civil Engineering and Technology	43
5	Alam M.M. et al.	The Islamic Shariah principles for investment in stock market	2017	Qualitative Research in Financial Markets	36
6	Retsikas K.	Reconceptualizing zakat in Indonesia: Worship, philanthropy and rights	2014	Indonesia and the Malay World	36
7	Tahiri Jouti A.	An integrated approach for building sustainable Islamic…	2019	ISRA Intl. J. of Islamic Finance	35
8	Nomran N.M.; Haron R.	Dual board governance structure and multi-bank performance….	2019	Corporate Governance (Bingley)	35
9	Goby V.P.; Nickerson C.	Conceptualization of CSR among Muslim consumers in Dubai…	2016	J. of Business Ethics	34
10	Andam A.C.; Osman A.Z.	Determinants of intention to give zakat on employment income…	2019	J. of Islamic Accounting and Business Research	33
11	Djaghballou C.-E.; Djaghballou M.; Larbani M.; Mohamad A.	Efficiency and productivity performance of zakat funds in Algeria	2018	Intl. J. of Islamic and Middle Eastern Finance and Management	33
12	Zaman R. et al.	Religiosity and corporate social responsibility practices…	2018	Social Responsibility Journal	32
13	Shaheen Al Ahwal M. et al.	Religious beliefs, practices, and health in colorectal cancer patients in Saudi Arabia	2016	Psycho-Oncology	29
14	Kailani N.; Slama M.	Accelerating Islamic charities in Indonesia: Zakat, sedekah and…	2020	South East Asia Research	28

(continued)

Table 2 (continued)

No	Authors	Title	Year	Source title	Cited by
15	Owoyemi M.Y.	Zakat management: The crisis of confidence in zakat agencies...	2020	J. of Islamic Accounting and Business Research	28
16	Farouk A.U.; Md Idris K.; Saad R.A.J.B.	Moderating role of religiosity on zakat compliance behavior in Nigeria	2018	Intl. J. of Islamic and Middle Eastern Finance and Management	28
17	Abdullah N. et al.	The effectiveness of zakat in alleviating poverty and inequalities a measurement...	2015	Humanomics	27
18	Wulandari P.; Kassim S.	Issues and challenges in financing the poor: Case of Baitul Maal Wa Tamwil in Indonesia	2016	Intl. J. of Bank Marketing	26
19	Ismail C.Z. et al.	Administration and management of waqf land in Malaysia: Issues and solutions	2015	Mediterranean J. of Social Sciences	26
20	Sohag K. et al.	Can zakat system alleviate rural poverty in Bangladesh? ...	2015	J. of Poverty	26

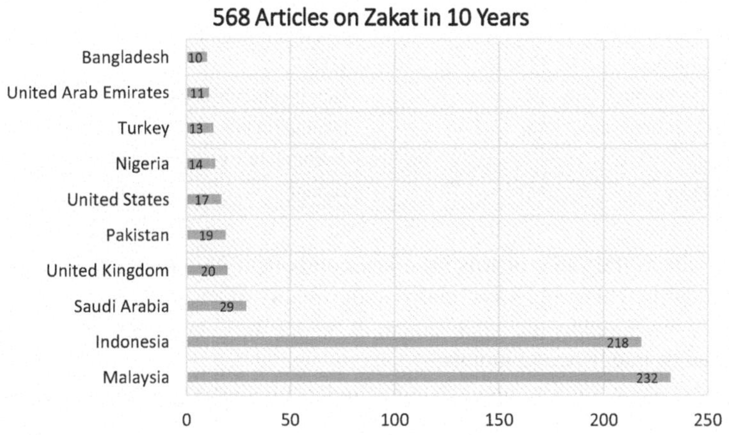

Fig. 2 The published articles on zakat by countries

studies, there are four types of integrity problems that befell Amil, namely breach of trust, and not submitting invoices within the time period for cross-border fraud and unauthorized fraud (Ab Rahman & Abd Shakor, 2024). All these integrity problems do not occur not only involving amil who work in zakat institutions but also involving appointed amil who work outside the institution such as mosque amil, professional amil, company amil, and assistant amil.

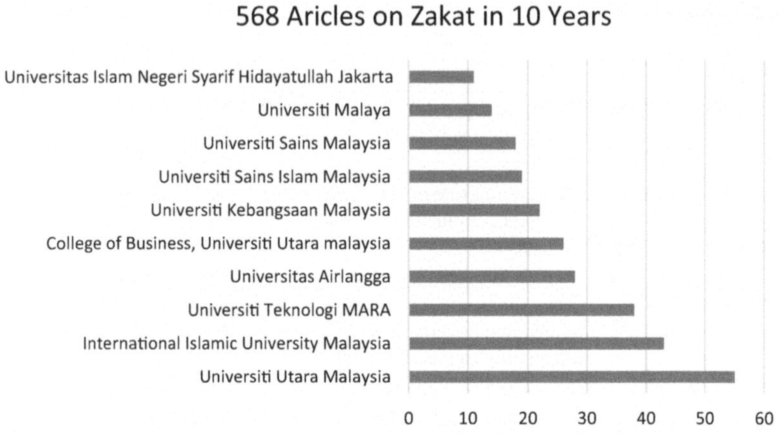

Fig. 3 The published articles on zakat by affiliations

Table 3 Top 20 most emerged keywords in 568 articles on Zakat in 10 years

Keywords	Number of Articles	Keywords	Number of articles
Zakat	200	Zakat management	12
Malaysia	37	Zakat institutions	12
Indonesia	30	Shariah	12
Poverty	26	COVID-19	12
Islam	22	Zakat distribution	11
Zakat institution	21	Trust	11
Poverty alleviation	18	Islamic philanthropy	11
Islamic social finance	18	Islamic finance	11
Waqf	17	Accountability	11
Islamic banks	13	Human	10

Islam places the value of trust as a key component in human resource development standards. This problem is related to the quality of an amil's personality, which is not only measured by the achievement of the results obtained but also by looking at the level of integrity and morals.

In Amil's view, Amil's career requires high professionalism because he needs to have specific skills, master zakat issues, be dedicated, honest, and trustworthy. This is supported by Adnan (2017) because the quality of amil's professionalism is formed from careful zakat management skills based on comprehensive training and amil's own experience. However, according to Romli and Daud (2023), today's amil skills need to be more sustainable and in line with the passage of time, otherwise they will decline and be limited to the elements of traditional zakat management only. The study supports the existence of a professional code of ethics as a guideline for Amil's ethics in carrying out his duties as stated by Md Hairi et al. (2012).

2. *Amil's Professionalism Through the Application of Digital Skills and Innovation*

According to Muharom (2010), the problem of employees who are not creative and have no initiative has hampered efforts to defend the fate of asnaf. So, it is not surprising that the lack of creative and innovative values is what gives rise to the problem of excessive bureaucracy which hampers the quality of amil services. Amil's expertise in reducing bureaucracy is the main factor influencing people's self-confidence in the aspect of zakat distribution (Hairunnizam & Sanep, 2014). This view is supported by Mohd Fisul et al. (2017) who state that public trust is based on amil's obedience in implementing standard collection procedures set by zakat institutions in Malaysia. Meanwhile, according to Pg et al. (2017), bureaucratic stigma, such as the long distribution process and the need for various supporting documents, is the cause of the perception that amil is less efficient and effective. Therefore, it is very important for amil to improve the distribution procedures adopted so that the distribution process to asnaf can be carried out easily and quickly.

This practice shows that amil is still immersed in the old ways that burden asnaf. In fact, the use of digital technology in the distribution aspect is also considered to be still lagging behind compared to advances in zakat collection technology through various sophisticated online payment platforms. The imbalance between the application of collection and distribution technology shows the amil's indifference in defending the fate of asnaf compared to the increase in the number of collections every year.

Amil's expertise in providing an efficient zakat implementation system is closely related to the ability to collect asnaf data which is an important part of their duties. However, according to Yamaludin et al. (2023), there have been many cases of asnaf data leaks, especially involving ASNAF in higher education institutions. This problem was exacerbated when many asnaf complained that they faced strict application procedures and difficulties in obtaining application forms due to the Movement Control Order (PKK). Therefore, it is time for Amil to introduce a method of implementing zakat on a path that is more practical and easier to use to avoid the issue of breaking up asnaf in the future.

There are also several studies that focus their discussion on innovation practices at the level of zakat institutions, mosques, and higher learning institutions. Muneeza et al.'s study (2022), for example, have shared several forms of collection and distribution innovations carried out by amil in Malaysia, such as introducing rice ATM machines, telephone applications, zakat payments via websites, and financial technology (fintech). While Rendy et al. (2020) also reviewed the use of smartphone applications in increasing the effectiveness of zakat collection where the frequency of zakat payments via telephone applications did not show a significant increase even though zakat collection increased. This is because there are amil innovations that do not consider the suitability of the educational background, economy, and way of life of local communities before introducing them to the community.

Amil's expertise in utilizing the latest technology in daily tasks can increase the efficiency of zakat management (Salleh and Chowdhury (2020). According to Laldin and Furqani (2019), the use of technology can reduce operational costs by saving human energy and reducing zakat advertising costs. In effect, management can avoid further leakage of internal expenditure by utilizing available allocations to cover other, more important operational costs (Ibrahim et al., 2021). Even more importantly, it can reduce the risk of financial negligence due to dependence on a fully human workforce (Mohd Nor et al., 2021).

According to Saad et al. (2023), valuable zakat management innovation does not lie in technological sophistication alone, but in the production of new ideas and concepts that can facilitate zakat payments and improve the welfare of asnaf. For example, the innovation carried out by Amil Sultan Idris Education University was to improve the interview management system and perfect it. The effectiveness of delivering information to be more effective, cheaper, and customer friendly is an innovation in zakat management that has provided benefits for many asnaf and zakat payers in higher learning institutions (Rosliza et al., 2015).

Therefore, zakat consultation practices that are efficient, friendly, and customer-oriented need to be implemented and become the core of amil's career. The slogans "the customer is always right" and "customer first" describe Amil's friendly nature in treating customers regardless of position and rank, even asnaf and zakat payers come from various backgrounds and levels of society. This is in accordance with the objectives outlined by the Selangor Zakat Council (LZS) in improving the quality and quality of amil services in assisting asnaf and zakat payers in the country (Firdaus et al., 2021).

Amil's new innovations can continue the continuity of zakat services and make them relevant to current developments. However, the majority of amil are still comfortable implementing traditional management systems with only a few appearing to be serious about exploring innovation and new technology.

3. *Practice Amil Competencies*

Practicing an appropriate competency culture is very important in producing qualified and competitive employees. According to Muhsin (2020), a competent amil will be held in high esteem by the community, which will then influence the achievement of collection and the quality of distribution at a zakat institution. The research above. Previous research found that there are five types of scientific competency practices that are often discussed, namely personality competency practices, scientific competency practices, and skills competency practices.

The practice of amil skills competency has been touched upon by several researchers who tend to examine the types of zakat management skills separately, such as collection skills, distribution skills, accounting skills, and zakat technology. For example, Nurhadi's research (2014) which examined the problem of income zakat collection skills found that skills in managing data on zakat payers were an important aspect. Things that need to be considered because there are still some of them who do not have the skills to manage data systematically, making it difficult to make references when necessary.

In addition, several studies focus on the nature of integrity in preparing zakat financial reports (Norfaiezah et al. 2017). Complete and comprehensive zakat reporting practices reflect Amil's personality qualities of being transparent and thorough in carrying out his duties (Naz'aina, 2015). According to Muhsin (2014), amil must reject the mentality that preparing zakat reports is only the role of zakat institutions. However, on the other hand, parties working in the field also play a role in preparing archives or reports on collection and distribution, submitting reports to zakat institutions on a regular basis, as well as carrying out independent audits, because they are always faced with the risk of public distrust.

A study by Naimi et al. (2019) has touched on amil's scientific competence. According to him, Amil's career requires individuals who have basic knowledge about zakat, including the matters of collecting and distributing zakat and knowledge related to financial management. Even though amil are not necessarily experts in various matters of religion or hold important positions in local organizations, they need to know in depth about the law and the correct way to calculate zakat, be able to communicate in depth about the background of the people in their area (Som et al. (2021). According to Nasir & Rahman (2020), a knowledgeable amil is an amil who knows the law of zakat and tries to add to the knowledge of other fardu ain. Apart from that, amil's responsibility as an educator is also heavy for them to understand the latest zakat issues and fatwas related to zakat through discussion activities, forums, workshops, or seminars which are often held by zakat institutions and higher education institutions (Muhsin, 2014).

Spiritual practice and integrity are two important elements that shape amil's personality. Research conducted by Taha et al. (2017) found that spiritual practices such as prayer, fasting, and Hajj have a significant relationship to the level of amil transparency which in turn can influence the performance of zakat institutions in Malaysia. This is supported by research by Abrar et al. (2021) who stated that these practices, Islamic morals that form the character of trust, such as prayer and fasting, should be used as a benchmark in preaching, evaluating the level of competence and professionalism of amil every year.

Apart from collecting and distributing zakat, Amil also acts as a preacher who gives wise warnings to increase awareness of paying zakat among the community (Muhsin, 2014). According to Azman (2014), amil not only needs to be good at managing zakat but also needs to equip himself with various communication skills such as verbal and non-verbal mass communication skills and skills to become an effective zakat preacher. Apart from preaching against zakat payers, amil also plays a role in shifting the asnaf paradigm from weak thinking and unable to think positively, stubbornness and competitive (Wahid & Zolkiaply, 2018). This is supported by Teh et al. (2016) who found that local amyl was able to change.

Asnaf's thinking pattern through the right approach then becomes an evaluation indicator to improve the quality of life of asnaf in the city. In summary, the discussion in this section has highlighted the views that have been expressed by previous researchers regarding the role and practice of amil professionalism, amil professionalism through aspects of the application of digital skills and innovation, as well as amil competency practices in the aspects of personality, knowledge, and skills.

5 Conclusion

The chapter elaborates on the development and potential of zakat management particularly on the professionalism and competency of amil in Malaysia and Indonesia. The research on this topic was crucial to be explored and studied in order to get great economic impact of zakat, i.e., poverty alleviation in certain Muslim countries, i.e., Bangladesh, Pakistan, Nigeria, Chad, and Yemen. Each country has uniqueness in managing zakat fund, however, through effective management and professional human resources of amil, zakat potential could be perceived.

Amil zakat plays a crucial role in the collection and distribution of zakat funds, which are used to alleviate poverty and provide assistance to those in need. The education system has not provided a clear pathway for training professionals in zakat management, which is a complex process that requires specialized knowledge and skills. Meanwhile, the government commitment, rules and regulations, and facilities and infrastructure should support the professionalism of amil zakat. Understanding the zakat management with an adoption of technology is crucial to do acceleration in developing national economy. With that, issues related to amil zakat can be eliminated such as low reporting, transparency, accountability, and effectiveness of zakat institutions, low public awareness, integration with state revenue and political issues.

The conclusion from the search results is that professionalism and competency of amil zakat has potential to significantly improve the management and transparency of zakat collection and distribution. This can increase public trust, awareness, and provide a more comprehensive system of managing zakat funds. This can lead to a more effective distribution of zakat funds and ultimately contribute to economic growth and poverty reduction. This study is expected to give valuable information and insights for future research in mapping out a more comprehensive framework mainly on zakat management and its human resources.

References

Al-Bawwab, R. A. (2023). Zakat: Changing the framework of giving. *Islamic Economic Studies, 30*(2), 86–103. https://doi.org/10.1108/IES-08-2021-0026

Ab Rahman, A., & Abd Shakor, M. F. (2024). Cabaran Pengurusan Amil Zakat di Institut Pengajian Tinggi: The Challenges of Amil Zakat in Higher Education Institutions. *Journal of Fatwa Management and Research, 29*(1), 142–152.

Abrar, R. N. M., Dewi, T. C., Dewantari, W., Rosyidah, S. F., & Prastowo, A. (2021). Analisis Kemampuan Kognitif Siswa terhadap Materi Zakat melalui Pembelajaran Fikih Madrasah Ibtidaiyah Kurikulum 2013. *Jurnal Pendidikan Islam, 5*(1), 51–60.

Adnan, M. A. (2017). The need of establishment of professional amil zakat to enhance the future zakat development. *International Journal of Zakat, 2*(1), 71–79.

Alshater, M. M., Saad, R. A. J., Abd Wahab, N., & Saba, I. (2021). What do we know about zakat literature? A bibliometric review. *Journal of Islamic Accounting and Business Research, 12*(4), 544–563.

Akhyar Adnan, M., & Barizah Abu Bakar, N. (2009). Accounting treatment for corporate zakat: A critical review. *International Journal of Islamic and Middle Eastern Finance and Management, 2*(1), 32–45.

Ardani, M., & Pujiyono, A. (2021). The priority problems and solutions in formulating strategies to optimize zakat collection in Indonesia. *International Journal of Zakat, 6*(3), 69–86.

Azman Ab Rahman. (2014). *Modul Pengurusan Amil Zakat Malaysia*. Penerbit USIM.

Bin-Nashwan, S. A., Abdul-Jabbar, H., Aziz, S. A., & Ismail, S. S. H. (2020a). Challenges of zakah management in muslim developing countries. *Journal website: journal. zakatkedah. com. my, 2*(1).

Bin-Nashwan, S. A., Abdul-Jabbar, H., Aziz, S. A., & Haladu, A. (2020b). Zakah compliance behavior among entrepreneurs: Economic factors approach. *International Journal of Ethics and Systems, 36*(2), 285–302.

Firdaus, Y. E., Refnita, D., & Saputra, A. (2021). The mustahik zakat in various dimensions of fiqh in era society 5.0. *El-Mashlahah, 11*(1), 13–28.

Hairunnizam, W., & Sanep, A. (2014). Faktor mempengaruhi tahap keyakinan agihan zakat: Kajian terhadap masyarakat Islam di Selangor. *Jurnal Ekonomi Malaysia, 48*(2), 41–50.

Hudaefi, F. A., Caraka, R. E., & Wahid, H. (2022). Zakat administration in times of COVID-19 pandemic in Indonesia: A knowledge discovery via text mining. *International Journal of Islamic and Middle Eastern Finance and Management, 15*(2), 271–286.

Ibrahim, A. Z., Zainol, Z., Wahab, N. A., & Samsi, A. (2021). The relationship between livelihood assets, zakat aid and food security achievement among zakat recipient in Kedah, Malaysia. *Cent Asia Caucasus, 22*, 373–384.

Kamal, S., Berakon, I., Hamid, A., & Muttaqin, Z. (2024). How do muzakki pay professional zakat? (the qualitative inquiries using the bloom model). *Journal of Islamic Marketing, 15*(3), 866–885.

Laldin, M. A., & Furqani, H. (2019). Fintech and Islamic finance: Setting the Sharī 'ah parameters. In *Fintech in Islamic finance* (pp. 113–119). Routledge.

Mawardi, I., Widiastuti, T., Al Mustofa, M. U., & Hakimi, F. (2023). Analyzing the impact of productive zakat on the welfare of zakat recipients. *Journal of Islamic Accounting and Business Research, 14*(1), 118–140.

Md Hairi Md Hussain, Kamil Md. Idris & Ram Al Jaffri Saad. (2012). Ketelusan di dalam tadbir urus institusi zakat (Prosiding Seminar Isu-isu kontemporari Zakat di Malaysia), 63–74.

Mohammed, M. O., Robbana, A., & Bedoui, H. (2021). *Zakat digital management techniques* (pp. 299–317). Insights and Solutions.

Mohd Fisul, I., Wahid, H., & Noor, M. A. M. (2017). *Pengurusan zakat di Malaysia: Suatu pendekatan analisis gelagat* (pp. 67–82). Pusat Penyelidikan Ekonomi dan Kewangan Islam UKM.

Mohd Nor, S., Abdul-Majid, M., & Esrati, S. N. (2021). The role of blockchain technology in enhancing Islamic social finance: The case of Zakah management in Malaysia. *Foresight, 23*(5), 509–527.

Muharom, F. (2010). Model-Model Kreatif Distribusi Zakat Berbasis Masyarakat. *Ijtihad: Jurnal Wacana Hukum Islam dan Kemanusiaan, 10*(1), 35–51.

Muhsin N. P. (2014). "Amalan Dakwah zakat Di Wilayah Persekutuan Malaysia: Satu Pemerhatian" dalam proceeding of the international conference on Arabic studies and Islamic civilization iCasic 2014, 4–5 march 2014,. Kuala Lumpur.

Muhsin Nor Paizin. (2020). Tingkah Laku Asertif dan Tahap Kepuasan Bekerja Dalam kalangan Kakitangan Institusi Zakat di Malaysia. *Journal on Technical and Vocational Education., 5*(2), 29.

Mutamimah, M., Alifah, S., Gunawan, G., & Adnjani, M. D. (2021). ICT-based collaborative framework for improving the performance of zakat management organisations in Indonesia. *Journal of Islamic Accounting and Business Research, 12*(6), 887–903.

Muneeza, A., Bin-Nashwan, S. A., Moshin, M. I. A., Mohamed, I., & Al-Saadi, A. (2022). Zakat payment from cryptocurrencies and crypto assets. *International Journal of Islamic and Middle Eastern Finance and Management, 16*(3), 482–497.

Norfaiezah, S., Aziz, N. M. A., & Ram al Jaffari Saad. (2017). Zakat accountability practices: Case study of zakat institution. *International Journal of Economic Research., 14*(16), 223–232.

Naimi, M., Nor, M., & Abdul, D. (2019). Fraud Dan Mekanisme Pencegahan Di Institusi zakat (fraud and prevention mechanism at zakat institution). Proceedings of the 1st Kedah international zakat conference 2019 (KEIZAC 2019) pp. 233–243, 2019 (Keizac), 233–243.

Nasir, M. S., & Rahman, F. U. (2020, November). Analysis of the recipients of zakat and the current situation. In Indonesian conference of zakat-proceedings (pp. 331–338).

Naz'aina. (2015). The effect of internal control system and Amil competence on the financial reporting quality at zakat management institution active member of zakat forum in special Capital City region Jakarta and West Java provinces. *Procedia - Social and Behavioral Sciences, 211*, 753–760. https://doi.org/10.1016/j.sbspro.2015.11.100

Owoyemi, M. Y. (2020). Zakat management: The crisis of confidence in zakat agencies and the legality of giving zakat directly to the poor. *Journal of Islamic Accounting and Business Research, 11*(2), 498–510. https://doi.org/10.1108/JIABR-07-2017-0097

Pg, M. F., & F.A.O., Hairunnizam, W. & Mohd Ali, M.N. (2017). Kecekapan Pengurusan Kewangan Dan Agihan Zakat: Kajian Di Majlis Ugama Islam Sabah (MUIS). *Jurnal Syariah, 25*(3), 415–414.

Rendy, N. A. S. (2020). *Pemahaman Masyarakat Mengenai Mekanisme Pembagian Zakat Fitrah Secara Merata Dalam Perspektif Hukum Islam (Studi Kasus Mushola Baiturrahman Kelurahan Metro Kecamatan Metro Pusat Kota Metro)*. (Doctoral dissertation,. IAIN Metro.

Romli, A. B., & Daud, M. N. (2023). Konsep Kompetensi Keperibadian Amil Zakat. *AZKA International Journal of Zakat & Social Finance*, 57–67.

Rosliza, R., & @ R., Ibrahim, H., Sanep, A., Abdul Basir, H.H. & Shofian, H.A. (2015). Inovasi dalam Pengurusan Zakat di UPSI: Kesannya Terhadap Kepuasan Asnaf dalam Kalangan Pelajar. *Jurnal Perspektif, 7*(1), 101–112.

Saad, R. A. J., Ahmi, A., Sawandi, N., & Abdul Aziz, N. M. (2023). Zakat administration reformation towards an effective and efficient zakat revenue generation. *Journal of Islamic Accounting and Business Research, 14*(8), 1232–1260.

Salleh, M. C. M., & Chowdhury, M. A. M. (2020, October). *Technology adoption among zakat institutions in Malaysia*. In Indonesian Conference of Zakat-Proceedings (pp. 1-14).

Sarif, S., Ali, N. A., & Kamri, N. A. (2024). Zakat for generating sustainable income: An emerging mechanism of productive distribution. *Cogent Business & Management, 11*(1), 2312598.

Som, H. M., Paizin, M. N., Mokhar, D. R., & Sungit, D. F. (2021). Kesan penswastaan terhadap pengurusan sumber manusia di institusi zakat.

Teh, S. T., Azman, A. S., & Shamsuddin, N. (2016). Impak agihan zakat terhadap kualiti asnaf di Selangor. In *Proceeding 2nd international conference on Economics & Banking* (pp. 226–237).

Taha, R., Adam, F., Ali, N. N. M., & Ariff, A. M. (2017). Religiosity and transparency in the management of zakat institutions. *J. Legal Ethical & Regul. Isses, 20*, 1.

Wahid, H., & Zolkiaply, N. (2018). *Kepuasan Asnaf Terhadap Bantuan Sewa Rumah Oleh Lembaga Zakat*. Peranan Penolong Amil Kariah.

Yamaludin, S. M., Syed Alwi, S. F., Rosman, R., & Khamis, M. R. (2023). The role of zakat distribution on the sustainability of gharimin (genuine debtors) in Islamic financial institutions in Malaysia. *Journal of Islamic Accounting and Business Research.*

Zhao, D., & Strotmann, A. (2015). *Analysis and visualization of citation networks*. Morgan & Claypool Publishers.

Azman Ab Rahman is a Professor at the Faculty of Shariah and Law, University of Islamic Sciences Malaysia (USIM) and also holds a position as Director at the Institute of Fatwa and Halal (IFFAH). He holds a Bachelor's Degree from the University of al Al-Bayt, Mafraq, Jordan in the field of Fiqh and Usul. He obtained a Master and Doctor of Philosophy from the International Islamic University of Malaysia (UIAM) in the field of Fiqh and Usul Fiqh. He has published more than 100 articles in indexed and non-indexed journals and also produced more than 50 academic books in the field of zakat and contemporary fiqh. He has led 50 research grants and member of more than 90 research projects including research grants awarded by USIM, Ministry of Education, other state agencies, and private institutions. His outstanding outputs in research were evident from his receiving distinguished awards at national and international levels.

Ahmad Rafiki is currently an Associate Professor in the Faculty of Economics and Business, the Dean of Faculty of Economics and Business, and the Head of Islamic Management and Halal

Industry Center at Universitas Medan Area, Indonesia. He obtained his Bachelor's Degree in Business Administration with a major in Marketing from MARA University of Technology (UiTM) and a Masters in Management from the International Islamic University of Malaysia (IIUM). He completed his thesis for the award of Doctor of Philosophy from the Islamic Science University of Malaysia (USIM). He holds a Certified Islamic Marketer (CIMA) from the International Islamic Marketing Association (IIMA). His research interest is in the Islamic entrepreneurship, management, marketing, and leadership. He has published chapters and papers in international publishers such as IBA-MacMillan, IGI Global, Emerald Publishing, Springer, IntechOpen, and Routledge. He also became the editorial advisory board and reviewer in reputable publishers of Emerald and Elsevier.

Hery Syahrial is an assistant professor at the Faculty of Business and Management, Universitas Medan Area. He graduated his Master of Science from Universitas Sumatera Utara and has been a lecturer for more than 20 years in the field of Management. He has extensive experience in academic affairs and was Deputy Dean for Academics at the Faculty of Business and Management for about 16 years. His research field is in human resource management and now actively publishing research articles and book chapters on that field as well as attended several international conferences in Malaysia, Turkey, and Macedonia.

Fitriani Tobing is a lecturer at the Faculty of Business and Management, Universitas Medan Area. She is currently completing her doctoral program in the field of Management Science at Universitas Syiah Kuala. She graduated her Master of Science from Universitas Padjadjaran and has been serving as a lecturer in management for 6 years. Her research area focuses on marketing management, and she is actively engaged in publishing research articles and book chapters in the field of Islamic marketing. She has also participated in several international conferences in Malaysia.

Perception of Customers Toward the Introduction of Islamic Banking to Ghana

Jibril Yahuza and Omar Kachkar

Abstract This study seeks to determine customers' impressions of Islamic Banking (IB) in Ghana. As a non-Islamic Sub-Saharan African country, the study's goal is to investigate Ghana's behavioral goals regarding IB implementation. The recommended approach is based on the *"theory of planned behaviour (TPB),"* which is widely utilized in customer behavior evaluation. This theory consists of belief structures related to behavior, norms, and control that influence consumers' intentions. The questionnaire was designed to be self-administered. The results of the self-administered structured questionnaire were employed to gather information from 312 respondents to investigate banking customers' perceptions of embracing IB in Ghana. SPSS Amos and partial least squares structural equation modeling (PLS-SEM) were utilized in conducting the descriptive technique. The technique of the model shows an acceptable high goodness-of-fit index, and the data is consistent with the statistical model; the measurement model had a good enough validity and reliability, and the Standardized Mean Square Values (SRMR) were acceptable because it was below the threshold. All three predictors were found positive and significant; the attitude of banking customers in Ghana toward IB significantly impacted their behavioral intention to use it when introduced. Similarly, subjective norms were found to be significant and positive, and lastly, perceived behavior control was found to be positive and effective from all three components. Attitude is the most important factor determining a consumer's desire to use IB, and it has the greatest performance value, followed by subjective norms and perceived behavior control. Nevertheless, perceived behavioral control outperformed subjective norms, which means more awareness must be created to motivate more banking consumers to use IB when introduced.

J. Yahuza · O. Kachkar (✉)
Department of Management, Ibn Haldun University, Istanbul, Türkiye
e-mail: omar.kachkar@ihu.edu.tr

A. Rafiki et al. (eds.), *Strategic Islamic Business and Management*, Contributions to Management Science,
https://doi.org/10.1007/978-3-031-61778-2_16

1 Introduction

The introduction of the Shari'a-compliant Islamic business can be traced back to 1975 when the Islamic Development Bank (IsDB) of Saudi Arabia and the Islamic Bank of Dubai took the initiative (Hanieh, 2020). Since the 1960s, the global market has been dominated by the five sub-sectors of the Islamic financial operations including Islamic banking (IB), takaful, and Islamic funds, Islamic bonds, and other Islamic financial organizations (Tijjani et al., 2021). For the past decades, the Islamic financial system has witnessed much development, with its assets reaching USD 2.4 trillion in 2017, marking a growth of more than 40% since 2012. Furthermore, the consistent growth pattern of the Islamic financial industry led to a total market worth of 2.19 trillion dollars as of 2018 (IFSB, 2019). The industry was predicted to experience noteworthy expansion in the overall worth. It was projected by the Islamic Financial Services Industry (IFSI) forecasts that the industry will expand by USD 3.8 trillion by 2023, with an annual compound growth rate of 6% (Amin, 2020).

According to the International Monetary Fund (IMF), the United Nations Development Programme (UNDP), and the World Bank (WB), Islamic finance is founded on Islamic principles and ideals that uphold permissible standards, oppose interest, and embody equitable earnings, assets, and production practices (Mohamad & Borhan, 2017; Tijjani et al., 2021). Islamic finance prohibits transactions involving uncertainty (gharar) as well as immoral behavior such as insider trading, market manipulation, short-selling, and dealing in shares of unregistered firms. Islamic finance aims to achieve the eighth objective of sustainable economic growth, which includes comprehensive and productive employment based on Islamic principles (Hanieh, 2020).

Islamic financial institutions engage in competition with conventional banks through the provision of varied outlets of alternatives to conventionally grounded products and services. Scholars and professionals express a keen interest in researching various factors that potentially impact a client's attitude toward Islamic banking, encompassing elements such as traditions, cultural and religious perspectives, societal influences, lending decisions made by the bank, and other economic variables. His interest has been stimulated by the expanding prevalence of Islamic banking globally (Junaidi et al., 2022; Abid & Jie, 2022). These studies have presumed evidence suggesting that non-economic factors may exert an influence on the financial choices made in Ghana, encompassing aspects such as culture, religion, family, tradition, society, and political reliance, the roles played by different social classes, individuals, and the government, as well as the presence of economic duality within individual nations.

Traditional banking services primarily characterize the Ghanaian Banking Industry (GBI). All 30 certified financial institutions in Ghana currently engage in interest-based banking activities for nearly all financing activities. The regulatory framework governing the GBI consists of vital legislative acts such as the Bank of Ghana (BoG) Act 2002 (Act 612), the Non-Bank Financial Sectors Act 2008 (Act

774), the Banking Act 2004 (Act 673), the Companies Code Act 179, 1963, and numerous BoG, Directives, Circulars, and Regulations, and Notices. IB is experiencing rapid expansion and is capable of attracting financial resources and investment from domestic and foreign origins. Many banks in Ghana, including those in rural areas, provide E-banking activities to their clients, ranging from Automated Teller Machine (ATM) cards to Internet banking services. These banks provide investment banking and insurance activities in addition to typical current and savings accounts.

The acceptance process of the Islamic financial operation in some marketing geographies is a complex undertaking due to its status as a comparatively new financing system. This process is closely intertwined with the instruction language, terminology, and frameworks sourced from the teachings of Islam. The resemblance between certain Islamic banking services and their traditional counterparts sometimes sparks debates and discussions among institution participants (Ezeh & Nkamnebe, 2020). Moreover, the expansion of Islamic financing without a parallel enhancement of the Islamic economic technique has hindered the acceptance of this new method (Khurshid & Mahmood, 2020).

Islamic financing is rooted in the beliefs of Islamic principles and is commonly chosen by adherents of the religion due to their spiritual obligations. It differs from conventional banking by adhering to Sharia-based financial practices built on Islamic law, which prohibits usury of certain types of gambling and uncertain transactions. Islamic banking offers alternative products such as Muzaraha, Ijarah, and Musharakah to address these prohibited transactions. These products aim to fill the gaps caused by these forbidden transactions. Home goods rely on shared earnings or losses depending on performance, the things backing them up and selling and buying items (Butt et al., 2018; Lujj et al., 2016).

Islam practices require followers to consistently apply spiritual guidelines regarding their bond with God and how they relate to nature and people. Therefore, Muslims always follow moral and ethical jurisdictions and follow private and public codes (Bukhari et al., 2020). When purchasing anything, Muslims must ensure that every item used is halal, even in economic and monetary matters. Islam recognizes monetary management while employing a specific, fair, and just technique for all parties involved (Lujja et al., 2016; Al Balushi et al., 2019).

Islamic banking has seen substantial development globally in many years because of assistance from Islamic and non-Islamic nations (Abdullah & Dusuki, 2006; Ahmed et al., 2021; Mbawuni & Nimako, 2018). In Indonesia, home to the world's greatest Muslim population, IB is broadening more quickly than conventional banks. Assets held by Islamic banks leaped a remarkable 13.94% in 2021—higher than the 9.86% increase traditional banks saw. What is more, development in IB deposits reached a noteworthy 15.30% yearly, evidently surpassing the 11.93% progress traditional financial organizations posit accomplished. In this respect, Islamic bank financing increased at a more rapid rate than conventional banks and grew by 6.90% as compared to 5.11% in OCT 2021 and IBS 2021. In this respect, Islamic bank financing increased more rapidly than conventional banks and grew by 6.90% compared to 5.11% in OCT 2021 and IBS 2021.

Islamic finance growth research was significantly linked to the spread of Islamic banking, mainly whenever it revolves around the research's emphasis on Islamic banking and conduct intention (Erol and El-Bdour, 1989; Hoque & Awang, 2019). Previous studies on the intents of Islamic banks concerning services have shown varying outcomes depending on the conditions. Prior research, however, has primarily focused on specific groups of Muslims: Millennials Muslims pursuing Islamic colleges with a concentration on advanced study (Aji et al., 2020; Shome et al., 2018). In general, Muslim university students are better aware, particularly riba. The result will probably be a vital impetus for ordinary people to switch to Islamic financial management procedures.

Globally, Islamic Banking (IB) is being implemented in more non-Muslim and Muslim nations (Bananuka et al., 2019; Charag et al., 2020; Bananuka et al., 2020a; Junaidi, 2021). The reasoning behind been built on the significant fundamentals of Islamic Banking, such as rejecting gharar, limiting interest rates and Altima, promoting trade, and investing in halal businesses (Junaidi, 2021; Charag et al., 2020; Ali et al., 2018; Aziz & Afaq, 2018; Mahdzan et al., 2017). As stated in Surat Baqara verse 274 of the Holy Quran and Deuteronomy 23 verse 19 of the Holy Bible (King James Version), the holy Qur'an both prohibits extortion and interest rates from a religious standpoint. In the words of Bananuka et al. (2020a, 2020b, 2020c), Islamic Banking regards money is both a store of value and a vehicle of trade. The overall worth of IB assets went up US$1765.8 trillion in 2019 as opposed to US$1571.3 trillion in 2018 (Financial et al. Services, 2020). Nevertheless, Africa continues to have a minor fraction of worldwide banking assets—1.6%—while the Gulf Cooperation Council (GCC) area controls the majority of global IB reserves (45.4%) (Islamic Financial Services Board, 2020).

In many countries, especially emerging ones, financial services have been crucial to the growth of their economy. However, traditional banking practices are based on principles that are not according to the teachings of Islam, and this has led to concerns about their legitimacy. To address this issue, an alternative to traditional banking is Islamic financial management, which is built on Islamic fundamentals. According to Abdullah and Dusuki (2006), Islamic banking is proliferating and becoming increasingly popular in both non-Muslim and Muslim nations (Iqbal & Molyneux, 2006).

Furthermore, Islamic banks have grown to offer a diverse variety of Shari'a-compliant current accounts, savings accounts, mortgage loans, credit cards, products, and other services (Amin et al., 2014; Butt et al., 2018). Increasing competition among Islamic and conventional banks translates to more inventiveness in goods and services. Therefore, Islamic financial institutions are pushed to produce the services and products that their clientele demand the most. Customers will use the services of Islamic banks if their offerings are of excellent quality.

Nonetheless, only a few of these kinds of research investigations consider client religiosity while deciding whether to utilize Islamic banking. Some notable exemptions include Aziz and Afaq (2018), Kaabachi and Obeid (2016), and Echchabi et al. (2016). Consequently, the model used in the study was created by Aziz and Afaq (2018) to describe why Indonesian customers prefer to utilize Islamic banking services. Unlike previous research, this study emphasizes religion as a factor affecting customers' intention to accept the services of Islamic finance. The survey is also unusual regarding the types, amount, and where the studies were conducted.

Traditional banking operates in a way that is fundamentally different from Islamic banking. Instead of using interest rates, Islamic banking relies on profit/loss sharing to facilitate borrowing and lending. This approach is built on Shariah law and ensures that all financial operations are in line with the rules and regulations of Islamic Finance. Customers can get a range of financial operations using Islamic instruments designed to comply with Shariah law. In contrast, traditional banking is based on interest, and religious laws do not influence its operations. These key distinctions make Islamic banking a unique and appealing option for individuals and businesses seeking an alternative to traditional banking. The details differences between conventional and Islamic banks are shown in Table 1 beneath.

Islamic banking is observed as a potential competitor to conventional banks in Ghana due to its free interest characteristic. The spring up of Islamic Financing, the

Table 1 Differences between conventional banks and Islamic banks

Conventional system	Islamic system
Involves payment or receipt of interest as a fundamental component of loans and deposits.	Prohibits receipt and payment of interest (riba) as it is a way of exploiting people.
Primarily relies on fixed interest-based returns to depositors and interest charges to borrowers.	It operates based on profit and loss sharing, where profits and losses are shared between the financial institution and the clients.
It may involve unsecured loans, and the backing of loans is not always tied to specific assets.	Require transactions to be supported by physical assets or services, promoting actual economic activity.
Places more emphasis on risk transfer, often involving collateral and insurance.	Emphasizes risk-sharing of risk among all partners in the transaction of finance.
Subject to regulatory frameworks, may engage in speculative activities, such as derivatives trading.	Avoids speculative transactions and uncertainties (gharar) in financial dealings.
May invest in various industries without specific ethical considerations.	Encourages socially responsible investments, avoiding industries like gambling, alcohol, and pork.
Operates under the regulatory framework set by central banks and financial authorities.	Operates with guidance from Sharia boards and may be subject to Islamic financial regulations.
Often emphasizes the value of collateral without significant ethical considerations.	It may require collateral but tends to prioritize the nature of the asset and the ethical considerations.
Engages in short- and long-term transactions based on the nature of financial products.	Encourages shorter-term transactions to reduce uncertainty and ensure adherence to Sharia principles.
Operates on a fractional reserve system where banks can create money through lending beyond their actual reserves.	Critiques the conventional system for the fractional reserve banking model, where money creation is not necessarily linked to actual productive activities.

Source: Developed by authors

expansion of the population of Muslims, and the funding need in Ghana are some of the reasons behind this belief. However, according to the Oxford Business Group, there is still more to be done for both Islamic and conventional services in the Ghanaian marketing setting due to the growing middle class, expanding middle class, and substantial long-term economic outlook of the country. Ghana Islamic Microfinance, a microfinance company established in 2010, is currently using an Islamic banking and finance strategy to offer fierce and ethically adaptable interest-free microfinance products to Muslim clients in Ghana. Moreover, the Central Bank of Ghana has received two requests in order to form Islamic banks, steps are being taken to revise the Banking Act, 2004 (Act 673) to include laws governing Islamic financing in the nation, according to reports from March 2022. Given the paucity of empirical surveys on Islamic Banking in non-Islamic developing nations, the potential and distinctiveness of Islamic Banking in Ghana, the survey aims to investigate customers' perspectives toward implementing Islamic Banking in Ghana. The study hopes to expand our knowledge of consumer behavior toward Islamic banks' services by advancing the discussion on the significance of Islamic banks generally and in non-Islamic nations.

Although Islamic banking looks promising, there are many barriers to the implementation of Islamic financial systems in other countries due to a lack of understanding amongst the banking customers, especially in countries where the non-Muslim population is disproportionately higher than the Muslim population, like Ghana. Lack of knowledge of the people can lead non-Muslims and some Muslims to believe that Islamic banking is only available to Muslims mistakenly. The main issue is a misunderstanding of the actual variations between the activities of conventional and Islamic banks regarding financing, deposits and investment methods, agreements, leasing, and similar matters.

The difficulties impeding the implementation of Islamic banking in Ghana were identified in this study as significant and huge, necessitating the search for solutions. Even though the issues discussed in the paper are entirely religious and are invariably associated with the objection to Islamic banking, these issues may well be related to ignorance and misperception on the part of certain non-Muslims and even their religious organizations, which have a history of criticizing the system and shunning it for a variety of reasons.

Moreover, even the Muslim populace refrained from using Islamic banking because of insufficient awareness, comprehension, and knowledge of its fundamental principles. Based on these challenges, the study examines customers' perceptions regarding the commencement of Islamic banking in Ghana.

Research Objectives

The following are the research objectives:

- To identify the effect of the perception and attitude of bank customers on their intention toward Islamic banking.
- To determine the impact of subjective norms of bank customers on their intention toward Islamic Banking.
- To examine the effect of perceived behavioral control of bank customers on their intention toward Islamic Banking.

Research Questions
The following are the research objectives:

- What is the effect of the perception and attitude of bank customers on their intention toward Islamic banking?
- What is the effect of subjective norms of bank customers on their intention toward Islamic Banking?
- What is the effect of perceived behavioral control of bank customers on their intention toward Islamic Banking?

2 Literature Review

The Theory of Planned Behavior (TPB) is a useful paradigm for describing human behavior in social psychology. It is founded on the Theory of Reasoned Action (TRA), introduced by Fishbein and Ajzen in 1975, which links attitude and behavior and includes social elements in this relationship. According to Kocagöz (2010, p. 140).

However, TRA has come under fire because of some restrictions that resulted in the TPB being incorporated into consumer behavior. TPB is regarded as the preeminent conceptual framework for researching human behavior, building upon TRA (Fishbein & Ajzen, 1977) (Ajzen, 2001). Its foundation lies in social psychology, which addresses the relationship between beliefs and behavior (Ajzen, 1991). TPB combines behaviors influenced by non-voluntary external variables and is a modification of TRA (Ajzen, 1985). It is made up of four parts: subjective norms, attitude, behavioral control, and behavioral intents (Fig. 1). Among these four components, the desire to implement Islamic banking in Ghana is identified by subjective norms, attitude, and perceived behavioral control are the key components of the Theory of Planned Behavior (TPB). Additionally, attitudes and subjective standards were added to expand on TRA by incorporating perceived behavioral control, which considers volitional elements for assessing the likelihood of behavior among people (Armitage & Comer, 1998).

2.1 Attitude and Intentions

People's attitudes toward socially relevant things, people, events, or symbols are frequently constant and may be described by a series of thoughts, feelings, and behavioral patterns. This can result in a liking or aversion to specific objects, according to psychology (Vaughan & Hogg, 2005; Eagly & Chaiken, 1993). Suki (2010) defines attitude as a measure of how someone feels about something, which may assist anticipate their intentions and behaviors (Phau and Teah, 2009). Attitudes have also been shown to influence behavior (Table 2).

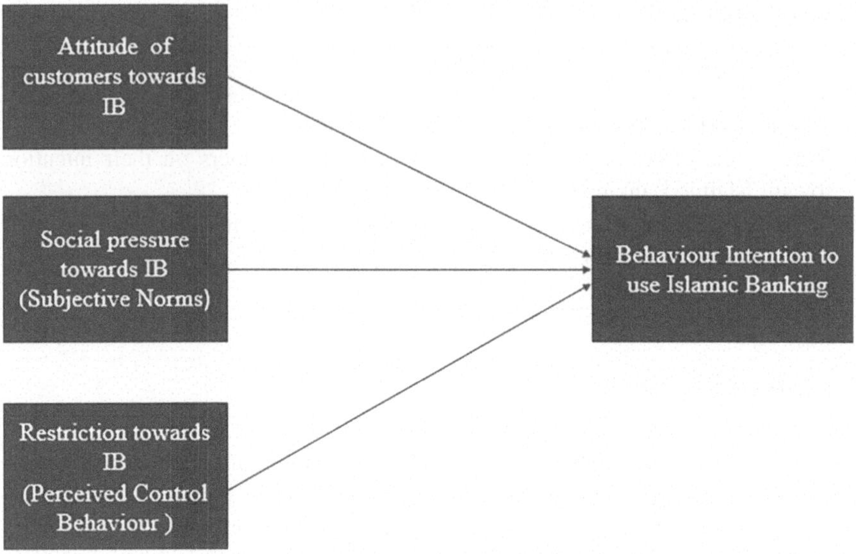

Fig. 1 Conceptual framework based on TPB

There is much research on intention and attitude, specifically how it affects behavioral intention. Research has been done on how technology and customer behavior interact, especially with Islamic financial services. Numerous research studies have demonstrated that a favorable attitude is positively correlated with a stronger behavioral intention to purchase; yet, in specific domains, such as life insurance (Britto & Amer, 2007) and being physically active (Armitage, 2005), the relationship between attitude and intention is negatively correlated. Overall, the existing literature suggests the need for more research on this theme, specifically in the backdrop of Islamic banking. To achieve this, the following hypothesis is put forth:

H1. Consumer attitude significantly influences an individual intention to accept Islamic banking.

2.2 Subjective Norms and Intention

The effect of social elements, like social pressure to exhibit a particular behavior, is called subjective norms (Fishbein & Ajzen, 1977). It has been suggested that social pressures can sometimes control attitudes toward specific behavior (Fishbein & Ajzen, 1977). The subjective norms combine the motivation to uphold normative beliefs and the fundamental belief structures known as normative beliefs. Normative opinions refer to a person's belief that a particular group referent motivates them to act, while the significance of those reference groups drives an individual to comply.

Table 2 Some literature studies

Author	Country	Study Scope	Sample (N)	Theory	Findings	Link to the study
Al Umar and Setyono, (2023)	Indonesia	Islamic banking	275	TPB	Attitude → intention ($\beta = 0.246$) subjective norm → intention ($\beta = 0.203$) PBC → intention ($\beta = 0.226$) Islamic branding → intention ($\beta = 0.254$) Hijra intention → intention ($\beta = 0.025$)	https://ejournal.iaisyarifuddin.ac.id/index.php/iqtishoduna/article/view/2243/729
Si and Chin, (2023)	Malaysia	Islamic banking	303	TPB	Attitude → intention ($\beta = 0.235$) awareness → intention ($\beta = 0.025$) PBC → intention ($\beta = 0.392$) subjective norm → intention ($\beta = 0.155$) religiosity → intention ($\beta = 0.140$)	https://journal.ump.edu.my/jgi/article/view/9736/2991
Sulistiowati et al., (2023)	Indonesia	Islamic Finance	200	TPB	PBC(KP) → intention ($\beta = 0.434$) attitude (SK) → intention ($\beta = 0.232$) PBC → usage intention ($\beta = 0.388$)	https://journal.ipb.ac.id/index.php/ijbe/article/view/42655/24971
Nurjanah et al., (2023)	Indonesia	Islamic Finance	140	TPB	Religiosity → intention ($\beta = 0.455$) social environment → intention ($\beta = 0.343$) government support → intention ($\beta = 0.076$)	https://ije3.esc-id.org/index.php/home/article/view/147/153
Mindra, Bananuka, Kaawaase, Mindra et al., (2022)	Uganda	Islamic banking	300	TRA	The results of this study show that attitude is positively associated with utilizing Islamic banking.	File:///C:/Users/user/Downloads/2022_Mindra_Juma_etal2022_ModerationattitudeandIB%20(1).pdf
Alzadjal et al., (2022)	Oman	Islamic Finance	638	TPB	Attitude → intention ($\beta = 0.471$) subjective norm → intention ($\beta = 0.406$) PBC → intention ($\beta = 0.557$)	https://www.emerald.com/insight/content/doi/10.1108/JIMA-05-2020-0150/full/html#sec019
Ali et al., (2021)	Pakistan	Islamic banking	311	TPB	PBC → acceptance of Islamic banking ($\beta = 0.273$) SN → acceptance of Islamic banking ($\beta = 0.234$) financial literacy → acceptance of Islamic banking ($\beta = 0.106$) risk perception → acceptance of Islamic banking ($\beta = 0.061$)	File:///C:/Users/user/Downloads/1929.pdf

(continued)

Table 2 (continued)

Author	Country	Study Scope	Sample (N)	Theory	Findings	Link to the study
Maryam et al., (2021)	Pakistan	Islamic banking	300	TPB	Social influence → attitude → acceptance intentions (β =0.077) religious obligations → attitude → acceptance intentions ($\beta = 0.066$) reputation →attitude → acceptance intentions ($\beta = 0.111$) support for business→ attitude → acceptance intentions ($\beta = 0.022$)	File:///C:/Users/user/Downloads/10-1108_JIMA-03-2021-0059.pdf
Naseri & Sharofiddin, (2021)	Afghanistan	Islamic finance	334	DIT	The multiple regression analysis finding indicated that product knowledge, relative advantage, and religiosity significantly and positively influenced the acceptance of Islamic banking.	https://tujise.org/content/6-issues/17-2/5-a169/a169.pdf
Smith (2021)	Malaysia	Islamic banking	149	TPB	Attitude → intention ($\beta = 0.01$) subjective norm → intention (β =0.11) PBC → intention ($\beta = 0.229$) intention → religiosity ($\beta = 0.00$)	https://journal.unisza.edu.my/jimk/index.php/jimk/article/view/575/418

The Theory of Planned Behaviors (TPB), The Diffusion of Innovation Theory (DIT), The Theory of Reasoned Action (TRA)

Subjective norms are the total number of beliefs and norms plus the worth assigned to them.

Subjective norms have a substantial influence on behavioral intention in a variety of industries, including Islamic banking, real estate mortgages, and insurance. Subjective norms are important in defining behavioral intention, according to studies, however, some studies contradict this conclusion, emphasizing the need for more study (Amin, 2012; Juwaheer et al., 2012; Ali & Jama, 2016). Moreover, as previously done by Amin (2012) and Omar & Owusu-Frimpong (2007). Hence:

H2. Subjective norms significantly and positively influence an individual's intention to accept Islamic banking.

2.3 Perceived Behavioral Control and Intentions

According to the Theory of Planned Behavior (TPB), Perceived Behavioral Control (PBC) is defined by Ajzen and Madden (1986) as an individual's impression of the difficulty of an action. PBC refers to an individual's perceived level of control over a given behavior or objective. In other words, it is the individual's belief in their capacity to do a behavior effectively. PBC can be impacted by a number of circumstances, including previous experiences, accessible resources, and social support. PBC, attitudes, and subjective norms impact our behavioral intentions. In simpler terms, the more positive our attitude, subjective norms, and PBC are concerning a particular behavior, the more we will carry it out with solid intentions.

The literature has reported varied findings from earlier studies examining the connection between behavioral intention and PBC. Many studies have found a link between behavioral intentions and PBC in a variety of domains, including Islamic banking (Echchabi & Abdul-Aziz, 2012; Md Husin and Ab Rahman, 2016), Islamic insurance (Ali & Jama, 2016), online banking activities (AL-Ajam & Nor, 2013), budgeting and scheduling (Echchabi & Abdul-Aziz, 2012; Saeed et al., 2012), and credit card use by community-university members (Krishna et al., 2011). This implies that having a sense of control over one's behavior is critical in getting the desired results. Individuals can improve their chances of effectively converting their intentions into action by identifying and resolving issues that impact PBC.

However, many studies on introducing Islamic banking have shown a positive correlation between PCB and intention. This is particularly beneficial in the Ghanaian situation, where Islamic banking is still in the early stages, as hypothesized as follows:

H3. Perceived behavior control significantly affects an individual intention to accept Islamic banking.

3 Research Design

The suggested model's structural linkages were tested in the research using partial least squares structural equation modeling (PLS-SEM) with SmartPLS 4.2 software (Magno et al., 2024). PLS-SEM is a valuable tool for the presentation or validation of models while illuminating numerous statistical correlations simultaneously. Due to its distribution-insensitive supposition, anticipatory concentration, and descriptive building of models' method, PLS-SEM was established in the research to be ideal for appreciating the components that impact IB adoption (Chin, 2010). The total number of participants for PLS-SEM was determined using the act of ten, as described by Hair et al. As per the guidelines, the smallest possible sample for every latent construct should be tenfold the most significant number of structural routes (Hair et al., p. 144). A helpful sample of survey respondents will be retrieved to find out the intention of customers toward IB, and the survey comprises 312 questionnaires that were distributed for research purposes. As a result, our 312-respondent sample size is sufficient.

3.1 Respondents Demographics Profile

The questionnaires of the survey design included a thorough description of each respondent's demographic profile, based on the theory that these characteristics might impact how banking customers view the commencement of Islamic banking in Ghana.

According to Cooper and Schindler (2014), it is essential to screen the data gathered from research tools to ensure that there are no missing data, outliers, and normalcy. This step is crucial in maintaining the accuracy and validity of the study's results. Pairwise exclusion has been employed in the research to deal with missing data. According to Toka and Çetin (2016), this approach allows researchers to examine data, get each of the observed components of parameters independently, and connect the results. Furthermore, the boxplot test was used to evaluate univariate outliers. After missing information and outliers were examined, Kurtosis and Skewness were used to perform the normality test.

After data screening, the SPSS and Amos software programs were used to do inferential and descriptive analyses of 312 survey responses. Five demographic attributes were examined in this study. These comprise the respondent's age, gender, occupation, degree of education, and religion. The overall outcomes of the demographic profile of the respondents are displayed in Table 3. The findings indicate that, among the 312 respondents, the proportion of men and women was 50.3% and 49.7%, respectively. This ratio is regarded as impartial and fair.

Regarding religion, the ratio remained the same. Specifically, 50% of respondents identified as Muslims, and the remaining 50% did not. They were carried out to provide religious balance and prevent bias in the findings about the respondents' ages. The majority of them were aged between 31 and 40 years. The other age ranges are displayed in Table 3.

Table 3 Respondents' demographic profile

Demographic information		Respondents no.	Percent	Cumulative %
Gender	Male	155	49.7	49.7
	Female	157	50.3	100
Age	< 20 years	11	3.5	3.5
	20–30 years	98	31.5	34.9
	31–40 years	115	36.9	71.8
	41–50 years	76	24.4	96.2
	51 years >	12	3.8	100
Religion	Muslim	156	50	50
	Non–Muslim	156	50	100
Marital status	Single	67	21.5	21.5
	Married	171	54.8	76.3
	Widowed	32	10.3	86.5
	Divorced	42	13.5	100
Education level	High school	11	3.5	3.5
	Diploma	115	36.9	40.4
	Degree	113	36.2	76.6
	Master	49	15.7	92.3
	PhD	24	7.7	100
Occupation	Government employee	102	32.7	32.7
	Private employee	84	26.9	59.6
	Self-employed	79	26.9	84.9
	Farmer	7	2.2	87.2
	Housewife	8	2.6	89.7
	Student	30	9.6	99.4
	Unemployed	2	0.6	100
Monthly income	<$ 1000	248	79.5	79.5
	$ 1001—2000	49	15.7	95.2
	$ 2001—3000	12	3.8	99
	$ 3001 >	3	1	100

The sample's demographic breakdown ($n = 312$)

During the survey, respondents' marital status was taken into consideration. According to the study, over half of the participants are married, followed by 21.5% of single people, 10.3% of widowed people, and 13.5% of divorced people. Regarding education, the findings show that 3.5% of respondents have completed a higher education program, followed by 36.9% with a diploma, 36.2% with a degree, 15.7% with a master's degree, and 7.7% with a doctorate. In terms of the respondents' employment status (i.e., occupation), over 59.6% worked for the public sector (32.7%) or the private sector (26.9%). The remaining 40.4% work as independent contractors, farmers, housewives, students, or unemployed. Table 3 shows that most respondents, or 79.5%, earn less than $1000 per month, while the remaining respondents, or 20.5%, make more than that amount.

Table 4 Numerical

Construct	Variables	Standardized loading	Cronbach Alpha	CR	AVE	MSV
Attitudes	ATT2	0.921	0.923	0.953	0.837	0.411
	ATT4	0.926				
	ATT6	0.923				
	ATT8	0.887				
Subjective norms	SN2	0.875	0.920	0.944	0.807	0.046
	SN4	0.918				
	SN6	0.911				
	SN8	0.888				
Perceived behavior control	PC2	0.893	0.918	0.942	0.803	0.251
	PC4	0.901				
	PC6	0.909				
	PC8	0.881				
Intention to use IB	INT2	0.800	0.909	0.929	0.687	0.420
	INT4	0.856				
	INT5	0.811				
	INT6	0.868				
	INT7	0.785				
	INT8	0.848				

Suggested composite reliability (CR) > 0.70*
Average variance extracted (AVE) > 0.50 **CR > AVE* AVE > MSV**

3.2 Using Confirmatory Factor Analysis to Validate the Overall Measurement Model

Each construct's variables were combined into a single model to assess how well they fit, discriminant validity, convergent validity, and reliability. The study's measurement framework was examined using statistical measures of fit such as Chi-square (X^2), Root Mean Square Error of Approximation (RMSEA), and Comparative Fit Index (CFI), as endorsed by several academics (Byrne, 2010; Hair et al., 2010; Kline, 2011).

3.3 Specification and Assessment of the Structural Model

Hair et al. (2010) state that testing the entire structural technique and the suggested research hypotheses comes next after the evaluation of the model has produced a satisfactory result and demonstrated goodness of fit. Using AMOS software to conduct Structural Equation Modeling (SEM) to accomplish this task. SEM was utilized in this study to evaluate the hypotheses of the research under investigation (Table 4).

Hair et al. (2010) claim that SEM is the most influential numerical tool to employ when examining the homogenizing effects of several external and internal variables in SEM. It uses the relationships between latent and measured variables to express a theory. After that, SEM will carefully assess the magnitude to which the theory fits the data that accurately depicts fact (Hair et al., 2010). Additionally, Kline (2011) noted that to minimize sampling error, the sample size in SEM tools ought to be high enough ($N > 200$). As a result, the ultimate data analysis used 312 survey responses.

The default structural model's first result, as seen in Fig. 2, indicates that the fit indices were scientifically insufficient (CFI = 0.337, RMSEA = 0.064). After examining modification indices, it was determined that the covariance of measurement error between certain items was required to satisfy the goodness-of-fit requirements of the model. The updated structured model's outcome showed respectable high goodness-of-fit indices.

At $p = 0.001$, the Chi-square (X2 = 329.460, df = 113) was statistically significant. However, given the sample size and complexity of the model, the chi-square test could be deceptive (Byrne, 2010). Therefore, the normed chi-square was used rather than using chi-square as advised by (Hair et al., 2010). The outcome showed that the currently proposed model's normed chi-square (CMIN/DF = 2.916) was less than the cut-off value of 5.0. Additionally, the comparative fit index (CFI = 0.954) indicated a good fit of the model, within the acceptable range of exceeding or equal to 0.90. The approximation's Root Mean Square Error

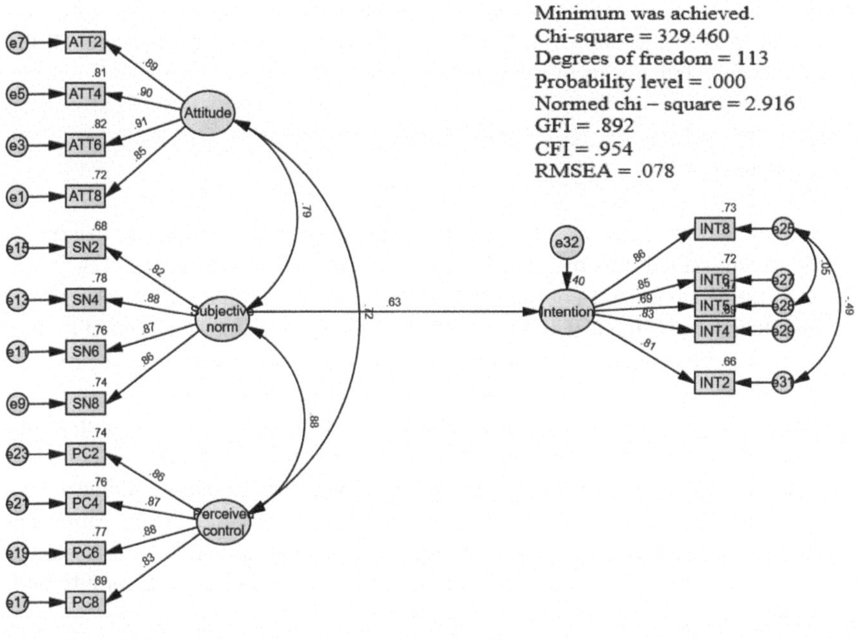

Fig. 2. The researcher's updated the structural framework

Table 5 Hypothesized model estimated values

Structural path	Hypothesized relationship	Std. reg. weight (b)	CR	S E	*p*
INT–ATT	H1	0.847	9.165	0.044	***
INT–SN	H2	0.860	9.376	0.040	***
INT–PC	H3	0.828	8.807	0.041	***
Statistics				Suggested	Obtained
X^2 significant				≥0.05	0.000
Normed x2 (CMIN/df) Comparative Fit Index (CFI)				≤5.00	2.91
Root mean error square of approximation (RMSEA)				≥ 0.90 ≤ 0.08	0.954 0.078

Note: *p* indicates the degree of significance for regression weight. ***significance of.001

(RMSE = 0.078) is less than the threshold of 0.08. Each of these demonstrated a good model fit.

Additionally, all the path coefficients were acceptable and highly supportive, showing that the investigation revealed essential correlations. Table 5 displays the projected value of the total structural parameters.

3.3.1 Hypotheses Testing

As previously stated, the present research produced three hypotheses centered around TPB's primary constructs. All hypotheses were determined to be highly significant at (p 0.05), as shown in Fig. 2. The analysis that follows presents all the outcomes of the hypothesis evaluation.

3.3.2 Research H1

H1. Consumer attitude positively impacts the intention to accept Islamic banking.

The findings about this hypothesis amply validate it. As illustrated in Fig. 2 and Table 5, the standardized coefficient is 0.044, and the *t*-values (critical ratio of regression weight (C.R.) is 9.165 and $p < 0.05$. The standardized coefficient's value indicates that if customers' attitudes increase by one standard deviation, their intention to use I.B. in the model increases by 0.847 standard deviations.

Overall, the findings of this investigation are in line with earlier research (Gopi & Ramayah, 2007; Alam Choudhury & Nurul Alam, 2012; Knabe, 2012; Schulze & Wittmann, 2003) that came to similar findings about the beneficial influence that people's attitudes have on their intention to behave. The study from Amin and Chong (2011) that used the TRA to predict respondents' intention to use Islamic pawnshops also found that subjective norms and attitudes correlate favorably with intention.

3.3.3 Research H2

H2. Subjective norms significantly influence an individual's intention to accept Islamic banking.

The model's results support this theory because of the positive relationship between the variables. As a result, there is a considerable relationship between consumers' S.N. and their inclination to accept I.B. when introduced.

As shown in Fig. 2 and Table 5, the standardized coefficient is 0.040, and the t-values (critical ratio of regression weight (C.R.)) is 0.860 and $p < 0.000$. The standardized coefficient reveals that if a consumer's perception increases by one standard deviation, their desire to use I.B. increases by 0.040 standard deviations.

This study's findings support previous research that found a favorable association between people's behavioral intentions and subjective norms (Gopi & Ramayah, 2007).

3.3.4 Research H3

H3. Perceived behavior control significantly and positively influences individual intention to accept Islamic banking.

The findings unequivocally and strongly concur with this hypothesis. Arguably, a customer's P.C. and capacity and the intent to use I.B. are strongly correlated. The standardized coefficient is 0.828, and the t-values (critical ratio of regression weight, or C.R.) are 8.807 and $p < 0.000$, as shown in Table 5. According to the standardized coefficient's value, customers' intention to use I.B. can increase by up to 0.828 standard deviations for every standard deviation their P.C. increases. Thus, the hypothesis is strongly supported by the results of the analysis.

3.4 The Structural Equation Modeling

The PLS-SEM procedure was utilized for data analysis, as it can handle non-normal data and clarify the variance among primary variables. Smart-PLS version 4.2 software was employed for this purpose. The PLS-SEM process has two stages. The first phase evaluates the outer model, which is the measurement model, and verifies the reliability and validity of the data. Once the measurement model is examined, the structural model is evaluated for hypothesis testing. Figure 3 depicts the constructed path model.

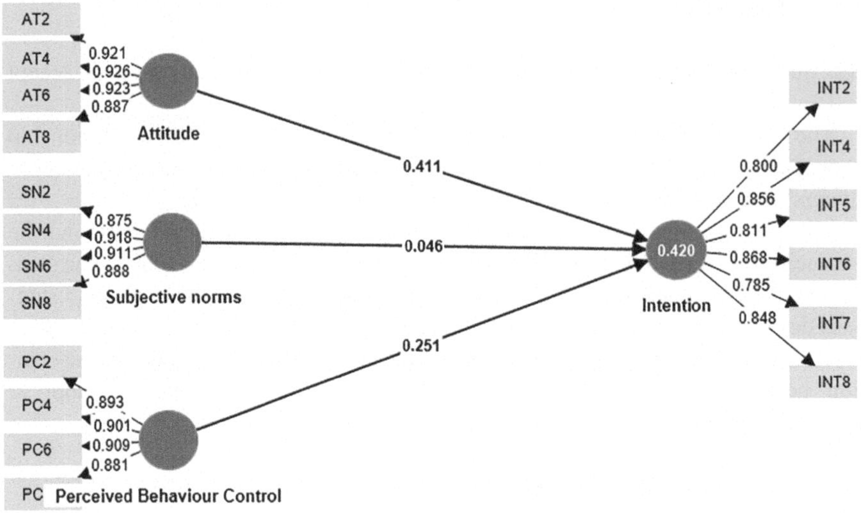

Fig. 3 Path model in SmartPLS

Table 6 Discriminant validity

	Attitude	Intention	Perceived behavior control	Subjective norms
Attitude	0.915			
Intention	0.615	0.829		
Perceived behavior control	0.678	0.567	0.896	
Subjective norms	0.726	0.547	0.807	0.898

The diagonal numbers reflect the square root of the AVE, while the off-diagonals indicate correlations

3.5 *Model Measurement*

The process of structural equation modeling begins with the evaluation of the measurement results framework. It makes clear how the variable and its items relate to one another. It can be verified by evaluating factor loadings, also called composite reliability, internal reliability, which computes internal consistency, discriminant validity determined by the Fornell and Lacker criterion, construct validity (AVE), and HTMT ratio. The smartPLS algorithm was run to obtain the measurement results (Table 6).

The discriminant validity was also assessed using the Fornell and Larcker (1981) criteria (Hair et al., 2014). Fornell and Larcker (1981) relate the AVE of each construct to the relationships for the rest of the constructions. Each construct's square root of AVE for the current study was more significant than its relationship with other constructs. In addition, as Table 7 illustrates, HTMT establishes discriminant validity using a cut limit value of less than 0.90. This study's measurement model guaranteed a sufficient degree of discriminant validity. Thus, the structural model

Table 7 HTMT

	Attitudes	Intention	Perceived behavior control	Subjective norms
Attitude				
Intention	0.665			
Perceived behavior control	0.732	0.616		
Subjective norms	0.784	0.594	0.876	

Table 8 Predictive relevance and goodness of fit

	R^2	Q^2
Intention	0.420	0.408

Table 9 Hypothesis testing

Hypothesis	Relationship	Beta	Std Error	*t*-value	Decision	F^2
H1	Att- > intention	0.411	0.056	7.393	Accepted	0.130
H2	SN- > intention	0.251	0.071	3.515	Accepted	0.036
H3	PBC- > intention	0.046	0.080	0.576	Accepted	0.001

$p^* < 0.05$, $p^{**} < 0.01$

analysis may proceed since the measurement model revealed appropriate reliability and validity.

3.6 Structural Model

The structural model was examined after the outer model was inspected. It involves the interplay between the reflecting constructions and the theories' predictive significance. For that purpose, the R^2 and Q^2 parameters were calculated. Table 8 reveals that the R^2 for Intention was 42%, indicating that the model had sufficient explanatory power. Values exceeding zero indicate appropriate predictive relevance, and the Q^2 value for Intention is 0.408, as shown in Table 8.

Another measure for assessing the structural model is effect size (f^2). F^2 examines the unique result of each outside variable about the internal variable. Cohen (1988) defined f^2 values of 0.01, 0.036, and 0.130 as small, medium, and significant, respectively. As a result, the impact size (f^2) of attitude toward Islamic banking acceptance is significant. In contrast, the effect size (f^2) of perceived behavioral control and subjective norms is minor, as indicated in Table 9. As a result, the study model had appropriate predictive relevance.

In order to test the hypotheses, the PLS-SEM algorithm was used to obtain approximate values for path coefficients. Bootstrapping was performed for 5000 samples. H1 depicted the relationship between attitude and Intention to use Islamic banking, which was backed by $B = 0.411$, $p < 0.01$ and t statistics 7.393. H2 represented the relationship between subjective norms and Intention to use Islamic

Table 10 Model fit value

Criterion	Value
SRMR	0.051

Table 11 IMPA Results

Latent variables	Intention to use	
	Total effect (Importance)	Index value (performance)
Attitude	0.411	65.272
Subjective norms	0.251	50.950
Perceived behavioral control	0.046	54.605

banking, which was backed by $B = 0.251$, $p < 0.01$ and t statistics 3.515. H3 revealed the relation between perceived behavioral control and Intention to use Islamic banking, which is also supported by $B = 0.046$, $p < 0.01$, and t statistics 0.576. The current research's analysis of data confirmed all three hypotheses. This means a significant relationship exists between the dependent variables and the Intention to use IB when introduced in Ghana.

3.7 Model Fit

The Smart-PLS algorithm is an effective tool in modeling data, which uses Standardized Mean Square Values (SRMR) to assess the goodness of fit of the model. In general, the model fit cut-off should be smaller than 0.080, which indicates an acceptable level of fit. For this specific study, the SRMR value was calculated to be 0.051, which is well within the acceptable range as it is less than the maximum threshold of 0.080. This suggests that the model fits the data well and the results obtained from it are reliable (Table 10).

3.8 Importance of Performance Map Analysis

According to the importance-performance map study, attitude was found to be the most important factor in determining the willingness to use Islamic banking services, and it also had the highest performance value. Perceived behavioral control and subjective norms were next in line, but the performance value of perceived behavioral control did not meet that of subjective norms. The results are displayed in Table 11 and Fig 4.

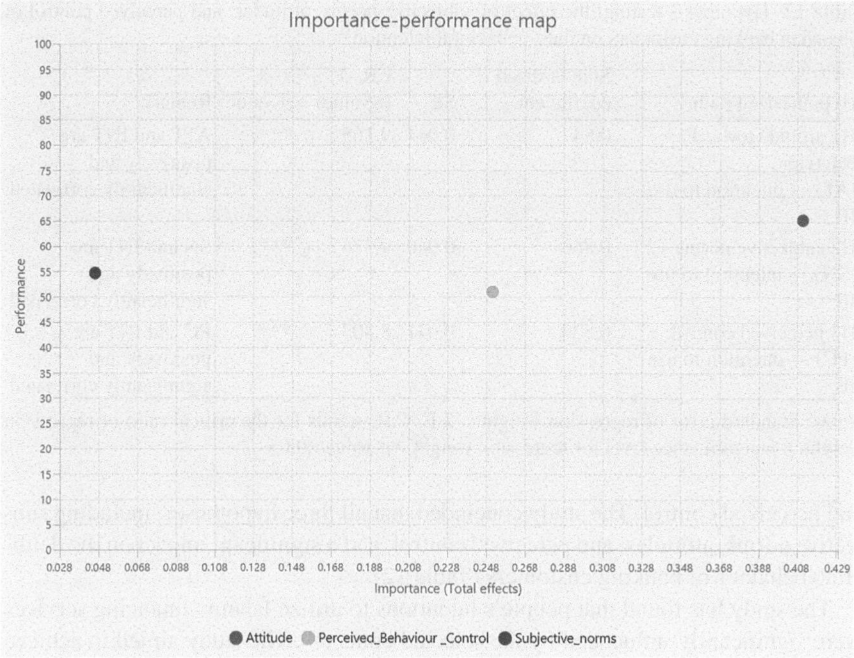

Fig. 4 Importance of performance map analysis

3.9 Measures

In order to assess Ghanaian banking customers' behavioral intention to use I.B., 32 items were added to the questionnaire, which was distributed based on the primary TPB constructs. In the discussion that follows, those constructs are emphasized.

Intention: Eight variables were employed to evaluate the intention of Ghanaian banking customers to accept Islamic Banking products.

Attitude: Eight variables have assessed the attitude of Ghanaian banking customers to accepting Islamic banking.

Subjective norms: The subjective norms were assessed employing eight variables.

Perceived control: Perceived control was assessed by employing eight variables.

4 Discussion

The study is based on the TPB framework, which was used to create survey questions about the behavior of banking customers toward Islamic banking participation. The three main hypotheses of TPB were examined in the study, which included the positive relationship between intended actions and subjective norms, attitudes,

Table 12 Hypothesis testing: the effect of subjective norms, attitudes, and perceived control of Ghanaian banking customers on their behavioral intention

Hypothesized path	Standardized coefficient	SE	CR (*t*-value)	*p*-value	Remark
H1 attitude toward behavior (AT) → intention to use IB	0.847	0.044	9.165	***	ATT and INT are positively and significantly correlated
H2 subjective norms (SN) → intention to use IB	0.860	0.040	9.376	***	SN and INT are positively and significantly correlated
H3 perceived control (PC) → intention to use IB	0.828	0.041	8.807	***	PC and INT are positively and significantly correlated

Notes: Standard error of regression weight = S.E. C.R. stands for the critical ratio of regression weight. p = significance level for regression weight. *p-value <0.05

and perceived control. The study concluded that all three hypotheses, including subjective norms, attitudes, and perceived control, had a significant impact on the deliberate behavior of banking customers (Table 12).

The study has found that people's intentions to utilize Islamic financing services were significantly influenced by the selected concepts. The study aimed to achieve its goals by developing three hypotheses. Thus, it was postulated that the implementation of Islamic financing was influenced by mindset. This confirmed hypothesis is supported by the study's findings. The study's outcomes indicate that a customer will be more likely to offer Islamic banking services if he has a positive attitude about accepting Islamic banking. The results align with those of earlier research (Ali & Jama, 2016; Amin et al., 2014). It was found that one's thinking is the most significant indicator. This implies that a consumer's attitude can successfully promote the formation of good intentions (Gopi & Ramayah, 2007). This favorable attitude also implies that Islamic bank clients are more likely to use the product. The study was investigated whether personal beliefs and opinions had a positive impact on the intention to use Islamic financing. According to the results, there was indeed a positive correlation between subjective norms and the willingness to engage in Islamic financing practices, which was consistent with earlier studies (Ali & Jama, 2016; Amin et al., 2014).

This suggests that fostering an atmosphere conducive to the acceptance of Islamic financing services involves considering the perspective of people close to the potential consumer. It is important to understand the cultural and religious beliefs that influence the decision-making process when it comes to financial matters. By taking into account the opinions and perspectives of those close to the potential consumer, Islamic financing services can better tailor their products and services to meet the needs and preferences of their target audience. Anyone who needs more awareness and expertise must rely on the group's source for advice and aid. Subjective norms may grow if the consumer's closest companions encourage employing the Islamic financing system.

5 Conclusion

The study has demonstrated that three critical components, attitude, subjective standards, and perceived behavioral control, can be used to determine if Islamic financing is accepted in any community.

This study evaluated consumer attitude, subjective standards, perceived behavioral control, and intention as critical characteristics of the I.B. introduction to Ghana. A conceptual model built on I.B. research was created and assessed using information from an online poll of 312 Ghanaian customers. Study discovered a correlation between consumers' attitudes, perceived behavior control, and subjective norms regarding Islamic Banking in Ghana, and their intentions to adopt it. This could have significant implications for the future of the industry and its growth in the region.

According to the proposed framework, Muslim groups are more likely groups to intend to accept Islamic financing than non-Muslims. Although this study offers essential theoretical views to the literature on Islamic finance in non-Islamic Sub-Saharan African countries, it also has several shortcomings that point to topics for additional research. First, because the study only included Ghanaian consumers, the findings are limited in their relevance to other Sub-Saharan African situations. More research is required to adopt the recommended paradigm in other Sub-Saharan Africa to strengthen the study's external validity. Furthermore, because Islamic Banking still needs to acquire a complete operating license in Ghana, clients may be unable to ascertain the level of service they got from Islamic Banking institutions, which is why service quality characteristics were excluded from this study.

In order to increase the predicted worth of the theoretical framework of variables that influence Islamic Banking acceptance proposed in the study, future research should expand the suggested framework by adding quality of service factors.

References

Abdullah, N. I., & Dusuki, A. W. (2006). Customers perceptions of Islamic hire-purchase facility in Malaysia: An empirical analysis. *IIUM Journal of Economics and Management, 14*(2), 177–204.

Abid, A., & Jie, S. (2022). Understanding farmers' decision-making to use Islamic finance through the lens of the theory of planned behavior. *Journal of Islamic Marketing., 14*, 1084. https://doi.org/10.1108/JIMA-10-2020-0324

Ahmed, Z., Zhang, B., & Cary, M. (2021). Linking economic globalization, economic growth, financial development, and ecological footprint: Evidence from symmetric and asymmetric ARDL. *Ecological Indicators, 121*, 107060. https://doi.org/10.1016/j.ecolind.2020.107060

Aji, H. M., Berakon, I., & Riza, A. F. (2020). The effects of subjective norm and knowledge about riba on intention to use e-money in Indonesia. *Journal of Islamic Marketing*, ahead-of-print(ahead-of-print). https://doi.org/10.1108/JIMA-10-2019-0203

Ajzen, I. (1985). From intentions to actions: A theory of planned behavior. In *Action control* (pp. 11–39). Springer. https://doi.org/10.1007/978-3-642-69746-3_2

Ajzen, I. (1991). The theory of planned behavior. *Organizational Behavior and Human Decision Processes, 50*(2), 179–211. https://doi.org/10.1016/0749-5978(91)90020-T

Ajzen, I. (2001). Nature and operation of attitudes. *Annual Review of Psychology, 52*, 27–58. https://doi.org/10.1146/annurev.psych.52.1.27

Ajzen, I., & Madden, T. J. (1986). Prediction of goal-directed behavior: Attitudes, intentions, and perceived behavioral control. *Journal of Experimental Social Psychology, 22*, 453–474.

Al-Ajam, A. S., & Nor, M. K. (2013). Internet banking adoption: Integrating technology acceptance model and trust. *European Journal of Business Management, 5*, 17–20.

Alam Choudhury, M., & Nurul Alam, M. (2012). Corporate governance in Islamic perspective. *International Journal of Islamic and Middle Eastern Finance and Management, 6*(3), 180–199.

Ali, A. Y. S., & Jama, A. K. A. (2016). Determinants of Islamic insurance acceptance: Empirical evidence from Somalia. *European Journal of Business and Management, 8*(15), 102–108.

Ali, F., Rasoolimanesh, S. M., Sarstedt, M., Ringle, C. M., & Ryu, K. (2018). An assessment of the use of partial least squares structural equation modeling (PLS-SEM) in hospitality research. *International Journal of Contemporary Hospitality Management, 30*(1), 514–538.

Ali, B. J., Saleh, Akoi, S., Abdulrahman, A. A., Muhamed, A. S., Noori, H. N., & Anwar, G. (2021). Impact of service quality on the customer satisfaction: Case study at online meeting platforms. *International journal of Engineering, Business and Management, 5*(2), 65–77.

Alzadjal, M. A. J., Abu-Hussin, M. F., Md Husin, M., & Mohd Hussin, M. Y. (2022). Moderating the role of religiosity on potential customer intention to deal with Islamic banks in Oman. Journal of Islamic Marketing, 13(11), 2378–2402.

Amin, H. (2012). Patronage factors of Malaysian local customers toward Islamic credit cards. *Management Research Review, 35*(6), 512–530. https://doi.org/10.1108/01409171211238271

Amin, H. (2020). Critical success factors for the receptiveness of Islamic home financing in Malaysia. *International Journal of Emerging Markets, 15*(5), 849–873. https://doi.org/10.1108/IJOEM-04-2018-0187

Amin, H., Abdul-Rahman, A. R., & Abdul Razak, D. (2014). Theory of Islamic consumer behavior: An empirical study of consumer behavior of Islamic mortgage in Malaysia. *Journal of Islamic Marketing, 5*(2), 273–301.

Amin, H., & Chong, R. (2011). Determinants for Ar-Rahnu usage intention: An empirical investigation. *Africal Journal Business and Management, 5*(20), 8181–8191.

Al Balushi, Y., Locke, S., & Boulanouar, Z. (2019). Omani SME perceptions towards Islamic financing systems. *Qualitative Research in Financial Markets, 11*(4), 369–386. https://doi.org/10.1108/QRFM-06-2018-0078

Al Umar, A. U. A., & Setyono, J. (2023). Hijrah Intention and Customer Decisions to Use Islamic Banks: Development of Theory of Planned Behavior. IQTISHODUNA: Jurnal Ekonomi Islam, 12(1), 301–320.

Armitage, C. J. (2005). Can the theory of planned behavior predict the maintenance of physical activity? *Health Psychology, 24*(3), 235–245. https://doi.org/10.1037/0278-6133.24.3.235

Armitage, C. J., & Comer, M. (1998). Meta-analysis of the theory of planned behavior. Manuscript submitted for publication.

Aziz, S., & Afaq, Z. (2018). Adoption of Islamic banking a behavioral perspective. *Cogent Business and Management, 5*(1), 1–18. https://doi.org/10.1080/23311975.2018.1548050

Bananuka, J., Kaawaase, T. K., Kasera, M., & Nalukenge, I. (2019). Determinants of the intention to adopt Islamic banking in a non-Islamic developing country: The case of Uganda. *ISRA International Journal of Islamic Finance, 11*(2), 166–186.

Bananuka, J., Kasera, M., Najjemba, G. M., Musimenta, D., Ssekiziyivu, B., & Kimuli, S. N. L. (2020a). Attitude: Mediator of subjective norm, religiosity and intention to adopt Islamic banking. *Journal of Islamic Marketing, 11*(1), 81–96.

Bananuka, J., Katamba, D., Nalukenge, I., Kabuye, F., & Sendawula, K. (2020b). Adoption of Islamic banking in a non-Islamic country: Evidence from Uganda. *Journal of Islamic Accounting and Business Research, 11*(5), 989–1007.

Bananuka, J., Mukyala, V., Tumwebaze, Z., Ssekakubo, J., Kasera, M., & Najjuma, M. S. (2020c). The intention to adopt Islamic financing in emerging economies: Evidence from Uganda. *Journal of Islamic Accounting and Business Research, 11*(3), 610–628.

Britto, P. R., & Amer, M. M. (2007). An exploration of cultural identity patterns and the family context among Arab Muslim young adults in America. *Applied Developmental Science, 11*(3), 137–150. https://doi.org/10.1080/10888690701454633

Butt, I., Ahmad, N., Naveed, A., & Ahmed, Z. (2018). Determinants of low adoption of Islamic banking in Pakistan. *Journal of Islamic Marketing, 9*(3), 655–672. https://doi.org/10.1108/JIMA-01-2017-0002

Bukhari, S. A. A., Hashim, F., & Amran, A. (2020). Determinants of green banking adoption: A theoretical framework. *KnE Social Sciences*, 1–14. https://doi.org/10.1177/2277977920905306

Byrne, B. M. (2010). *Structural equation modeling with AMOS: Basic concepts, applications, and programming*. Taylor and Francis Group.

Charag, A. H., Fazili, A. I., & Bashir, I. (2020). Determinants of consumer's readiness to adopt Islamic banking in Kashmir. *Journal of Islamic Marketing, 11*(5), 1125–1154. https://doi.org/10.1108/JIMA-10-2018-0182

Chin, W. W. (2010). How to write up and report PLS analyses. In V. Esposito Vinzi, W. W. Chin, J. Henseler, & H. Wang (Eds.), *Handbook of partial least squares: Concepts, methods and applications* (pp. 655–690). Springer. https://doi.org/10.1007/978-3-540-32827-8_29

Cohen, J. (1988). *Statistical power analysis for the behavioral sciences* (2nd ed.). Lawrence Erlbaum Associates, Publishers.

Cooper, D. R., & Schindler, P. S. (2014). *Business research methods* (12th ed.). McGraw-Hill Irwin.

Eagly, A., & Chaiken, S. (1993). *The psychology of attitudes*. Harcourt Brace Jovanovich College Publishers.

Echchabi, A., & Abdul-Aziz, H. (2012). Empirical investigation of customers' perception and acceptance towards Islamic banking services in Morocco. *Middle East Journal of Scientific Research, 12*(6), 849–858.

Echchabi, A., Azouzi, D., & Aziz, H. A. (2016). The prospects of Islamic banking in Tunisia: An empirical survey. *EuroMed Journal of Business, 11*(1), 119–131.

Erol, C., & El-Bdour, R. (1989). Attitudes, behaviour, and patronage factors of Bank customers towards Islamic banks. *International Journal of Bank Marketing, 7*(6), 31–37. https://doi.org/10.1108/02652328910132060

Ezeh, P. C., & Nkamnebe, A. D. (2020). A conceptual framework for the adoption of Islamic banking in a pluralistic-secular nation: Nigerian perspective. *Journal of Islamic Marketing, 9*(4), 951–964. https://doi.org/10.1108/JIMA-03-2017-0022

Fishbein, M., & Ajzen, I. (1977). Belief, attitude, intention, and behavior: An introduction to theory and research. *Journal of Business Venturing, 5*, 177–189.

Fornell, C., & Larcker, D. F. (1981). Evaluating structural equation models with unobservable variables and measurement error. *Journal of Marketing Research, 18*(1), 39–50.

Hanieh, A. (2020). New geographies of financial power: Global Islamic finance and the Gulf. *Third World Quarterly, 41*(3), 525–546. https://doi.org/10.1080/01436597.2019.1675505

Hair, J.F., Black, W.C., Babin, B.J. and Anderson, R.E. (2010), Multivariate data analysis.

Hair, J. F., Jr., Sarstedt, M., Hopkins, L., & Kuppelwieser, V. G. (2014). Partial least squares structural equation modeling (PLS-SEM): An emerging tool in business research. *European Business Review, 26*(2), 106–121. https://doi.org/10.1108/EBR-10-2013-0128. Management & Data Systems, 121(1), 5–11.

Hoque, A. S. M. M., & Awang, Z. (2019). Does gender difference play moderating role in the relationship between entrepreneurial marketing and Bangladeshi SME performance? *Accounting, 5*(1), 35–52.

IBS. (2021). *From materials to application of Lithium-ion battery in renewable energy and electric vehicle*. Binus University Graduate Program.

IFSB. (2019). *Islamic financial service board*. Kuala Lumpur.

Gopi, M., & Ramayah, T. (2007). Applicability of theory of planned behavior in predicting intention to trade online: Some evidence from a developing country. *International Journal of Emerging Markets, 2*(4), 348–360. https://doi.org/10.1108/17468800710824509

Iqbal, M., & Molyneux, P. (2006). Thirty years of islamic banking. *History, Performance and Prospects, Bankhistorisches Archiv, 32*(2), 155.

Islamic Financial Services Board. (2020). *Islamic financial services industry stability report 2020*, available at: www.ifsb.org/download.php?id=5724&lang=English&pg=/sec03.php (accessed 15 November 2023).

Junaidi, J. (2021). The awareness and attitude of Muslim consumer preference: The role of religiosity. *Journal of Islamic Accounting and Business Research, 12*(6), 919–938. https://doi.org/10.1108/JIABR-08-2020-0250

Junaidi, J., Anwar, S. M., Alam, R., Lantara, N. F., & Wicaksono, R. (2022). Determinants to adopt conventional and Islamic banking: Evidence from Indonesia. *Journal of Islamic Marketing., 14*, 892. https://doi.org/10.1108/JIMA-03-2021-0067

Juwaheer, T. D., Pudaruth, S., & Noyaux, M. M. E. (2012). Analysing the impact of green marketing strategies on consumer purchasing patterns in Mauritius. *World Journal of Entrepreneurship, Management and Sustainable Development, 8*(1), 36–59. https://doi.org/10.1108/20425961211221615

Kaabachi, S., & Obeid, H. (2016). Determinants of Islamic banking adoption in Tunisia: Empirical analysis. *International Journal of Bank Marketing, 34*(7), 1069–1091. https://doi.org/10.1108/IJBM-02-2015-0020

Kline, R. B. (2011). *Principles and practice of structural equation modeling*. Guilford Press.

Knabe, A. (2012). *Applying Ajzen's theory of planned behavior to a study of online course adoption in public relations education*. Marqutte University Publication.

Kocagöz, E. D. (2010). *Kadınların Makyaj Malzemelerini Satın Alma Davranışlarının İncelenmesi: Planlanmış Davranış Teorisinin Bir Uygulaması*. Erciyes Üniversitesi.

Krishna, A., Sari, M., & Rofaida, R. (2011). Analisis Tingkat Literasi Keuangan di Kalangan Mahasiswa dan Faktor-Faktor yang Mempengaruhinya: Survey pada Mahasiswa Universitas Pendidikan Indonesia. In *Proceedings of the 4th international conference on teacher education; join conference UPI & UPSI Bandung, Indonesia, 8–10 November 2010*. Hal 552-560.

Mohamad, S., & Borhan, N. A. (2017). Islamic finance and social sustainability: Parameters for developing a model for social impact measurement. *Malaysian Journal of Sustainable Environment, 3*(2), 81. https://doi.org/10.24191/myse.v3i2.5596

Tijjani, B., Ashiq, M., Siddique, N., Khan, M. A., & Rasul, A. (2021). A bibliometric analysis of quality research papers in Islamic finance: Evidence from web of science. *ISRA International Journal of Islamic Finance, 13*(1), 84–101. https://doi.org/10.1108/IJIF-03-2020-0056

Khurshid, M., & Mahmood, M. R. (2020). A comparative study of adoption of conventional banking and Islamic banking in Pakistan. *Hamdard Islamicus, XLIII*(3).

Lujja, S., Rusni, M. O. M., & Hassan, R. (2016). Modelling public behavioral intention to adopt Islamic banking in Uganda: The theory of reasoned action. *International Journal of Islamic and Middle Eastern Finance and Management, 9*(4), 583. https://doi.org/10.1108/IMEFM-08-2015-0092

Magno, F., Cassia, F., & Ringle, C. M. (2024). A brief review of partial least squares structural equation modeling (PLS-SEM) use in quality management studies. *The TQM Journal, 36*(5), 1242–1251. https://doi.org/10.1108/TQM-06-2022-0197

Mahdzan, S. N., Zainudin, R., & Au, F. S. (2017). The acceptance of Islamic banking services in Malaysia. *Journal of Islamic Marketing, 8*(3), 496.

Maryam, S. Z., Mehmood, M. S., & Khaliq, C. A. (2021). Factors influencing the community behavioral intention for adoption of Islamic banking: Evidence from Pakistan. *International Journal of Islamic and Middle Eastern Finance and Management, 12*(4), 586–600.

Mbawuni, J., & Nimako, S. G. (2018). Muslim and non-Muslim consumers' perception towards introduction of Islamic banking in Ghana. *Journal of Islamic Accounting and Business Research, 9*(3), 353–377.

Md Husin, M., & Ab Rahman, A. (2016). Predicting intention to participate in family takaful scheme using decomposed theory of planned behaviour. *International Journal of Social Economics, 43*(12), 1351–1366. https://doi.org/10.1108/IJSE-03-2015-0074

Mindra, R., Bananuka, J., Kaawaase, T., Namaganda, R., & Teko, J. (2022). Attitude and Islamic banking adoption: Moderating effects of pricing of conventional bank products and social influence. *Journal of Islamic Accounting and Business Research, 13*(3), 534–567. https://doi.org/10.1108/JIABR-02-2021-0068

Naseri, K., & Sharofiddin, A. (2021). Empirical Study on Adoption of Islamic Banking System Using Quantitative Method A Case Study of Afghanistan. Turkish Journal of Islamic Economics, 8(2).

Nurjanah, R., Tabroni, I., Wang, Y., & Zou, G. (2023). Drill method to improve students' prayer movements. *Journal Emerging Technologies in Education, 1*(5), 297–307. https://doi.org/10.55849/jete.v1i5.499

Omar, O., & Owusu-Frimpong, N. (2007). Life insurance in Nigeria: An application of the theory of reasoned action to consumers' attitudes and purchase intention. *The Service Industries Journal, 27*(7), 963–976. https://doi.org/10.1080/02642060701570891

Phau, I., & Teah, M. (2009). Devil wears (counterfeit) Prada: A study of antecedents and outcomes of attitudes towards counterfeits of luxury brands. *Journal of Consumer Marketing, 26*(1), 15–27. https://doi.org/10.1108/07363760910927019

Saeed, I., Juell-Skielse, G., & Uppström, E. (2012). Cloud enterprise resource planning adoption: Motives & barriers. In *Advances in enterprise information systems II*. CRC Press. https://doi.org/10.1201/b12295-45

Schulze, R., & Wittmann, W. W. (2003). A meta-analysis of the theory of reasoned action and the theory of planned behavior: The principle of compatibility and multidimensionality of beliefs as moderators. In R. Schulze, H. Holling, & D. Böhning (Eds.), *Meta-analysis: New developments and applications in medical and social sciences* (pp. 219–250). Hogrefe & Huber Publishers.

Shome, A., Jabeen, F., & Rajaguru, R. (2018). What drives consumer choice of Islamic banking services in The United Arab Emirates? *International Journal of Islamic and Middle Eastern Finance and Management, 11*, 79–95.

Si, K. S., & Chin, P. N. (2023). Intention to Use Islamic Banking Products and Services: Evidence from Malaysia. Journal of Governance and Integrity, 6(2), 616–623.

Smith, M. (2021). infed.org: Education, community-building and change. https://www.infed.org

Suki, N. M. (2010). An empirical study of factors affecting the internet banking adoption among Malaysian consumers? *The Journal of Internet Banking and Commerce, 15*(2), 1–11.

Sulistiowati, L., Arisanti, K., & Lestari, W. (2023). Analisis Kesulitan Soal AKM dalam Pemecahan Masalah Matematika Ditinjau Dari Taksonomi Solo. *Jurnal Pendidikan Dan Konseling (JPDK), 4*(4), 702–708. https://doi.org/10.31004/jpdk.v4i4.5305

Toka, O., & Çetin, M. (2016). Imputation and deletion methods under the presence of missing values and outliers: A comparative study. *Gazi University Journal of Science, 29*(4), 799–809.

Vaughan, G. M., & Hogg, M. A. (2005). *Introduction to social psychology*. Pearson Education.

Jibril Yahuza is a dedicated master's student pursuing his master's degree in Management at Ibn Haldun University, Turkey, He had his first degree in Accounting and Finance at Asia Pacific University of Technology and Innovation, (APU) Malaysia. With a keen interest in Islamic Banking and Finance, he embarked on a journey of academic exploration and scholarly inquiry.

Omar Kachkar is a Trainer and Islamic finance expert with over 15 years of work experience as a consultant and academician in English and Arabic languages. Registered Shariah Advisor at Securities Commission Malaysia. Associate of Association of Shariah Advisors in Islamic Finance – ASAS. Certified Zakah Accountant, and Certified Shari'ah Advisor and Auditor (CSAA) of AAOIFI. Assistant professor at Ibn Haldun University, deputy director of Ibn Haldun University Center for Islamic Economics and Finance.

Cash Waqf Linked Deposit: A Proposed Waqf Model for Education Program and Poverty Alleviation in Indonesia

Mitra Sami Gultom and Muhammad Iman Sastra Mihajat

Abstract The objective of this paper is to examine the potential role of Cash Waqf Linked Deposit (CWLD) that could contribute to Indonesia's education system and efforts to reduce poverty. The study, which employed a qualitative approach and data from papers and scholarly works on the subject of Cash Waqf, included Cash Waqf Linked Sukuk (CWLS) and compared it with Cash Waqf Linked Deposit. CWLD shall consist of three instruments, namely temporary cash waqf, Islamic fixed deposit, and Islamic financing in the Islamic bank. The existing product offered by Islamic banks is whereby the customer places a deposit in an Islamic bank, while the return will be transferred to nazhir as a permanent waqf from the depositor. While in CWLD, the capital will become a temporary waqf, and the return will be a permanent waqf. The paper proposes the use of cash waqf from deposits for the program that may provide more benefits in the future for the citizen including education programs and poverty alleviation. This research shows that the return from CWLD is that it may be used for providing education programs such as building schools and traditional markets. The analysis of Cash Waqf Linked Deposit (CWLD) a new Islamic product that OJK launched in 2023 is valuable in drawing the attention of Islamic banking entities to develop and implement CWLD without jeopardizing their sustainability, competitiveness, or institutional survival. This is evident from the previous model of CLWS and the proposed model of CWLD to incorporate their social financing product initiatives.

M. S. Gultom · M. I. S. Mihajat (✉)
Department of Islamic Banking, Faculty of Islamic Studies, Universitas Muhammadiyah Prof. Dr. Hamka, Jakarta, Indonesia
e-mail: imansastra@uhamka.ac.id

A. Rafiki et al. (eds.), *Strategic Islamic Business and Management*, Contributions to Management Science,
https://doi.org/10.1007/978-3-031-61778-2_17

1 Introduction

Islamic banking and finance in Indonesia have been experiencing significant growth in the last two decades triggered by increasing Islamic awareness (Chong & Liu, 2009) even though it was hit by COVID-19. It is one of the fastest growing industries and spread to all corners of the globe and at the same time receiving wide acceptance (Iqbal & Molyneux, 2005) that is not only accepted in countries where Muslims are the majority, such as the Middle East, but is also developing in Muslim minority countries, for example, the UK, Germany, Thailand, and Singapore. This is because Islamic finance has an element of the social sector that could potentially encourage the business sector and has tremendous advantages in terms of societal welfare to the social sector. This mutually beneficial partnership has the ability to both advance Islamic financial development and serve as a stimulus for sustainable economic growth (Ismal et al., 2015).

Since one of the main objectives of establishing Islamic financial institutions that offer Islamic financial products and services is to empower low-income societies. This paper tries to examine a combined waqf structure with Islamic banking products namely Islamic fixed deposit and Islamic financing because these instruments have the ability to strengthen the aforementioned societies. The two primary instruments offered by Islamic banks that can support the structure are Islamic fixed deposits and Islamic financing. The former will provide a return on the waqf, while the latter will advance the project without requiring the Islamic fixed deposit to mature, say, for 3 years.

Second instrument is waqf. A number of countries have created waqf programs that can boost their economies and promote social welfare. According to Mohsin (2013), these countries including Indonesia, Malaysia, Kuwait, and the UK have developed waqf share schemes with the aim of raising funds from the public to support welfare, finance lawful goods and services that are required by the societies. While in Indonesia according to Timur (2001), Waqf's main objective is to fund different initiatives that benefit the Muslim community as a whole. This strategy was successful in providing funds for programs that promoted entrepreneurship, provided free healthcare to the impoverished and disadvantaged, established a number of educational projects, and reduced poverty.

Therefore, the main objective of this paper is to propose the three important elements in Islamic finance instruments which are waqf, Islamic fixed Deposit, and Islamic financing by collaborating with two important institutions which are waqf Institutions such as Badan Wakaf Indonesia (Indonesia Waqf Board) and Islamic Bank. The cooperation of these organizations will turn into a plan to address national problems like poverty and education. The Cash Waqf Link Deposit (CWLD) structure is a mechanism utilized in Indonesia for the purpose of poverty alleviation and education programs.

2 Research Methods

The research employs a qualitative descriptive research approach in order to perform an exploratory investigation of the literature pertaining to the CWLD idea and structure in this research. The study used secondary data for its data collection. Secondary data about Waqf, and Islamic banking's products is derived from literature in the form of scientific journals, published articles, and regulatory reports.

Meanwhile, this study included the following techniques for data collection: (1) Review of the literature, which entailed compiling data on the subject of the investigation from books and journals. To determine whether or not the new CWLD structure is workable as compared to Cash Waqf Linked Sukuk (CWLS), documentary activities based on reports, publications, and scientific journals are relevant to the problems discussed in this study (Mustafida, 2015).

3 Research Discussion

3.1 The Development of Waqf and Islamic Bank in Indonesia

Despite numerous obstacles and difficulties, the growth of the Islamic banking and financial sector is occasionally nevertheless promising in Indonesia and many other parts of the world. In Indonesia, based on data from the Indonesia Financial Services Authority (OJK) total Islamic Finance Asset in Indonesia as of June 2023 reached Rp 2.450 trillion (USD 163,03 bio) (excluding Islamic Stock). While Islamic banking industry has the second largest contributor after Islamic capital market, the total asset of Islamic banking Industry is Rp 822,26 trillion (USD 54 bio), increase of almost 19% compared to previous year amounting to Rp 693,8 trillion (OJK, 2023a). However, globally Islamic banks contribute 80% of the entire global Islamic financial assets (Aliyu et al., 2017; Hussain et al., 2015; Kammer et al., 2015).

This is supported by the conversion of one conventional bank owned by Riau Province to an Islamic Bank, Bank Riau Kepri (BRK) to become Bank Riau Kepri Syariah (BRKS). So as of March 2023, the total Islamic banks operating in Indonesia are 13 Islamic commercial banks, 20 Islamic windows, and 167 Islamic financing banks. This also makes Islamic banking market share compared to total banking market share in Indonesia increase from 6.62% to 7,31% compared to the previous year (see Table 1). However, the Islamic banking market share is dominated by Bank Syariah Indonesia, one largest Islamic bank in Indonesia owned 39,1% market share from total Islamic banking market share (Mihajat, 2023).

Following the epidemic, Indonesia's Islamic banking industry has recovered and has been expanding in step with the country's improving economic circumstances. As of the end of June 2023, Islamic banking institutions provided financing totaling IDR541.09 trillion, according to data from the Otoritas Jasa Keuangan (2023a). This represents a rise of 16.9% over the previous year or 7,95% of market share. By

Table 1 Islamic finance asset 2023

Industry	Total asset (in IDR and USD)				Market share (%)
	National	National	Sharia (Islamic)	Sharia (Islamic)	
	Rp T	USD Bn	Rp T	USD Bn	
Banking	11.258,65	749,28	822,53	54,74	7,31%
Non-Bank institutions	3.209,10	213,57	157,60	10,49	4,91%
Capital market	7.923,13	527,29	1470,42	97,86	18,56%
Total	**22.390,88**	**1.490,14**	**2.450,55**	**163,09**	**10,94%**

Source: OJK Islamic Finance Market Update, June 2023 (Otoritas Jasa Keuangan, 2023a)

the end of June this year, the total deposits held by Islamic banks and windows represented 7.64% of the market share, up 10.14% from the previous year to IDR 625.57 trillion.

The last instrument is cash waqf also known as *waqf al-Nuqud*, it is a type of waqf made in the form of cash by an individual, group of individuals, organization, or legal body. In the Islamic History documents waqf made a significant contribution to the welfare and educational advancement of Muslim societies (Shulthoni & Saad, 2018). Islamic history also has demonstrated that zakah, sadaqah, waqf, and qard al-hasan are redistributive tools of Islamic finance that can improve financial inclusion by reducing poverty (Suseno & Fitriyani, 2018). Therefore, it is required for Islamic finance and economics stakeholders to contribute by suggesting the optimal model with the goal to enhance the tool and increase its impact beyond what is typically seen.

In Indonesia, the development of cash waqf has been showing a tremendous increase since the last decade, which according to Sharia has also received permissibility from the Indonesian Council of Ulama based on its fatwa as issued in 2002 (DSN-MUI, 2002). Mohsin (2013) agrees that cash waqf is acceptable as long as it adheres to the same legal requirements as immovable waqf. Even in the present day, cash waqf can be a new source of funding. Therefore, cash waqf is permissible provided and utilized for items that are allowed by sharia.

Nowadays, the use of cash waqf as a productive form of waqf will give greater advantages than traditional waqf, which involves movable or immovable assets. In many societies, asset waqf is more identic to graveyards, mosques, and schools if it is deemed to be property waqf. Generally speaking, only those who have extra assets are able to contribute a waqf. However, the waqf assets that contribute and generate income are still in the form of small-size assets, because the traditional management of waqf (nazhir) still lacks of knowledge in modern management practices. In contrast to cash waqf, which is more accessible to a larger number of people, this can be contributed by everyone in the form of an amount of money that is more affordable for every Muslim (Al-Arif, 2012).

According to the Indonesian Waqf Board (BWI), however, only barely 10% of profitable waqf land is used productively by the community. This issue arises as a result of the conventional nature of the waqf administration paradigm. As a result, a new paradigm is required that combines commercial use of waqf land with sharia provisions and cash waqf in order to optimize the potential use of waqf assets. Cash Waqf Linked Deposit will positively contribute to the well-being of the people if waqf is managed productively like the government and *nazhir* have managed effectively Cash Waqf Linked Sukuk (CWLS) in the last 4 years.

3.2 Cash Waqf Linked Deposit Versus Cash Waqf Linked Sukuk

Musari (2021) states that the first financial innovation to integrate two Islamic financial instruments was sukuk, which was tied to waqf. Where the money raised from the Sukuk issuance was used to construct profitable enterprises on the waqf asset, and where waqf assets functioned as the foundation for Sukuk linked with waqf. Therefore, this instrument has two positive effects: Mauquf Alaih benefits from the monthly coupon from Sukuk by constructing beneficial facilities for low-income populations, and the State uses the proceeds from Sukuk to expand the Indonesian economy and support the state budget (Saptono, 2019).

Ismal et al. (2015) assert that to enhance the previous model, Bank Indonesia suggested a new transformation of the waqf and sukuk structure through the creation of Cash waqf-linked sukuk (CWLS). Finally, the Governor announced CWLS during the 2018 World Bank and International Monetary Fund (IMF) Annual Meetings, which were held from October 12 to 14 in Nusa Dua, Bali. The Indonesian Waqf Board (BWI), the Ministry of Religious Affairs, the Central Bank of Indonesia (BI), and the Ministry of Finance (MoF) then signed a memorandum of understanding (MoU) for the development of CWLS on November 1, 2018. The Islamic Sovereign Sukuk (SBSN) Act and the State Budget (APBN) Act would make Cash Waqf Linked Sukuks feasible.

Despite its inception in 2018, it took the Ministry of Finance about 2 years to issue the first CWLS. However, currently, CWLS is becoming a successful tool for waqf investment to finance the social program and social project. CWLS was finally first issued in March 2020 with series SW001 at a value of IDR 50.8 billion (MoF, 2020). Under its role as a nazhir (waqf manager), the Indonesian Waqf Agency (BWI) places cash waqf funds via the private placement technique. The coupon from Sukuk will be transferred to BWI which will be utilized for medical equipment purchases and renovations to help construct a new center at the Wakif Achmad Wardi Hospital in Serang, Banten Province. The coupon from Sukuk will support 2513 impoverished patients over the course of 5 years by purchasing ambulances and offering free cataract surgery. The coupons are paid on a monthly basis by the government. The funds from the Waqf Sukuk will be fully refunded to the Waqf holder SW001 series on the maturity date. The terms and conditions of the CWLS series from the date of issuance until the end of 2023 are displayed in Tables 2 and 3.

Table 2 Cash Waqf Linked Sukuk Issuance from 2020 to 2023

Explanation	SW001	SWR001	SWR002	SWR003	SWR004[a]
Issuance time	March 2020	November 2020	June 2021	July 2022	August 2023
Issuance amount	IDR50.849 billion (USD 3.272 mio)	IDR14.912 billion (USD 959 k)	IDR24.141 billion (USD 1.553 mio)	IDR 38.25 billion (USD 2.39 mio)	IDR 112.56 billion (USD 7,04 mio)
Tradability	Non-tradable	Non-tradable	Non-tradable	Non-tradable	Non-tradable
Aqad (Islamic contract)	*Wakala*	*Wakala*	*Wakala*	*Wakala*	*Wakala*
Rate/yield	6.15%	5.5%	5.57%	5.05%	5.85%
Profit/coupon rates p.a.	5.00%	5.5%	5.57%	5.05%	5.85%
Duration	5 years	2 years	2 years	2 years	2 years
Number of *waqif*	–	1041	591	688	709
Method of offering	Private placement	Book building	Book building	Book building	Book building

Source: DJPPR (2020a, 2020b, 2021a, 2021b, 2022), Rudya (2023) and Detik.com (2023), processed by author

[a]The offering period lasts from July 7 to August 31, 2023

Table 3 Features of CWLD

Features of CWLD	
The fixed deposit amount	The waqf amount
The coupon amount will be transferred to Nazhir (BWI)	For the benefit of mauquf alaih
Minimum fixed deposit investment	IDR 1 million
Maximum fixed deposit amount	As per the mini prospectus of the waqf project
Minimum tenor of CWLS	1–3 years
Profit	Will be transferred to Nazhir's account as sources of payment for financing

The development of CWLS issuance for the last 4 years has shown tremendous increase and promise in Indonesia despite many challenges and uncertainties due to COVID-19. Based on data from DJPPR (2020a, 2020b, 2021a, 2021b, 2022) total CWLS issuance in Indonesia as of October 2023 reached IDR 112.56 billion (USD 7.04 million) increase of almost 193% compared to previous year which was only IDR 38.25 billion (USD 2.39 million) which is the largest throughout the CWLS Retail issuance. Waqf certificate buyers came from individual wakifs amounting to IDR 24,323 billion and institutional wakifs amounting to IDR 88,240 billion. The number of waqf certificate holders also increased from time to time, in 2021, there were 591 CWLS certificate holders in 2021, which increase to 688 CWLS certificate holders in 2022 and increased to 709 CWLS holders in 2023. The remarkable growth of CWLS is a result of Indonesian Muslims' growing awareness of waqf,

waqf lifestyle among the societies and the CWLS literacy increase strategy is quite effective, which includes both Gen Z and older individuals (Rudya, 2023). According to Sulaeman and Zubaidah (2022), the government of Indonesia already prioritized financing the State Budget (APBN) through Sukuk since the issuance of Law No. 19 of 2008 concerning State Sharia Securities (SBSN), which was enacted in April 2008.

Finally, OJK (2023b) has recently introduced Cash Waqf Linked Deposit as one of the Islamic banking products that assist the waqf program, in response to the success of CWLS in supporting waqf through Sukuk. The primary goal of the initiative is to raise citizen education levels and support the efforts of the Waqf institutions to end poverty. By starting the initiative, it will be expected that the Islamic banking sector will contribute to Indonesia's efforts to address its poverty and economic difficulties. As a result, the article attempts to elucidate the potential of CLWD in assisting waqf institutions' efforts for educational attainment and poverty alleviation.

CWLD can be defined as a social investment in the form of temporary waqf with respect to deposit amounts and permanent waqf with respect to profit sharing from fixed deposits which cannot be broken or transferred or sold before the end of the agreed waqf period. The waqf amount and profit will be channeled to the waqf institution body in collaboration with Islamic banks both in managing deposits and in providing financing. The program can be accomplished at the beginning of the program without waiting for the agreed waqf period to end by obtaining financing from an Islamic bank whose installments are taken from profit sharing from waqif's fixed deposits.

3.3 Cash Waqf Linked Deposit for Education Program and Poverty Alleviation

In the previous two decades, the Indonesian Islamic banking sector has grown dramatically. The growth is apparent in terms of business volume, financial goods, and financial sectors. So far, the Islamic banking sector has taken the lead in delivering products and services to society. Indeed, the Islamic financial system has incorporated aspects that assist low-income societies in gaining access to financial services that are currently insufficient due to a lack of collateral and technical capacity. Waqf, an Islamic social sector, is one of these elements. According to the Indonesian Waqf Board's (BWI) most recent identification procedure, Indonesia has a lot of promise in terms of waqf assets. The BWI has a total of 5 billion m2 of waqf lands dispersed around the country. Unfortunately, those assets are underutilized due to a lack of competent management, making them less economically productive and having little economic impact, particularly on low-income society.

According to a contemporary interpretation of waqf, it is permissible to temporarily transfer assets in order to raise additional assets that may be put to charitable and social purposes. The kinds of objects that permit the categorization of financial assets as waqf assets that can be securitized for the issuance of waqf-based Deposits are also developing further. The concept of project-based deposit and waqf-linked sukuk are similar both aim to develop the real economy through the issuance of sukuk and fixed deposit. The following are two uses for the cash waqf linked sukuk and cash waqf lined deposit: The inception of government and Islamic bank or corporate-oriented commercial for the development of waqf land. This configuration has the potential to facilitate the mobilization of resources.

Therefore, the author tries to examine a combined instrument issued by OJK (2023b) such as waqf, fixed deposit, and financing from Islamic bank, the collaboration between two Islamic organization such as Indonesia Waqf Board (BWI) and Islamic Bank. We believe that the collaboration between these Islamic instruments and institutions will give a greater impact economically and socially on low-income societies. Whereby the purpose of CWLD issuance is used by each nazir to finance the program/project. The paper tries to explore the use of cash waqf from fixed deposits for the program that may provide more benefit in the future for the citizen including education program and poverty alleviation.

This research shows that the profit from CWLD is proven it may be used for providing education programs to Indonesian citizens through building schools and universities under waqf land. It is expected that after graduation from the education program, the students can contribute to developing the economic of the country as well as decreasing the national poverty level. The students may waqf one of his 7 days for people empowerment and educating the needy. The second one is for poverty alleviation, the government may build through nazhir (BWI) a traditional market under the land of waqf that can be used by local traditional traders by paying low rental rates, if the local traditional traders need financing for its business, the Islamic bank can finance them with low rate of financing. Please find the OJK-CWLD model below (Fig. 1):

Process Flow for CWLD Scheme

1. MoU (PKS) agreement between Nazhir and Islamic bank as LKS-PWU (Islamic banking entities who act as the recipient of cash waqf) providing waqf programs and waqf fund management.
2. Wakif opened fixed deposits Via LKS-PWU, the Wakif gave Nazir the bill of paper of fixed deposit as a waqf.
3. The Wakif obtained a duplicate deposit slip, a cash waqf certificate, and the waqf pledge deed.
4. Returns from fixed deposits are used for the waqf program and are sent to Nazir.
5. The financing installments are paid through the amount received on a monthly basis from profit sharing of fixed deposit.
6. Islamic bank will return the wakif's temporary waqf fund at the end of waqf period.

A mechanism for depositing waqf funds (cash waqf) in deposit instruments within a specific timeframe (at least 1 year) is provided by this temporary cash waqf product, which is designed to integrate the social and commercial functions of Islamic banks as Islamic Financial Institutions the recipient of Cash Waqf

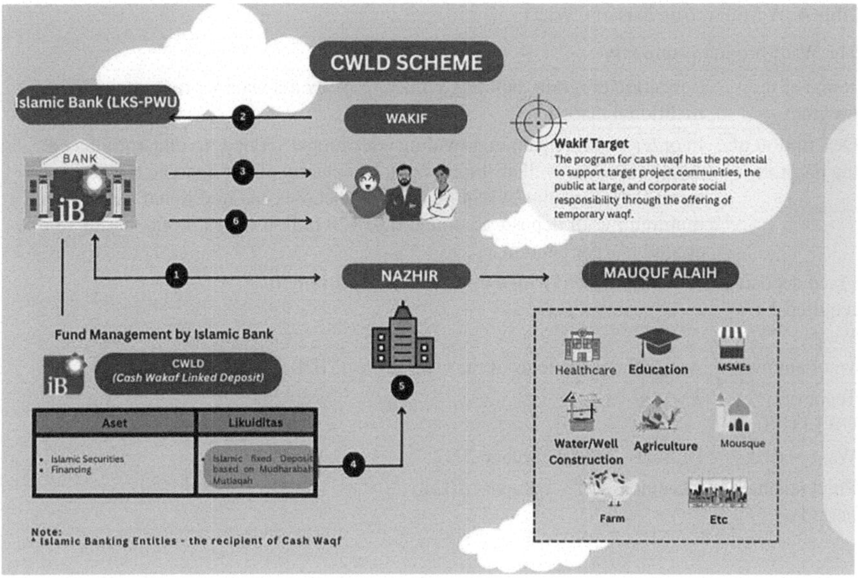

Fig. 1 Proposed CWLD Structure. Source: Adapted from Otoritas Jasa Keuangan (2023b) with modifications by author

(LKS-PWU). When the waqf period ends, the cash waqf funds in the form of deposits are returned to the wakif. The results can be used directly by the waqf recipients (mauquf 'alaih). Upon maturity, the cash waqf monies are returned to the wakif, just like with deposit products generally. The distinction is that, as a partner of an Islamic bank, Nazhir (a waqf asset management) distributes the profit sharing to mauquf 'alaih (beneficiaries) rather than depositors.

3.4 Features of the CWLS Product

CWLD uses Islamic bank's fixed deposit instrument to fund waqf project, where the profit from the fixed deposit will be transferred directly to Islamic bank as a monthly installment payment and notified nazhir (such as BWI) as waqf manager who manages the project. So that Mauquf "alaih" can benefit from the project. The Sukuk in the CWLS cannot be traded and it has to be held until maturity of 1–3 years. The minimum CWLS investment is IDR 1 million for the waqf certificate holder who wants to invest in the project as seen in Table 4. Islamic bank and nazhir both stand to benefit from CWLD, Islamic bank will boost its third-party fund, and nazhir will increase the waqf asset from the waqf amount.

To speed up the waqf project, Islamic bank will provide financing for this project. For example, the project needs IDR 20 billion, IDR 10 billion to build the traditional market under the waqf land and IDR 10 billion to build schools under the

Table 4 A Mini prospectus of CWLD

The Waqf program summary	
Name of the program	Education program, building schools/poverty alleviation program, building traditional market
Description of the program	In order to participate in providing educational services to underprivileged communities as well as improving the economy of low-income communities, about IDR 20 billion is required to build schools and traditional markets with a minimum waqf deposit of around IDR 120 billion with a profit-sharing rate of around 6% per annum
Fixed deposit required for Waqf	IDR 120 billion (with a minimum of Rp 1 billion)
Waqf amount	As per the prospectus of the waqf project (IDR 20 billion)
Tenor of CWLD	3 years
Waqf period	February 2024–February 2027
Fund raising period	December 2023–February 2024

Table 5 Coupon from fixed deposit and waqf project financing calculation

Profit from fixed deposit calculation								
Month	1	2	3	4	5	6	……	36
Sukuk proceed	IDR 120 bio							
Coupon amount	IDR 650 mio	IDR 650 mio	IDR 650 mio	IDR 650 mio	IDR 650 mio	IDR 650 mio	IDR 650 mio	IDR 650 mio
Sukuk coupon rate	6.5%							
Financing installment from Islamic Bank								
Financing amount	IDR 20 bio							
Installment amount	IDR 613 mio	IDR 613 mio	IDR 613 mio	IDR 613 mio	IDR 613 mio	IDR 613 mio	IDR 613 mio	IDR 613 mio
Financing rate	6.5%							

waqf land. The sources of payment from this financing will be coming from profit of the fixed deposit, the financing will be guaranteed by the sources of payment from profit of fixed deposit. BWI will act as nazhir to distribute the fund to mauquf alaih and manage the project as well as act as a financing recipient from Islamic bank. The monthly profit from fixed deposit will be used as a monthly installment that paid directly to Islamic bank from month 1 until month 36. Islamic banks will not have to worry about the payment capacity of the financing, because the sources of payment have zero risk in terms of payment capacity as seen in Table 5.

4 Conclusion

As a first move and issuer toward empowering low-income communities, reducing poverty, and improving the educational achievement of its people, Indonesia develops Cash Waqf Linked Sukuk and Cash Waqf Linked Deposit. CWLD is the first instrument to integrate waqf as social finance, fixed deposit, and financing from an Islamic bank as commercial finance. It works in tandem with social finance to take the lead and provide creative financing solutions for a range of social projects and initiatives. We think that the country can reduce its poverty rate by encouraging education. Because raising citizens' levels of education is expected, they can aid in the growth of the nation's economy and lower the rate of poverty there. The students may waqf one of his 7 days for people empowerment and educating the needy.

However, it is not that easy to adopt the CWLD structure, particularly from an Islamic bank. Typically, Islamic banks will need financing security for regular financing as a safety net and assurance in the event that financing encounters difficulties down the road. Because the bank must receive a security from the financing supplied that is equivalent to or greater than 100–1200% (some Islamic banks require 140–150%) of the financing value in order to conduct normal financing, as specified by the Islamic bank's policy and procedures. For non-profit organizations with social program finance as their goal, Islamic banks must create unique financing policies and procedures. Unlike traditional finance product development, the social product development program shall be facilitated by the Indonesia Financial Services Authority (OJK) through the use of simple regulations for this particular project.

References

Al-Arif, M. N. R. (2012). Wakaf Uang dan Pengaruhnya Terhadap Pengentasan Kemiskinan. *Indo Islamika, 2*(1), 2012.

Aliyu, S., Hassan, M. K., Mohd Yusof, R., & Naiimi, N. (2017). Islamic banking sustainability: A review of literature and directions for future research. *Emerging Markets Finance and Trade, 1–31*, 1. https://doi.org/10.1080/1540496X.2016.1262761

Chong, B. S., & Liu, M. H. (2009). Islamic banking: Interest-free or interest-based? *Pacific-Basin Finance Journal, 17*(1), 125–144.

Detik. (2023). *Bank Mega Syariah Cetak Rekor Penjualan CWLS Ritel SWR004 Rp 84.78 Milyar*. Accessed Nov 8, 2023, from https://finance.detik.com/moneter/d-6965288/bank-mega-syariah-cetak-rekor-penjualan-cwls-ritel-swr-004-rp-84-79-m

DJPPR. (2020a, March 10). *Press release: Issuance of waqf sukuk (cash waqf linked sukuk–CWLS) series SW001 on march 10, 2020 by private placement*. Directorate General of Budget Financing and Risk Management (DJPPR) Ministry of Finance (MoF).

DJPPR. (2020b, November 24). *Press release: "CWLS retail SWR001 series remains attracted by investors amid pandemic conditions"*. Directorate of Islamic Financing (DPS) Directorate General of Budget Financing and Risk Management (DJPPR) Ministry of Finance (MoF).

DJPPR. (2021a, April 9). *Press release: "The retail CWLS SWR002 series brings blessings to the economy of the ummah"*. Directorate of Islamic Financing (DPS) Directorate General of Budget Financing and Risk Management (DJPPR) Ministry of Finance (MoF).

DJPPR. (2021b, June 7). *Press release: "Amid pandemic conditions, CWLS retail series SWR002 is extremely incredible, the targets of retail green sukuk-sukuk tabungan ST008 series achieved before the offer ends"*. Directorate of Islamic Financing (DPS) Directorate General of Budget Financing and Risk Management (DJPPR) Ministry of Finance (MoF).

DJPPR. (2022, April 7). *Press release: "Offering plan for CWLS retail series SWR003"*. Directorate of Islamic Financing (DPS) Directorate General of Budget Financing and Risk Management (DJPPR) Ministry of Finance (MoF).

DSN-MUI. (2002). *Himpunan Fatwa Majelis Ulama Indonesia*.

Hussain, M., Shahmoradi, A., Turk, R. (2015). *An overview of Islamic finance* (No. WP/15/120). IMF Working Papers, IMF Working Paper Series.

Iqbal, M., & Molyneux, P. (2005). *Thirty years of Islamic banking: History, performance and prospects*. Palgrave Macmillan.

Ismal, R., Muljawan, D., Chalid, M. R., Kashoogie, J., & Sastrosuwito, S. (2015). *Awqaf linked Sukuk to support the economic development*. Occasional Paper.

Kammer, A., Norat, M., Pinón, M., Prasad, A., & Towe, C., 2015. *Islamic finance: Opportunities, challenges, and policy options* (No. SDN/15/05). Islamic finance, IMF Staff Discussion Note.

Mihajat, M. I. S. (2023, October 9). *Toward global standardization of fatwa in Islamic Bank in Indonesia*, Islamic Finance News.

Mohsin, M. I. A. (2013). Financing through cash-waqf: A revitalization to finance different needs. *International Journal of Islamic and Middle Eastern Finance and Management, 6*(4), 304–321.

MoF. (2020, October 9). The government of the republic of Indonesia, information memorandum of sukuk wakaf (cash waqf linked sukuk) SWR001 Series. Ministry of Finance (MoF) Republic of Indonesia

Musari, K. (2021). Green sukuk, Islamic green financing: A lesson learned from Indonesia. In O. M. Olarewaju & I. O. Ganiyu (Eds.), *Climate change and the sustainable financial sector* (pp. 1–16). IGI Global. https://doi.org/10.4018/978-1-7998-7967-1.ch001

Mustafida, R. (2015). *Optimization of waqf asset trough sukuk: Toward an inclusive economic development*. Forum Riset Ekonomi dan Keuangan Syariah IV.

Otoritas Jasa Keuangan. (2023a). *Market Update Keuangan Syariah Bulan Juni 2023*. Accessed Nov 6, 2023, from https://www.ojk.go.id/id/kanal/syariah/data-dan-statistik/reksa-dana-syariah/Default.aspx

Otoritas Jasa Keuangan. (2023b, October 25). *Pengembangan Cash Waqf Linked Deposit (Cwld) Di Bank Syariah*. CWLD Talkshow by OJK Islamic Banking Department in 10th Indonesia Sharia Economic Festival (ISEF).

Rudya. (2023, September 6). *SWR004 Capai Hasil Pemesanan Terbesar Sepanjang Penerbitan Sukuk Wakaf Ritel*. Accessed Nov 6, 2023, from https://mindcommonline.com/2023/09/06/swr004-capai-hasil-pemesanan-trbesar-sepanjang-penerbitan-sukuk-wakaf-ritel/

Saptono, I. T. (2019, November 12). *Waqf linked Sukuk for SDGs and zakat for micro hydro designing Islamic social finance instruments for development and humanitarian needs*. OIC Forum on Islamic Social Finance.

Shulthoni, M., & Saad, N. M. (2018). Waqf fundraising management: A conceptual comparison between traditional and modern methods in the waqf institutions. *Indonesian Journal of Islam and Muslim Societies, 8*(1), 57–86. https://doi.org/10.18326/ijims.v8i1.57-86

Sulaeman, S., Zubaidah, S., & Nur'atikah, T. (2022). Optimization of cash Waqf linked Sukuk (CWLS) for supporting Indonesia's economic recovery post-Covid-19 era. *Review of Islamic Social Finance and Entrepreneurship, 1*(2), 121–130. https://doi.org/10.20885/RISFE.vol1.iss2.art4

Suseno, P., & Fitriyani, Y. (2018). Role of Islamic finance development to financial inclusion: Empirical study in Islamic banking countries. *Jurnal Ekonomi & Keuangan Islam, 4*(1), 1–8. https://doi.org/10.20885/jeki.vol4.iss1.art1

Timur, K. (2001). The provision of public goods under Islamic law: Origins, impact, and limitations of the Waqf system. *Law and Society Review, 35*(4), 841–898.

Mitra Sami Gultom is a Lecturer in Islamic banking Department, Faculty of Islamic Studies, Universitas Muhammadiyah Prof. Dr. Hamka, Indonesia. He holds a bachelor's degree in Islamic Banking (IAIN North Of Sumatera, Indonesia), a master's in Islamic Economics (UIN North Of Sumatera, Indonesia), and a PhD candidate in Islamic Economics and Business (UIN Syarif Hidayatullah, Indonesia).

Muhammad Iman Sastra Mihajat is a Lecturer in Islamic banking Department, Faculty of Islamic Studies, Universitas Muhammadiyah Prof. Dr. Hamka, Indonesia, former Senior Vice President, Head of Compliance Division, Bank Mega Syariah and former Head of Sharia, Oman Arab Bank and Alizz Islamic Bank, Oman. He was an Expert/Advisor for Islamic Banking Products and Policies in PwC for the Saudi National Bank, Saudi Arabia. He was also an Advisor for Middle East Investor Relations for BPKH and Telkom Indonesia. He is a well-known Shari'a expert in Indonesia. He has experience not only in the Islamic banking industry but also in Islamic capital markets, investment advisory, private equity and venture capital, Islamic insurance (Takaful), and Shari'a advisory. He teaches in several well-known universities such as Sultan Qaboos University, Universitas Prof Dr. Hamka, Universitas Indonesia, Padjajaran University, Trisakti University, Islamic University of Syarif Hidayatullah, Al Azhar Indonesia University, and Azzahra Islamic University. He have presented papers in many international conferences such as Sudan, Azerbaijan, Italia, Saudi Arabia, Iran, Malaysia, Indonesia and Pakistan. He has published several Books and Journals on Islamic Banking and Finance related to highly reputed global Institutions. Dr. Muhammad Iman holds a LLB (Shari'a) from Al Azhar University, Cairo, Egypt; Postgraduate Diploma in Islamic Banking and Finance (IIiBF); Master in Finance, and PhD in Islamic Banking and Finance (Institute of Islamic Banking and Finance, IIiBF) from International Islamic University Malaysia (IIUM).

The manufacturer's authorised representative in the EU is Springer Nature Customer Service Centre GmbH, Europaplatz 3, 69115 Heidelberg, Germany. If you have any concerns regarding our products, please contact ProductSafety@springernature.com

Printed and bound by CPI Group (UK) Ltd, Croydon, CR0 4YY
01/10/2026
02225897-0010